Citroën C15 Van
Service and Repair Manual

Michael Gascoigne Bsc, MS, CEng, MIChemE, MISTC

Models covered
Citroën C15 vans with four-cylinder engines;
954 cc and 1124 cc petrol engines
1769 cc diesel engines

Does not cover 1118 cc, 1294 cc, 1360 cc petrol engines or Electrique

(3509 - 288)

© Haynes Group Limited 1999

A book in the **Haynes Service and Repair Manual Series**

ISBN **978 1 85960 509 7**

British Library Cataloguing in Publication Data
A catalogue record for this book is available from the British Library.

Haynes Group Limited
Haynes North America, Inc

www.haynes.com

Disclaimer

Contents

LIVING WITH YOUR CITROEN C15 VAN

Roadside repairs

Weekly checks

Tyre pressures

Lubricants and fluids

MAINTENANCE

Routine maintenance and servicing

Contents

The Citroën C15 Van covered by this Manual was introduced to the UK in 1989, except for the 954 cc carburettor petrol version which was only available in France. The front of the vehicle is based on the Visa GTi and the rear can be either a standard load compartment van, a combined passenger/load compartment with rear seats known as the Familiale, or else a platform cab which can be fitted with specialised units such as campers. This Manual covers only the standard load compartment van, but is applicable to the other vehicles since they all have the same basic design except for variations in the load compartment.

Citroën C15 Van

The standard van is available with payload capacities of 475 kg, 600 kg and 765 kg. The 475 kg version is available in petrol models only. All the vans have the same bodywork dimensions and the variation in payload capacity is achieved by the suspension components, so that the heavier payload vans are higher off the ground. There were minor bodywork variations in 1992, including side-rubbing strips and the vehicle was designated the "Champ" van.

Both petrol and diesel versions of the C15 van have been available in Europe since 1985. There was a major re-launch in July 1988 when new 954 cc and 1124 cc carburettor petrol engines were introduced, replacing previous versions.

In 1990, the Bosch Monopoint A2.2 petrol injection system was introduced on the 1124 cc engine, to meet the emission requirements of some European countries. This became available in the UK from June 1994 until January 1996 and replaced the 1124 cc carburettor model which was discontinued in November 1994.

The highly popular 1769 cc diesel models were introduced to the UK in 1989, alongside the petrol models, eventually to replace them and become the only models available.

All models have front-wheel drive and fully-independent front and rear suspension. Power-assisted steering was available on diesel models from December 1993 onwards and in the UK it was fitted as standard to the 765 kg payload Champ vans.

All engines are derived from well-proven designs which have appeared in many Citroën vehicles. Both the petrol and diesel engines are four-cylinder overhead camshaft design, mounted transversely, with manual transmission on the left-hand side. There are no automatic transmissions. Petrol models and early diesel models were fitted with either four or five-speed transmissions, depending on the engine size and vehicle payload capacity. During 1990, four-speed transmissions were phased out from the diesel models, so that they were all fitted with five gears.

Provided that regular servicing is carried out in accordance with the manufacturer's recommendations, the Citroën C15 Van should prove a reliable and economical small utility vehicle. The engine compartment is well-designed and most of the items needing frequent attention are easily accessible.

Project vehicles

This manual is based on the dismantling and reassembly of two C15 diesel models, before and after the fuel system was redesigned in February 1993. Work was also carried out on a petrol modell.

Your Citroën C15 manual

The aim of this manual is to help you get the best value from your vehicle. It can do so in several ways. It can help you decide what work must be done (even should you choose to get it done by a garage). It will also provide information on routine maintenance and servicing and give a logical course of action and diagnosis when random faults occur. However, it is hoped that you will use the manual by tackling the work yourself. On simpler jobs it may even be quicker than booking the vehicle into a garage and going there twice, to leave and collect it. Perhaps most important, a lot of money can be saved by avoiding the costs a garage must charge to cover its labour and overheads.

The manual has drawings and descriptions to show the function of the various components so that their layout can be understood. Tasks are described and photographed in a clear step-by-step sequence. The illustrations are numbered by the Section number and paragraph number to which they relate - if there is more than one illustration per paragraph, the sequence is denoted alphabetically.

References to the "left" or "right" of the vehicle are in the sense of a person in the driver's seat, facing forwards.

Acknowledgements

Thanks are due to Champion Spark Plug, who supplied the illustrations showing spark plug conditions and to Duckhams Oils, who provided lubrication data. Thanks are also due to Draper Tools Limited, who provided some of the workshop tools and to all those people at Sparkford who helped in the production of this manual.

We take great pride in the accuracy of information given in this manual, but vehicle manufacturers make alterations and design changes during the production run of a particular vehicle of which they do not inform us. No liability can be accepted by the authors or publishers for loss, damage or injury caused by any errors in, or omissions from, the information given.

The Citroën C15 Team

Haynes manuals are produced by dedicated and enthusiastic people working in close co-operation. The team responsible for the creation of this book included:

Author	**Michael Gascoigne**
Technical editor	**Martynn Randall**
Sub-editor	**Sophie Yar**
Editor & Page Make-up	**Steve Churchill**
Workshop manager	**Paul Buckland**
Photo Scans	**John Martin**
Cover illustration & Line Art	**Roger Healing**
Wiring diagrams	**Matthew Marke**

We hope the book will help you to get the maximum enjoyment from your car. By carrying out routine maintenance as described you will ensure your car's reliability and preserve its resale value.

Working on your car can be dangerous. This page shows just some of the potential risks and hazards, with the aim of creating a safety-conscious attitude.

General hazards

Scalding

• Don't remove the radiator or expansion tank cap while the engine is hot.
• Engine oil, automatic transmission fluid or power steering fluid may also be dangerously hot if the engine has recently been running.

Burning

• Beware of burns from the exhaust system and from any part of the engine. Brake discs and drums can also be extremely hot immediately after use.

Crushing

• When working under or near a raised vehicle, always supplement the jack with axle stands, or use drive-on ramps. *Never venture under a car which is only supported by a jack.*
• Take care if loosening or tightening high-torque nuts when the vehicle is on stands. Initial loosening and final tightening should be done with the wheels on the ground.

Fire

• Fuel is highly flammable; fuel vapour is explosive.
• Don't let fuel spill onto a hot engine.
• Do not smoke or allow naked lights (including pilot lights) anywhere near a vehicle being worked on. Also beware of creating sparks (electrically or by use of tools).
• Fuel vapour is heavier than air, so don't work on the fuel system with the vehicle over an inspection pit.
• Another cause of fire is an electrical overload or short-circuit. Take care when repairing or modifying the vehicle wiring.
• Keep a fire extinguisher handy, of a type suitable for use on fuel and electrical fires.

Electric shock

• Ignition HT voltage can be dangerous, especially to people with heart problems or a pacemaker. Don't work on or near the ignition system with the engine running or the ignition switched on.

• Mains voltage is also dangerous. Make sure that any mains-operated equipment is correctly earthed. Mains power points should be protected by a residual current device (RCD) circuit breaker.

Fume or gas intoxication

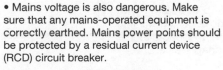

• Exhaust fumes are poisonous; they often contain carbon monoxide, which is rapidly fatal if inhaled. Never run the engine in a confined space such as a garage with the doors shut.
• Fuel vapour is also poisonous, as are the vapours from some cleaning solvents and paint thinners.

Poisonous or irritant substances

• Avoid skin contact with battery acid and with any fuel, fluid or lubricant, especially antifreeze, brake hydraulic fluid and Diesel fuel. Don't syphon them by mouth. If such a substance is swallowed or gets into the eyes, seek medical advice.
• Prolonged contact with used engine oil can cause skin cancer. Wear gloves or use a barrier cream if necessary. Change out of oil-soaked clothes and do not keep oily rags in your pocket.
• Air conditioning refrigerant forms a poisonous gas if exposed to a naked flame (including a cigarette). It can also cause skin burns on contact.

Asbestos

• Asbestos dust can cause cancer if inhaled or swallowed. Asbestos may be found in gaskets and in brake and clutch linings. When dealing with such components it is safest to assume that they contain asbestos.

Special hazards

Hydrofluoric acid

• This extremely corrosive acid is formed when certain types of synthetic rubber, found in some O-rings, oil seals, fuel hoses etc, are exposed to temperatures above 400°C. The rubber changes into a charred or sticky substance containing the acid. *Once formed, the acid remains dangerous for years. If it gets onto the skin, it may be necessary to amputate the limb concerned.*
• When dealing with a vehicle which has suffered a fire, or with components salvaged from such a vehicle, wear protective gloves and discard them after use.

The battery

• Batteries contain sulphuric acid, which attacks clothing, eyes and skin. Take care when topping-up or carrying the battery.
• The hydrogen gas given off by the battery is highly explosive. Never cause a spark or allow a naked light nearby. Be careful when connecting and disconnecting battery chargers or jump leads.

Air bags

• Air bags can cause injury if they go off accidentally. Take care when removing the steering wheel and/or facia. Special storage instructions may apply.

Diesel injection equipment

• Diesel injection pumps supply fuel at very high pressure. Take care when working on the fuel injectors and fuel pipes.

⚠️ *Warning: Never expose the hands, face or any other part of the body to injector spray; the fuel can penetrate the skin with potentially fatal results.*

Remember...

DO

• Do use eye protection when using power tools, and when working under the vehicle.

• Do wear gloves or use barrier cream to protect your hands when necessary.

• Do get someone to check periodically that all is well when working alone on the vehicle.

• Do keep loose clothing and long hair well out of the way of moving mechanical parts.

• Do remove rings, wristwatch etc, before working on the vehicle – especially the electrical system.

• Do ensure that any lifting or jacking equipment has a safe working load rating adequate for the job.

DON'T

• Don't attempt to lift a heavy component which may be beyond your capability – get assistance.

• Don't rush to finish a job, or take unverified short cuts.

• Don't use ill-fitting tools which may slip and cause injury.

• Don't leave tools or parts lying around where someone can trip over them. Mop up oil and fuel spills at once.

• Don't allow children or pets to play in or near a vehicle being worked on.

The following pages are intended to help in dealing with common roadside emergencies and breakdowns. You will find more detailed fault finding information at the back of the manual, and repair information in the main chapters.

If your vehicle won't start and the starter motor doesn't turn

☐ If it's a model with automatic transmission, make sure the selector is in 'P' or 'N'.

☐ Open the bonnet and make sure that the battery terminals are clean and tight.

☐ Switch on the headlights and try to start the engine. If the headlights go very dim when you're trying to start, the battery is probably flat. Get out of trouble by jump starting (see next page) using a friend's car.

If your vehicle won't start even though the starter motor turns as normal

☐ Is there fuel in the tank?

☐ Is there moisture on electrical components under the bonnet? Switch off the ignition, then wipe off any obvious dampness with a dry cloth. Spray a water-repellent aerosol product (WD-40 or equivalent) on ignition and fuel system electrical connectors like those shown in the photos. Pay special attention to the ignition coil wiring connector and HT leads. (Note that Diesel engines don't normally suffer from damp.)

A Check the security and condition of the battery terminals.

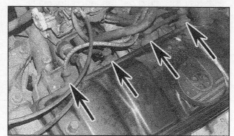

B Check that the four HT leads are securely connected to the spark plugs (arrowed). Trace the wires back and make sure they are securely connected to the distributor (petrol engine models).

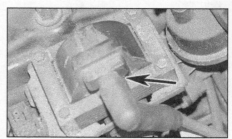

C Check that the HT lead (arrowed) is securely connected to the ignition coil. Trace the wire back and make sure it is securely connected to the distributor. Also check the LT wiring multi-plugs connecting the ignition coil to the distributor (petrol engine models).

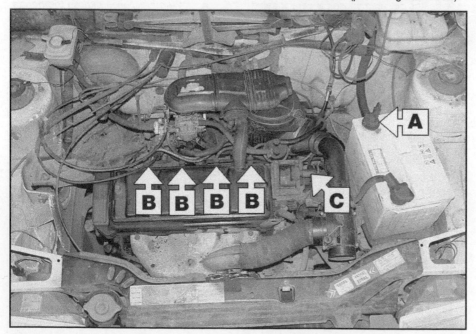

Check that electrical connections are secure (with the ignition switched off) and spray them with a water dispersant spray like WD40 if you suspect a problem due to damp

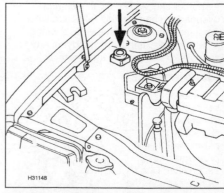

D Reset the inertia switch (where fitted) by depressing the button on top of the unit.

Jump starting

Jump starting will get you out of trouble, but you must correct whatever made the battery go flat in the first place. There are three possibilities:

1 *The battery has been drained by repeated attempts to start, or by leaving the lights on.*

2 *The charging system is not working properly (alternator drivebelt slack or broken, alternator wiring fault or alternator itself faulty).*

3 *The battery itself is at fault (electrolyte low, or battery worn out).*

When jump-starting a car using a booster battery, observe the following precautions:

✔ Before connecting the booster battery, make sure that the ignition is switched off.

✔ Ensure that all electrical equipment (lights, heater, wipers, etc) is switched off.

✔ Take note of any special precautions printed on the battery case.

✔ Make sure that the booster battery is the same voltage as the discharged one in the vehicle.

✔ If the battery is being jump-started from the battery in another vehicle, the two vehicles MUST NOT TOUCH each other.

✔ Make sure that the transmission is in neutral (or PARK, in the case of automatic transmission).

1 Connect one end of the red jump lead to the positive (+) terminal of the flat battery

2 Connect the other end of the red lead to the positive (+) terminal of the booster battery.

3 Connect one end of the black jump lead to the negative (-) terminal of the booster battery

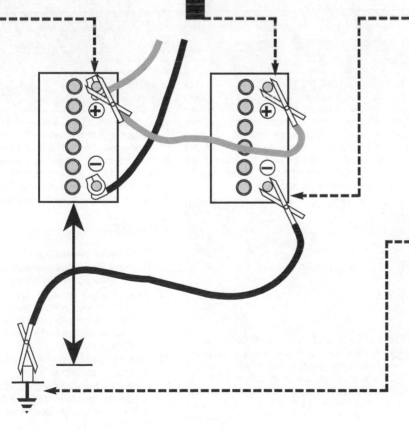

4 Connect the other end of the black jump lead to a bolt or bracket on the engine block, well away from the battery, on the vehicle to be started.

5 Make sure that the jump leads will not come into contact with the fan, drive-belts or other moving parts of the engine.

6 Start the engine using the booster battery and run it at idle speed. Switch on the lights, rear window demister and heater blower motor, then disconnect the jump leads in the reverse order of connection. Turn off the lights etc.

Wheel changing

 Warning: *Do not change a wheel in a situation where you risk being hit by another vehicle. On busy roads, try to stop in a lay-by or a gateway. Be wary of passing traffic while changing the wheel - it is easy to become distracted by the job in hand.*

Preparation

☐ When a puncture occurs, stop as soon as it is safe to do so.

☐ Park on firm level ground, if possible, and well out of the way of other traffic.

☐ Use hazard warning lights if necessary.

☐ If you have one, use a warning triangle to alert other drivers of your presence.

☐ Apply the handbrake and engage first or reverse gear

☐ Chock the wheel diagonally opposite the one being removed – a couple of large stones will do for this.

☐ If the ground is soft, use a flat piece of wood to spread the load under the jack.

Changing the wheel

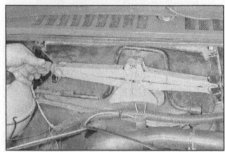

1 Undo the strap and remove the jack and wheelbrace from the engine compartment bulkhead.

2 Use the wheelbrace to slacken the spare wheel carrier securing bolt at the rear of the load compartment.

3 Lift the carrier upwards and unhook it, then lower it to the ground and remove the spare wheel. Hook the carrier back up while the wheel is removed.

4 Place the spare wheel under the sill, to protect the vehicle in case it falls off the jack.

5 Remove the wheel trim, where applicable. Slacken each wheel bolt by a half turn, using the wheelbrace, while the vehicle is still on the ground. If the bolts are too tight, DON'T stand on the wheelbrace to undo them - call for assistance.

6 Locate the jack below the reinforced jacking point nearest the wheel being changed. Place the jack head between the two horizontal flanges in the lower edge of the sill (don't jack the vehicle at any other point of the sill). Turn the jack handle clockwise until the wheel is raised clear of the ground.

Finally...

☐ Remove the wheel chocks.

☐ Stow the jack, wheelbrace and punctured wheel in the correct locations.

☐ Check the tyre pressure on the wheel just fitted. If it is low, or if you don't have a pressure gauge with you, drive slowly to the nearest garage and inflate the tyre to the right pressure.

☐ Have the damaged tyre or wheel repaired as soon as possible.

☐ Have the wheel bolts slackened and retightened to the specified torque at the earliest possible opportunity (see Chapter 10 Specifications).

7 Unscrew the wheel bolts and remove the wheel. Recover the centre trim, where applicable.

8 Fit the spare wheel and screw in the bolts. Ensure the centre trim, if applicable, is held securely under all four bolt heads. Lightly tighten the bolts with the wheelbrace, then lower the vehicle to the ground. Securely tighten the wheel bolts in a diagonal sequence, then refit the wheel trim where applicable.

Identifying leaks

Puddles on the garage floor or drive, or obvious wetness under the bonnet or underneath the car, suggest a leak that needs investigating. It can sometimes be difficult to decide where the leak is coming from, especially if the engine bay is very dirty already. Leaking oil or fluid can also be blown rearwards by the passage of air under the car, giving a false impression of where the problem lies.

⚠️ *Warning: Most automotive oils and fluids are poisonous. Wash them off skin, and change out of contaminated clothing, without delay.*

HAYNES HiNT *The smell of a fluid leaking from the car may provide a clue to what's leaking. Some fluids are distictively coloured. It may help to clean the car carefully and to park it over some clean paper overnight as an aid to locating the source of the leak.*
Remember that some leaks may only occur while the engine is running.

Sump oil

Engine oil may leak from the drain plug...

Oil from filter

...or from the base of the oil filter.

Gearbox oil

Gearbox oil can leak from the seals at the inboard ends of the driveshafts.

Antifreeze

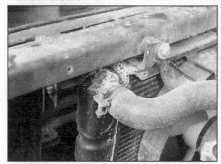

Leaking antifreeze often leaves a crystalline deposit like this.

Brake fluid

A leak occurring at a wheel is almost certainly brake fluid.

Power steering fluid

Power steering fluid may leak from the pipe connectors on the steering rack.

Towing

When all else fails, you may find yourself having to get a tow home – or of course you may be helping somebody else. Long-distance recovery should only be done by a garage or breakdown service. For shorter distances, DIY towing using another car is easy enough, but observe the following points:
□ Use a proper tow-rope – they are not expensive. The vehicle being towed must display an 'ON TOW' sign in its rear window.
□ Always turn the ignition key to position 'M' when the vehicle is being towed, so that the steering lock is released, and that the direction indicator and brake lights will work.
□ Towing eyes are provided below the bumpers, one at the front **(see illustration)** and two at the rear.

□ Before being towed, release the handbrake and select neutral on the transmission.
□ Note that greater-than-usual pedal pressure will be required to operate the brakes, since the vacuum servo unit is only operational with the engine running.
□ On models with power steering, greater-than-usual steering effort will also be required.
□ The driver of the car being towed must keep the tow-rope taut at all times to avoid snatching.
□ Make sure that both drivers know the route before setting off.
□ Only drive at moderate speeds and keep the distance towed to a minimum. Drive smoothly and allow plenty of time for slowing down at junctions.

Front towing eye (arrowed)

Introduction

There are some very simple checks which need only take a few minutes to carry out, but which could save you a lot of inconvenience and expense.

These "Weekly checks" require no great skill or special tools, and the small amount of time they take to perform could prove to be very well spent, for example;

☐ Keeping an eye on tyre condition and pressures, will not only help to stop them wearing out prematurely, but could also save your life.

☐ Many breakdowns are caused by electrical problems. Battery-related faults are particularly common, and a quick check on a regular basis will often prevent the majority of these.

☐ If your car develops a brake fluid leak, the first time you might know about it is when your brakes don't work properly. Checking the level regularly will give advance warning of this kind of problem.

☐ If the oil or coolant levels run low, the cost of repairing any engine damage will be far greater than fixing the leak, for example.

Underbonnet check points

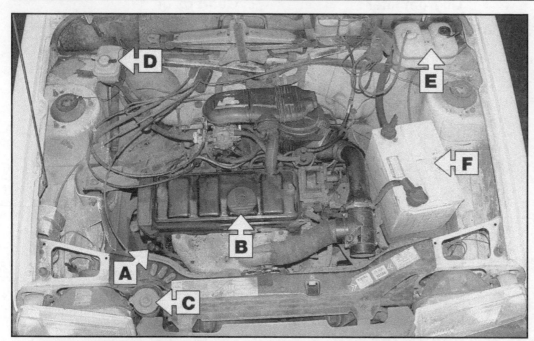

◀ **Carburettor petrol model**

A *Engine oil level dipstick*

B *Engine oil filler cap*

C *Radiator filler cap*

D *Brake fluid reservoir*

E *Screen washer fluid reservoir*

F *Battery*

Note: *On fuel-injection models, the underbonnet check points are the same except that the brake fluid reservoir is on the left.*

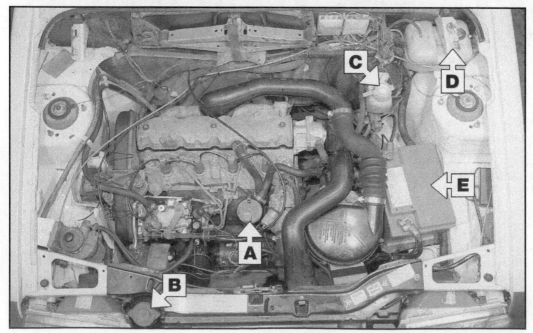

◀ **Diesel model, up to February 1993**

A *Engine oil level dipstick/filler cap*

B *Radiator filler cap*

C *Brake fluid reservoir*

D *Screen washer fluid reservoir*

E *Battery*

Note: *In February 1993 the fuel system was redesigned but the underbonnet check points remained the same until January 1994.*

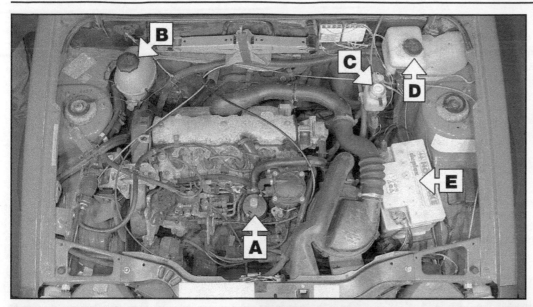

◀ **Diesel model, from January 1994**

A *Engine oil level dipstick/filler cap*

B *Coolant expansion tank*

C *Brake fluid reservoir*

D *Screen washer fluid reservoir*

E *Battery*

Engine oil level

Before you start
✔ Make sure that your car is on level ground.

✔ Check the oil level before the vehicle is driven, or at least 5 minutes after the engine has been switched off.

 HAYNES HiNT *If the oil is checked immediately after driving the vehicle, some of the oil will remain in the upper engine components, resulting in an inaccurate reading on the dipstick!*

The correct oil
Modern engines place great demands on their oil. It is very important that the correct oil for your vehicle is used (See "Lubricants and fluids" on Page 0•17).

Vehicle Care
● If you have to add oil frequently, you should check whether you have any oil leaks. Place some clean paper under the vehicle overnight, and check for stains in the morning. If there are no leaks, the engine may be burning oil (see "Fault Finding").

● Always maintain the level between the upper and lower dipstick marks (see photo 3). If the level is too low severe engine damage may occur. Oil seal failure may result if the engine is overfilled by adding too much oil.

1 On petrol models, the dipstick is located at the front right-hand corner of the engine. On diesel models, it is in the filler cap on the oil filler tube located in front of the engine. It is brightly coloured for easy identification. See "Underbonnet check points" on pages 0•10 and 0•11 for the exact location.

2 Withdraw the dipstick and wipe off all the oil using a clean rag or paper towel. Insert the clean dipstick into the tube as far as it will go, then withdraw it again.

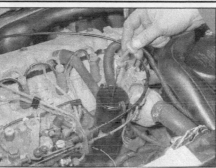

MIN   MAX

3 Note the oil level on the end of the dipstick, which should be between the upper "MAX" mark and the lower "MIN" mark. If the oil level is below the "MIN" mark, or only just above, topping-up is required. The amount of oil required to raise the level from the lower to upper mark is given in Chapter 1A or 1B Specifications.

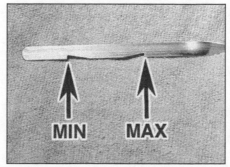

4 Oil is added through the filler cap. On petrol models, unscrew the cap, located on the cylinder head cover. On diesel models, release the clips and remove the cap from the oil filler tube, located in front of engine. Top-up the level, using a funnel if necessary to reduce spillage. Add the oil slowly, checking the level on the dipstick often and allowing time for the oil to fall to the sump. Add oil until the level is just up to the "MAX" mark on the dipstick. Do not overfill.

Coolant level

Warning: DO NOT attempt to remove the expansion tank pressure cap when the engine is hot, as there is a very great risk of scalding. Do not leave open containers of coolant about, as it is poisonous.

Vehicle Care

● With a sealed-type cooling system, adding coolant should not be necessary on a regular basis. If frequent topping-up is required, it is likely there is a leak. Check the radiator, all hoses and joint faces for signs of staining or wetness, and rectify as necessary.

● It is important that antifreeze is used in the cooling system all year round, not just during the winter months. Don't top-up with water alone, as the antifreeze will become too diluted.

1 The coolant level varies with the temperature of the engine. When the engine is hot, the level will rise slightly. On models with a coolant expansion tank integral with the radiator, the level should be visible when the filler cap is removed. On models with a separate expansion tank, the level should be above the plate which is visible on removing the cap. If "MAX" and "MIN" marks are provided, the level should be maintained between the marks and must be above the "MIN" mark at all times.

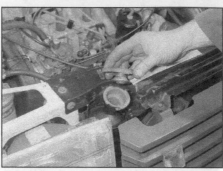

2 If topping up is necessary, **wait until the engine is cold**. Slowly unscrew the expansion tank cap, to release any pressure present in the cooling system and remove it.

3 Add a mixture of water and antifreeze to the expansion tank until the coolant level is correct. Refit the cap and tighten it securely.

Power steering fluid level

Note: *The majority of C15 vans in the UK are fitted with manual steering. Power steering became available in December 1993 and was fitted as standard to the 765 kg payload Champ vans.*

Before you start:

✔ Park the vehicle on level ground.

✔ Set the steering wheel straight-ahead.

✔ Ensure that the ignition is switched off (take out the ignition key).

HAYNES HiNT *For the check to be accurate, the steering must not be turned once the engine has been stopped.*

Safety First!

● The need for frequent topping-up indicates a leak, which should be investigated immediately.

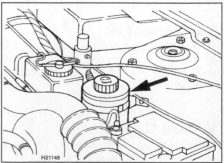

1 The power steering fluid reservoir is located at the left-hand side of the engine compartment, near the suspension strut. Wipe clean the area around the filler cap, then unscrew the cap and check that the fluid is level with the lugs at the centre of the filter base. If necessary, top-up the level, using only the specified type of fluid and do not overfill the reservoir. When the level is correct, securely refit the cap.

Brake fluid level

Warning:
● **Brake fluid can harm your eyes and damage painted surfaces, so use extreme caution when handling and pouring it.**

● **Do not use fluid that has been standing open for some time, as it absorbs moisture from the air, which can cause a dangerous loss of braking effectiveness.**

 • **Make sure that your vehicle is on level ground.**
• **The fluid level in the reservoir will drop slightly as the brake pads wear down, but the fluid level must never be allowed to drop below the "MIN" mark.**

Safety First!

● If the reservoir requires repeated topping-up this is an indication of a fluid leak somewhere in the system, which should be investigated immediately.

● If a leak is suspected, the vehicle should not be driven until the braking system has been checked. Never take any risks where brakes are concerned.

1 The brake master cylinder and fluid reservoir are mounted on the vacuum servo unit in the engine compartment on either the left or right-hand side of the bulkhead, according to model. The "MAXI" and "DANGER" marks are indicated on the side of the reservoir. The fluid level must be kept between the marks at all times.

3 Unscrew the cap and carefully lift it out of position, taking care not to damage the level switch float. Place the cap and float on a piece of clean rag. Inspect the reservoir; if the fluid is dirty, the hydraulic system should be drained and refilled (see Chapter 1A or 1B).

2 If topping-up is necessary, wipe the area around the filler cap with a clean rag before removing the cap.

4 Carefully add fluid, taking care not to spill it onto the surrounding components. Use only the specified hydraulic fluid; mixing different types of fluid can cause damage to the system and/or a loss of braking effectiveness. After filling to the correct level, refit the cap securely and wipe off any spilt fluid.

Screen washer fluid level

● Screenwash additives not only keep the winscreen clean during foul weather, they also prevent the washer system freezing in cold weather - which is when you are likely to need it most. Don't top up using plain water as the screenwash will become too diluted, and will freeze during cold weather.

● **On no account use coolant antifreeze in the washer system - this could discolour or damage paintwork.**

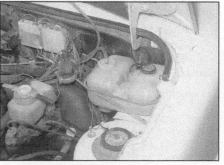

1 The screen washer fluid reservoir is located at the rear left-hand corner of the engine compartment. On models with rear window washers, an additional reservoir is located on a side panel in the load compartment.

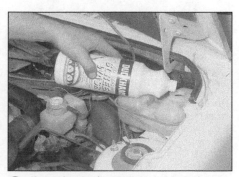

2 If topping up is necessary, open the cap and add water and a screenwash additive in the quantities recommended on the bottle.

Tyre condition and pressure

It is very important that tyres are in good condition, and at the correct pressure - having a tyre failure at any speed is highly dangerous. Tyre wear is influenced by driving style - harsh braking and acceleration, or fast cornering, will all produce more rapid tyre wear. As a general rule, the front tyres wear out faster than the rears. Interchanging the tyres from front to rear ("rotating" the tyres) may result in more even wear. However, if this is completely effective, you may have the expense of replacing all four tyres at once! Remove any nails or stones embedded in the tread before they penetrate the tyre to cause deflation. If removal of a nail does reveal that

the tyre has been punctured, refit the nail so that its point of penetration is marked. Then immediately change the wheel, and have the tyre repaired by a tyre dealer.

Regularly check the tyres for damage in the form of cuts or bulges, especially in the sidewalls. Periodically remove the wheels, and clean any dirt or mud from the inside and outside surfaces. Examine the wheel rims for signs of rusting, corrosion or other damage. Light alloy wheels are easily damaged by "kerbing" whilst parking; steel wheels may also become dented or buckled. A new wheel is very often the only way to overcome severe damage.

New tyres should be balanced when they are fitted, but it may become necessary to re-balance them as they wear, or if the balance weights fitted to the wheel rim should fall off. Unbalanced tyres will wear more quickly, as will the steering and suspension components. Wheel imbalance is normally signified by vibration, particularly at a certain speed (typically around 50 mph). If this vibration is felt only through the steering, then it is likely that just the front wheels need balancing. If, however, the vibration is felt through the whole car, the rear wheels could be out of balance. Wheel balancing should be carried out by a tyre dealer or garage.

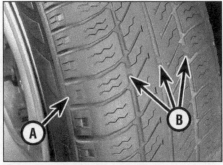

1 *Tread Depth - visual check*
The original tyres have tread wear safety bands (B), which will appear when the tread depth reaches approximately 1.6 mm. The band positions are indicated by a triangular mark on the tyre sidewall (A).

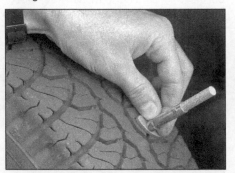

2 *Tread Depth - manual check*
Alternatively, tread wear can be monitored with a simple, inexpensive device known as a tread depth indicator gauge.

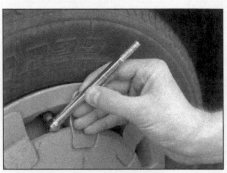

3 *Tyre Pressure Check*
Check the tyre pressures regularly with the tyres cold. Do not adjust the tyre pressures immediately after the vehicle has been used, or an inaccurate setting will result. Tyre pressures are shown on Page 0•16.

Tyre tread wear patterns

Shoulder Wear

Underinflation (wear on both sides)
Under-inflation will cause overheating of the tyre, because the tyre will flex too much, and the tread will not sit correctly on the road surface. This will cause a loss of grip and excessive wear, not to mention the danger of sudden tyre failure due to heat build-up.
Check and adjust pressures
Incorrect wheel camber (wear on one side)
Repair or renew suspension parts
Hard cornering
Reduce speed!

Centre Wear

Overinflation
Over-inflation will cause rapid wear of the centre part of the tyre tread, coupled with reduced grip, harsher ride, and the danger of shock damage occurring in the tyre casing.
Check and adjust pressures

If you sometimes have to inflate your car's tyres to the higher pressures specified for maximum load or sustained high speed, don't forget to reduce the pressures to normal afterwards.

Uneven Wear

Front tyres may wear unevenly as a result of wheel misalignment. Most tyre dealers and garages can check and adjust the wheel alignment (or "tracking") for a modest charge.
Incorrect camber or castor
Repair or renew suspension parts
Malfunctioning suspension
Repair or renew suspension parts
Unbalanced wheel
Balance tyres
Incorrect toe setting
Adjust front wheel alignment
Note: *The feathered edge of the tread which typifies toe wear is best checked by feel.*

Wiper blades

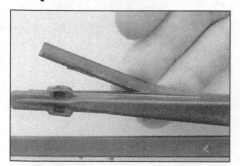

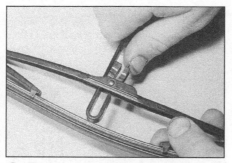

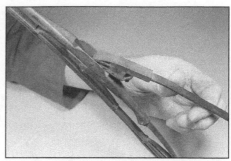

1 Check the condition of the wiper blades; if they are cracked or show any signs of deterioration, or if the glass swept area is smeared, renew them. For maximum clarity of vision, wiper blades should be renewed annually, regardless of their apparent condition. Don't forget the rear window wiper, if it is fitted.

2 The wiper arm may have a hooked end or a forked end. Make sure you obtain the correct type of blade to fit the wiper arm. To remove a wiper blade from an arm with a hooked end, pull the arm fully away from the glass until it locks. Swivel the blade through 90°, press the locking tab(s) with your fingers and slide the blade out of the hooked end of the arm. On refitting, ensure that the blade locks securely into the arm.

3 To remove a wiper blade from an arm with a forked end, pull the blade pivot from the fork until it releases. On refitting, push the blade pivot into the fork until it clicks home.

 HAYNES HINT *If smearing is still a problem despite fitting new wiper blades, try cleaning the windscreen with neat screen-wash additive or methylated spirit.*

Battery

Caution: *Before carrying out any work on the vehicle battery, read the precautions given in "Safety first" at the start of this manual.*

✔ Make sure that the battery tray is in good condition, and that the clamp is tight. Corrosion on the tray, retaining clamp and the battery itself can be removed with a solution of water and baking soda. Thoroughly rinse all cleaned areas with water. Any metal parts damaged by corrosion should be covered with a zinc-based primer, then painted.

✔ Periodically (approximately every three months), check the charge condition of the battery as described in Chapter 5A.

✔ On batteries which are not of the maintenance-free type, periodically check the electrolyte level in the battery. Remove the plugs from the top of the battery and check that the level is above the electrolyte plates.

✔ If the battery is flat, and you need to jump start your vehicle, see *Roadside Repairs*.

1 The battery is located on the left-hand side of the engine compartment. The exterior of the battery should be inspected periodically for damage such as a cracked case or cover.

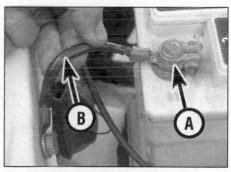

2 Check the tightness of battery clamps (A) to ensure good electrical connections. You should not be able to move them. Also check each cable (B) for cracks and frayed conductors.

HAYNES HINT

Battery corrosion can be kept to a minimum by applying a layer of petroleum jelly to the clamps and terminals after they are reconnected.

3 If corrosion (white, fluffy deposits) is evident, remove the cables from the battery terminals, clean them with a small wire brush, then refit them. Automotive stores sell a tool for cleaning the battery post . . .

4 . . . as well as the battery cable clamps

Electrical systems

✔ Check all external lights and the horn. Refer to the appropriate Sections of Chapter 12 for details if any of the circuits are found to be inoperative.

✔ Visually check all accessible wiring connectors, harnesses and retaining clips for security, and for signs of chafing or damage.

 If you need to check your brake lights and indicators unaided, back up to a wall or garage door and operate the lights. The reflected light should show if they are working properly.

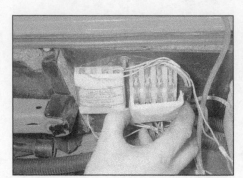

1 If a single light has failed, it is likely that a bulb has blown and will need to be replaced. Refer to Chapter 12 for details.

2 If more than one light has failed, it is likely that either a fuse has blown or there is a fault in the circuit (see Chapter 12). If both stop-lights have failed, it is possible that the switch on the brake pedal has failed (see Chapter 9, Section 18). The main fuses are located in two fuse panels in the engine compartment, mounted on the left-hand side of the scuttle panel. On some models, a supplementary fuse for the engine cooling fans is fitted to a fly-lead from the battery. To replace a blown fuse from either of the panels, pull off the cover . . .

3 . . . and remove the relevant fuse, then fit a new fuse of the correct rating (see Chapter 12). The fuse positions are colour-coded, according to a coloured dot on the fuseholder (not to be confused with the wire colours which may be different). If the fuse blows again, it is important that you find out why. A complete checking procedure is given in Chapter 12.

Tyre pressures (cold) - bar (psi)

Note: *On some models the tyre specifications and pressures are given on a sticker on the driver's side door pillar.*

Standard tyres (Michelin)*	Front	Rear
Petrol, 475 kg payload:		
135 R 13 MX (72 R) reinforced	2.7 (39)	3.5 (52)
Petrol, 600 kg payload:		
145 R 13 MX (78 R) reinforced	2.3 (33)	2.7 (39)
155 R 13 MX (78 R) reinforced	2.3 (33)	2.8 (41)
Petrol/diesel, 765 kg payload:		
155 R 13 XCA	2.7 (39)	4.5 (66)
Diesel, 600 kg payload:		
155 R 13 MX (78 S)	2.3 (33)	2.6 (38)

Inflate the spare wheel to 0.2 bar higher than the rear wheel, then deflate to the correct pressure when fitting.

Authorised fitting		
Petrol, 475 kg payload:		
145 R 13 (M + S)	2.7 (39)	3.5 (52)
Petrol, 600 kg payload:		
145 R 13 (M + S)	2.5 (36)	2.8 (41)
Petrol/diesel, 765 kg payload:		
155 R 13 XCA (M + S)	2.7 (39)	4.5 (66)
Diesel, 600 kg payload:		
155 R 13 (M + S)	2.3 (33)	2.6 (38)

Lubricants and fluids

Engine (petrol)	Multigrade engine oil, viscosity SAE 5W/40, 10W/40 or 15W/50, to API SG/SH *(Duckhams QXR Premium Petrol Engine Oil, or Duckhams Hypergrade Petrol Engine Oil)*
Engine (diesel)	Multigrade engine oil, viscosity SAE 15W/40 to API CD/CE *(Duckhams QXR Premium Diesel Engine Oil, or Duckhams Hypergrade Diesel Engine Oil)*
Cooling system	Ethylene glycol-based antifreeze with corrosion inhibitor *(Duckhams Antifreeze and Summer Coolant)*
Manual transmission	Total transmission BV 75W/80W *(Duckhams Hypoid Gear Oil 75W-80W GL-5)*
Brake hydraulic system	Brake fluid to DOT 4 on most models *(Duckhams Universal Brake & Clutch Fluid)*
Brake vacuum pump (diesel models)	SAE 10W/30 oil *(Duckhams QXR Premium Petrol Engine Oil)*
Power steering	Total ATX fluid *(Duckhams ATF Autotrans III)*
Steering rack	Multi-purpose lithium based grease *(Duckhams LB10)*
Hub/wheel bearings	Multi-purpose lithium based grease *(Duckhams LB10)*
Driveshaft CV joints	Special lubricant supplied in repair kit

Choosing your engine oil

Engines need oil, not only to lubricate moving parts and minimise wear, but also to maximise power output and to improve fuel economy. By introducing a simplified and improved range of engine oils, Duckhams has taken away the confusion and made it easier for you to choose the right oil for your engine.

HOW ENGINE OIL WORKS

• Beating friction

Without oil, the moving surfaces inside your engine will rub together, heat up and melt, quickly causing the engine to seize. Engine oil creates a film which separates these moving parts, preventing wear and heat build-up.

• Cooling hot-spots

Temperatures inside the engine can exceed 1000° C. The engine oil circulates and acts as a coolant, transferring heat from the hot-spots to the sump.

• Cleaning the engine internally

Good quality engine oils clean the inside of your engine, collecting and dispersing combustion deposits and controlling them until they are trapped by the oil filter or flushed out at oil change.

OIL CARE - FOLLOW THE CODE

To handle and dispose of used engine oil safely, always:

- **Avoid skin contact with used engine oil. Repeated or prolonged contact can be harmful.**
- **Dispose of used oil and empty packs in a responsible manner in an authorised disposal site. Call 0800 663366 to find the one nearest to you. Never tip oil down drains or onto the ground.**

OIL CARE
FOLLOW THE CODE
OIL BANK LINE
0800 66 33 66

Chapter 1 Part A:
Routine maintenance and servicing – petrol models

Contents

Degrees of difficulty

Easy, suitable for novice with little experience	**Fairly easy,** suitable for beginner with some experience	**Fairly difficult,** suitable for competent DIY mechanic	**Difficult,** suitable for experienced DIY mechanic	**Very difficult,** suitable for expert DIY or professional

Lubricants and fluids

Refer to "Weekly checks"

Capacities

Engine oil

Excluding filter .... 3.2 litres
Including filter .... 3.5 litres
Difference between MAX and MIN dipstick marks .... 1.4 litres
Cooling system (approximate .... 6.5 litres
Transmission (according to model*) .... 1.8 to 2.0 litres
Fill the transmission until oil runs out of the filler/level hole, then check the level as described in Section 23.
Fuel tank (all models) .... 47 litres
Braking system (approximate .... 0.25 litres
Washer reservoirs
Front .... 1.5 litres
Rear (if fitted) .... 1.5 litres

Engine

Oil filter type:
  954 cc carburettor engines .... Champion type not available
  1124 cc carburettor engines .... Champion F104
  1124 cc fuel-injected engines .... Champion type not available
Auxiliary drivebelt tension .... Approx 5.0 mm deflection midway between pulleys

Cooling system

Antifreeze mixture:
  28% antifreeze .... Protection down to -15°C
  50% antifreeze .... Protection down to -30°C
Note: *Refer to antifreeze manufacturer for latest recommendations.*

Fuel system

Air filter type:
  954 cc carburettor engines .... Champion type not available
  1124 cc carburettor engines .... Champion V402
  1124 cc fuel-injected engines .... Champion type not available
Fuel filter type:
  954 cc carburettor engines .... Champion type not available
  1124 cc carburettor engines .... Champion L101
  1124 cc fuel-injected engines .... Champion type not available
Idle speed:
  Carburettor models .... 750 ± 50rpm
  Fuel-injected models* .... 850 ± 50 rpm (not adjustable - controlled by ECU)
Idle mixture CO content:
  Carburettor models* .... 0.8 to 1.2 %
  Fuel-injected models* .... Less than 1.0 % (not adjustable - controlled by ECU)
See the relevant Part of Chapter 4 for further information.

Ignition system

Ignition timing .... Refer to Chapter 5B
Spark plugs:
  Carburettor engines .... Champion RC9YCC or RC9YC
  Fuel-injected engines .... Champion type not available
Spark plug electrode gap* .... 0.8 mm
Ignition HT lead resistance .... Approximately 600 ohms per 100 mm length
The spark plug gap quoted is that recommended by Champion for their specified plugs listed above. If spark plugs of any other type are to be fitted, refer to their manufacturer's recommendations.

Brakes

Front brake pad friction material minimum thickness .... 2.0 mm
Rear brake shoe friction material minimum thickness .... 1.0 mm

Torque wrench settings

	Nm	lbf ft
Spark plugs	25	18
Transmission filler/level and drain plugs	25	18
Roadwheel bolts:		
3 bolts (pre-launch models before July 1988)	70	52
4 bolts	80	59

The maintenance intervals in this manual are provided with the assumption that you, not the dealer, will be carrying out the work. These are the minimum maintenance intervals recommended by us for vehicles driven daily. If you wish to keep your vehicle in peak condition at all times, you may wish to perform some of these procedures more often. We encourage frequent maintenance, because it enhances the efficiency, performance and resale value of your vehicle.

If the vehicle is driven in dusty areas, used to tow a trailer, or driven frequently at slow speeds (idling in traffic) or on short journeys, more frequent maintenance intervals are recommended.

When the vehicle is new, it should be serviced by a factory-authorised dealer service department, in order to preserve the factory warranty.

Every 250 miles (400 km) or weekly
☐ Refer to "Weekly Checks"

Every 6000 miles / 10 000 km or 6 months, whichever comes first
In addition to the Weekly Checks, carry out the following:
☐ Renew the engine oil and filter (Section 3).
☐ Check all underbonnet components and hoses for fluid leaks (Section 4).
☐ Check the suspension and steering components for condition and security (Section 5).
☐ Check the condition of the driveshaft rubber gaiters (Section 6).
☐ Examine the exhaust system for corrosion and leakage (Section 7).
☐ Check the condition of the front brake pads, and renew if necessary (Section 8).
☐ Check the operation of the handbrake (Section 9).

Every 12 000 miles / 20 000 km or 12 months, whichever comes first
In addition to the 6000 mile service, carry out the following:
☐ Renew the spark plugs (Section 10).
☐ Check the ignition system and ignition timing (Section 11).
☐ Renew the fuel filter - carburettor models (Section 12).
☐ Check the idle speed and mixture adjustment (Section 13).
☐ Check the condition of the emission control system hoses and components (Section 14).
☐ Check the condition of the auxiliary drivebelt, and renew if necessary (Section 15).
☐ Check the clutch adjustment (Section 16).
☐ Lubricate the clutch control mechanism (Section 16).
☐ Check and adjust the valve clearances (Section 17).
☐ Check the condition of the rear brake shoes, and renew if necessary (Section 18).
☐ Carry out a road test (Section 19).

Every 18 000 miles / 30 000 km or 18 months, whichever comes first
In addition to the 6000 mile service, carry out the following:
☐ Renew the air filter (Section 20).
☐ Lubricate all hinges and locks (Section 21).

Every 24 000 miles / 40 000 km or 2 years, whichever comes first
In addition to the 12 000 mile service, carry out the following:
☐ Renew the brake fluid (Section 22).

Every 36 000 miles / 60 000 km or 3 years, whichever comes first
In addition to the 12 000 mile service, carry out the following:
☐ Renew the air filter (Section 20).
☐ Lubricate all hinges and locks (Section 21).
☐ Check the manual transmission oil level, and top-up if necessary (Section 23).

Every 48 000 miles / 80 000 km or 4 years, whichever comes first
In addition to the 24 000 mile service, carry out the following:
☐ Renew the fuel filter - fuel-injected models (Section 24).

Every 72 000 miles / 120 000 km or 6 years, whichever comes first
In addition to the 36 000 mile service, carry out the following:
☐ Renew the timing belt (Section 25).
☐ Renew the manual transmission oil (Section 26).
Note: *Citroën recommend that the timing belt renewal interval is halved to 36 000 miles (60 000 km) on vehicles which are subjected to intensive use, such as many short journeys or stop-start driving. The actual belt renewal interval is therefore very much up to the individual owner. That being said, it is highly recommended to err on the side of safety, and renew the belt at this earlier interval, bearing in mind the drastic consequences resulting from belt failure.*

Every 2 years, regardless of mileage
☐ Renew the coolant (Section 27).
☐ Lubricate all hinges and locks (Section 21).

Underbonnet view of a C15 carburettor petrol model

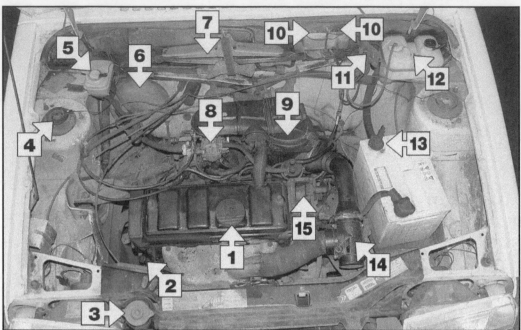

1 Engine oil filler cap
2 Engine oil dipstick
3 Radiator filler cap
4 Suspension strut upper mounting
5 Master cylinder/brake fluid reservoir
6 Brake vacuum servo unit
7 Wheel changing jack
8 Carburettor
9 Air cleaner housing
10 Fuseboxes
11 Washer pump
12 Washer fluid reservoir
13 Battery earth (negative) terminal
14 Air cleaner air temperature control valve
15 Ignition HT coil

Maintenance procedures - petrol models

1 General information

This Chapter is designed to help the home mechanic maintain his/her vehicle for safety, economy, long life and peak performance.

The Chapter contains a master maintenance schedule, followed by Sections dealing specifically with each task in the schedule. Visual checks, adjustments, component renewal and other helpful items are included. Refer to the accompanying illustrations of the engine compartment and the underside of the vehicle for the locations of the various components.

Servicing your vehicle in accordance with the mileage/time maintenance schedule and the following Sections will provide a planned maintenance programme, which should result in a long and reliable service life. This is a comprehensive plan, so maintaining some items but not others at the specified service intervals, will not produce the same results.

As you service your vehicle, you will discover that many of the procedures can - and should - be grouped together, because of the particular procedure being performed, or because of the close proximity of two otherwise-unrelated components to one another. For example, if the vehicle is raised for any reason, the exhaust can be inspected at the same time as the suspension and steering components.

The first step in this maintenance programme is to prepare yourself before the actual work begins. Read through all the

Sections relevant to the work to be carried out, then make a list and gather together all the parts and tools required. If a problem is encountered, seek advice from a parts specialist, or a dealer service department.

2 Intensive maintenance

If, from the time the vehicle is new, the routine maintenance schedule is followed closely, and frequent checks are made of fluid levels and high-wear items, as suggested throughout this manual, the engine will be kept in relatively good running condition, and the need for additional work will be minimised.

It is possible that there will be times when the engine is running poorly due to the lack of regular maintenance. This is even more likely if a used vehicle, which has not received regular and frequent maintenance checks, is purchased. In such cases, additional work may need to be carried out, outside of the regular maintenance intervals.

If engine wear is suspected, a compression test (refer to Chapter 2A) will provide valuable information regarding the overall performance of the main internal components. Such a test can be used as a basis to decide on the extent of the work to be carried out. If, for example, a compression test indicates serious internal engine wear, conventional maintenance as described in this Chapter will not greatly improve the performance of the engine, and may prove a waste of time and

money, unless extensive overhaul work is carried out first.

The following series of operations are those most often required to improve the performance of a generally poor-running engine:

Primary operations

a) Clean, inspect and test the battery (see "Weekly checks").
b) Check all the engine-related fluids (see "Weekly checks").
c) Check the condition and tension of the auxiliary drivebelt (Section 15).
d) Renew the spark plugs - (Section 10).
e) Inspect the distributor cap, rotor arm and HT leads (Section 11).
f) Check the condition of the air filter, and renew if necessary (Section 20).
g) Check the fuel filter (Section 12 or 24, as applicable).
h) Check the condition of all hoses, and check for fluid leaks (Section 4).
i) Check the idle speed and mixture settings (Section 13).

If the above operations do not prove fully effective, carry out the following secondary operations:

Secondary operations

All items listed under "Primary operations", plus the following:

a) Check the charging system (Chapter 5A).
b) Check the ignition system (Chapter 5B).
c) Check the fuel system (Chapter 4A or 4B).
d) Renew the distributor cap and rotor arm (Section 11).
e) Renew the ignition HT leads (Section 11).

Every 6000 miles / 10 000 km or 6 months

3 Engine oil and filter renewal

Note: *A suitable square-section wrench may be required to undo the sump drain plug on some models. These wrenches can be obtained from most motor factors, or from your Citroën dealer.*

1 Frequent oil and filter changes are the most important preventative maintenance procedures which can be undertaken by the DIY owner. As engine oil ages, it becomes diluted and contaminated, which leads to premature engine wear.

2 Before starting this procedure, gather together all the necessary tools and materials. Also make sure that you have plenty of clean rags and newspapers handy, to mop up any spills. Ideally, the engine oil should be warm, as it will drain better, and more built-up sludge will be removed with it.

Caution: Take care not to touch the exhaust or any other hot parts of the engine when working under the vehicle. To avoid any possibility of scalding, and to protect yourself from possible skin irritants and other harmful contaminants in used engine oils, it is advisable to wear gloves when carrying out this work.

3 Access to the underside of the vehicle will be greatly improved if it can be raised on a lift, driven onto ramps, or jacked up and supported on axle stands (see *"Jacking and Vehicle Support"*). Whichever method is chosen, make sure that the vehicle remains level, or if it is at an angle, that the drain plug is at the lowest point.

4 Slacken the drain plug about half a turn. On some models, a square-section wrench may be needed to slacken the plug **(see illustration)**. Position the draining container under the drain plug, then remove the plug completely. Recover the sealing ring from the drain plug.

> **HAYNES HINT**
> *Try to keep the drain plug pressed into the sump while unscrewing it by hand the last couple of turns. As the plug releases from the threads, move it away sharply, so that the stream of oil issuing from the sump runs into the container, not up your sleeve!*

5 Allow some time for the old oil to drain, noting that it may be necessary to reposition the container as the oil flow slows to a trickle.

6 After all the oil has drained, wipe off the drain plug with a clean rag, and fit a new sealing washer. Clean the area around the drain plug opening, and refit the plug. Tighten the plug securely.

7 Move the container into position under the oil filter, which is located on the front of the cylinder block, below the exhaust manifold.

8 Using an oil filter removal tool, slacken the filter initially, then unscrew it by hand the rest of the way **(see illustration)**. Empty the oil in the old filter into the container.

9 Use a clean rag to remove all oil, dirt and sludge from the filter sealing area on the engine.

> **HAYNES HINT**
> *Check the old filter to make sure that the rubber sealing ring hasn't stuck to the engine. If it has, carefully remove it.*

10 Apply a light coating of clean engine oil to the sealing ring on the new filter, then screw it into position on the engine. Tighten the filter firmly by hand only - **do not** use any tools.

11 Remove the old oil and all tools from under the vehicle, then lower the vehicle to the ground (if applicable).

12 Remove the dipstick, then unscrew the oil filler cap from the cylinder head cover. Fill the engine, using the correct grade and type of oil (refer to *"Weekly checks"* for details of topping up). A funnel, or an oil can with a spout may help to reduce spillage. Pour in half the specified quantity of oil first, then wait a few minutes for the oil to fall to the sump. Continue adding oil, a small quantity at a time, until the level is up to the lower mark on the dipstick. Then add a further quantity of oil, according to the Specifications, to bring the level up to the upper mark on the dipstick. Refit the filler cap.

13 Start the engine and run it for a few minutes. Check for leaks around the oil filter seal and the sump drain plug. Note that there may be a delay of a few seconds before the oil pressure warning light goes out when the engine is first started, as the oil circulates through the engine oil galleries and the new oil filter before the pressure builds up.

14 Switch off the engine, and wait a few minutes for the oil to settle in the sump once more. With the new oil circulated and the filter completely full, recheck the level on the dipstick, and add more oil as necessary.

15 Dispose of the used engine oil safely, with reference to *"General repair procedures"*.

OIL CARE
FOLLOW THE CODE
OIL BANK LINE
0800 66 33 66

Note: It is antisocial and illegal to dump oil down the drain. To find the location of your local oil recycling bank, call this number free.

4 Hose and fluid leak check

1 Visually inspect the engine joint faces, gaskets and seals for any signs of water or oil leaks. Pay particular attention to the areas around the camshaft cover, cylinder head, oil filter and sump joint faces. Bear in mind that, over a period of time, some very slight seepage from these areas is to be expected - what you are really looking for is any indication of a serious leak. Should a leak be found, renew the offending gasket or oil seal by referring to the appropriate Chapters in this manual.

2 Also check the security and condition of all the engine-related pipes and hoses. Ensure that all cable ties or securing clips are in place and in good condition. Clips which are broken or missing can lead to chafing of the hoses, pipes or wiring, which could cause more serious problems in the future.

3 Carefully check the radiator hoses and heater hoses along their entire length. Renew any hose which is cracked, swollen or deteriorated. Cracks will show up better if the hose is squeezed. Pay close attention to the hose clips that secure the hoses to the cooling system components. Hose clips can pinch and puncture hoses, resulting in cooling system leaks. If the original Citroën crimped-type hose clips are used, it may be a good idea to replace them with standard worm-drive clips.

4 Inspect all the cooling system components

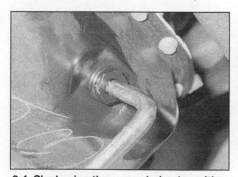

3.4 Slackening the sump drain plug with a square-section wrench

3.8 Using an oil filter removal tool (chain or strap wrench) to slacken the oil filter

(hoses, joint faces etc.) for leaks. A leak in the cooling system will usually show up as white or rust-coloured deposits on the area adjoining the leak. Where any problems of this nature are found on system components, renew the component or gasket with reference to Chapter 3.

5 With the vehicle raised, inspect the fuel tank and filler neck for punctures, cracks and other damage. The connection between the filler neck and tank is especially critical. Sometimes, a rubber filler neck or connecting hose will leak due to loose retaining clamps or deteriorated rubber.

6 Carefully check all rubber hoses and metal fuel lines leading away from the fuel tank. Check for loose connections, deteriorated hoses, crimped lines, and other damage. Pay particular attention to the vent pipes and hoses, which often loop up around the filler neck and can become blocked or crimped. Follow the lines to the front of the vehicle, carefully inspecting them all the way. Renew damaged sections as necessary. Similarly, while the vehicle is raised, take the opportunity to inspect all underbody brake fluid pipes and hoses.

7 From within the engine compartment, check the security of all fuel hose attachments and pipe unions, and inspect the fuel hoses and vacuum hoses for kinks, chafing and deterioration.

5 Suspension and steering check

Front suspension and steering

1 Apply the handbrake, jack up the front of the vehicle, and securely support it on axle stands (see "*Jacking and Vehicle Support*").

2 Visually inspect the balljoint dust covers and the steering rack-and-pinion gaiters for splits, chafing or deterioration **(see illustration)**. Any wear of these components will cause loss of lubricant, together with dirt and water entry, resulting in rapid deterioration of the balljoints or steering gear.

3 Grasp the roadwheel at the 12 o'clock and 6 o'clock positions, and try to rock it. Very slight free play may be felt, but if the movement is appreciable, further investigation is necessary to determine the source. Continue rocking the wheel while an assistant depresses the footbrake. If the movement is now eliminated or significantly reduced, it is likely that the hub bearings are at fault. If the free play is still evident with the footbrake depressed, then there is wear in the suspension joints or mountings.

4 Now grasp the wheel at the 9 o'clock and 3 o'clock positions, and try to rock it as before. Any movement felt now may again be caused by wear in the hub bearings or the steering track-rod balljoints. If the outer balljoint is worn, the visual movement will be obvious. If

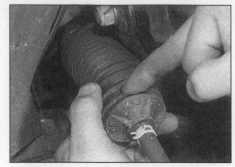

5.2 Checking a steering gear gaiter

the inner joint is suspect, it can be felt by placing a hand over the rack-and-pinion rubber gaiter and gripping the track rod. If the wheel is now rocked, movement will be felt at the inner joint if wear has taken place.

5 Using a large screwdriver or flat bar, check for wear in the suspension mounting bushes by levering between the relevant suspension component and its attachment point. Some movement is to be expected, as the mountings are made of rubber, but excessive wear should be obvious. Also check the condition of any visible rubber bushes, looking for splits, cracks or contamination of the rubber.

6 With the vehicle standing on its wheels, have an assistant turn the steering wheel back and forth about an eighth of a turn each way. There should be very little, if any, lost movement between the steering wheel and roadwheels. If this is not the case, closely observe the joints and mountings previously described, but in addition, check the steering column universal joints for wear, and the rack-and-pinion steering gear itself.

Rear suspension check

7 Chock the front wheels, then jack up the rear of the vehicle and support securely on axle stands (see "*Jacking and Vehicle Support*").

8 Working as described previously for the front suspension, check the rear hub bearings, and the suspension bushes and mountings for wear.

Shock absorber check

9 Check for any signs of fluid leakage around the front and rear shock absorbers. On the

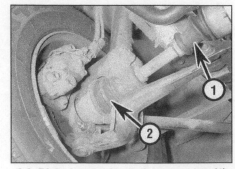

6.2 Right-hand driveshaft inner gaiter (1) and outer gaiter (2)

front shock absorber, check for fluid on the rubber gaiter around the piston rod. Should any fluid be noticed, the shock absorber is defective internally, and should be renewed. **Note:** *Shock absorbers should always be renewed in pairs on the same axle.*

10 The efficiency of the shock absorber may be checked by bouncing the vehicle at each corner. Generally speaking, the body will return to its normal position and stop after being depressed. If it rises and returns on a rebound, the shock absorber is probably suspect. Also examine the shock absorber mountings for signs of wear, and check that the rear shock absorber ball joints fit securely in their sockets.

6 Driveshaft and gaiter check

1 Apply the handbrake, jack up the front of the vehicle, and support it securely on axle stands (see "*Jacking and Vehicle Support*").

2 Turn the steering onto full lock, then slowly rotate the roadwheel. Inspect the condition of the inner and outer constant velocity (CV) joint rubber gaiters, squeezing the gaiters to open out the folds. Check for signs of cracking, splits or deterioration of the rubber, which may allow the grease to escape, and lead to water and grit entry into the joint. Also check the security and condition of the retaining clips **(see illustration)**. If any damage or deterioration is found, the gaiters should be renewed as described in Chapter 8.

3 At the same time, check the general condition of the CV joints themselves by first holding the driveshaft and attempting to rotate the wheel. Repeat this check by holding the inner joint and attempting to rotate the driveshaft. Any appreciable movement indicates wear in the joints, wear in the driveshaft splines, or a loose driveshaft retaining nut.

7 Exhaust system check

1 Chock the front wheels, then jack up the rear of the vehicle and support it securely on axle stands (see "*Jacking and Vehicle Support*").

2 With the engine cold (wait at least an hour after switching off the engine), check the complete exhaust system from the engine to the end of the tailpipe.

3 Check the exhaust pipes and connections for evidence of leaks, severe corrosion and damage. Make sure that all brackets and mountings are in good condition, and that all relevant nuts and bolts are tight. Leakage at any of the joints or in other parts of the system will usually show up as a black, sooty stain in the vicinity of the leak.

4 Proprietary pastes and bandages are available for repairing leaks. They work well in the short term, but eventually the exhaust section will need to be renewed as described in Chapter 4A or 4B.

5 Rattles and vibrations can often be traced to the exhaust system. Tap the silencer units with a soft mallet and listen for noises caused by corroded or displaced baffle material. *Caution: Do not strike the catalytic converter (where fitted), as this may damage the ceramic block inside.*

6 Carefully rock the pipes and silencers from side to side on their mountings. If the components are able to come into contact with the body or suspension parts, look for broken or worn rubber mountings.

7 Extra clearance can be gained by slackening the clamps between adjacent sections of the exhaust pipe to loosen the joints where possible, and twisting the pipes as necessary to provide the additional clearance.

8 Front brake pad check

1 Firmly apply the handbrake, then jack up the front of the vehicle and support it securely on axle stands (see "*Jacking and Vehicle Support*"). Remove the front roadwheels.

2 For a quick check, the thickness of friction material remaining on each brake pad can be measured from the front of the caliper **(see illustration)**. If any pad's friction material is worn to the specified thickness or less, *all four pads must be renewed as a set.*

3 For a comprehensive check, the brake pads should be removed and cleaned. The operation of the caliper can then also be checked, and the condition of the brake disc itself can be fully examined on both sides. Refer to Chapter 9 for further information.

9 Handbrake check and adjustment

1 Chock the front wheels, then jack up the rear of the vehicle, and support securely on axle stands (see "*Jacking and Vehicle Support*").

2 Apply the footbrake firmly several times to establish correct shoe-to-drum clearance, then apply and release the handbrake several times to ensure that the self-adjust mechanism has compensated fully for any wear in the linings.

3 Fully release the handbrake, and check that the rear wheels rotate freely, without binding. If not, check that all cables are routed correctly, and check that the cable components and levers move freely.

> **HAYNES HiNT** *If the handbrake mechanism fails to operate, or appears to be seized on one side of the vehicle only, remove the relevant brake drum (see Chapter 9) and check the handbrake lever pivot on the trailing brake shoe - it is possible for the lever to seize due to corrosion. If necessary, remove the lever, and clean the contact faces of the lever, brake shoe, and pivot.*

4 If all components are free to move, but the wheels still bind when rotated, adjustment is required as follows.

5 Pull up the boot from the handbrake to gain access to the equaliser and adjusting nut. The equaliser is a short metal bar at the rear of the lever, which lies transversely across the vehicle and pulls on both handbrake cables simultaneously.

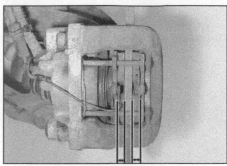

8.2 Brake pad friction material thickness (arrowed) can be checked with the pads in place

6 Apply the handbrake and set the lever on the third notch on the ratchet mechanism. With the lever in this position, slacken the locknut and turn the adjusting nut until only a slight drag can be felt when the rear wheels are turned.

7 Make sure there is equal tension on the left and right-hand cables. If the tensions are unequal, the equaliser bar will be inclined at an angle. If necessary, adjust the cables individually until the equaliser becomes transverse. This can only be done if the cables are fitted with in-line adjusters (normally underneath the vehicle). If the cables cannot be equalised they will need to be renewed (see Chapter 9).

8 Fully release the handbrake lever and check that the wheels rotate freely. Then apply the handbrake lever one notch at a time until both rear wheels are locked. The adjustment is correct if the wheels become locked at 5 notches. Turn the central adjustment nut on the equaliser bar to achieve the correct adjustment, then tighten the locknut.

9 Refit the boot to the handbrake and lower the vehicle to the ground.

Every 12 000 miles / 20 000 km or 12 months

10 Spark plug renewal

1 The correct functioning of the spark plugs is vital for the correct running and efficiency of the engine. It is essential that the plugs fitted are appropriate for the engine (a suitable type is specified at the beginning of this Chapter). If this type is used and the engine is in good condition, the spark plugs should not need attention between scheduled replacement intervals. Spark plug cleaning is rarely necessary, and should not be attempted unless specialised equipment is available, as damage can easily be caused to the firing ends.

2 If the marks on the original-equipment spark plug (HT) leads cannot be seen, mark the leads "1" to "4", to correspond to the cylinder the lead serves (No 1 cylinder is at the transmission end of the engine). Pull the leads from the plugs by gripping the end fitting, not the lead, otherwise the lead connection may be fractured.

3 It is advisable to remove the dirt from the spark plug recesses using a clean brush, vacuum cleaner or compressed air before removing the plugs, to prevent dirt dropping into the cylinders.

4 Unscrew the plugs using a spark plug spanner, suitable box spanner or a deep socket and extension bar **(see illustration)**. Keep the socket aligned with the spark plug - if it is forcibly moved to one side, the ceramic insulator may be broken off. As each plug is removed, examine it as follows.

5 Examination of the spark plugs will give a good indication of the condition of the engine.

If the insulator nose of the spark plug is clean and white, with no deposits, this indicates a weak mixture or too hot a plug (a hot plug transfers heat away from the electrode slowly, a cold plug transfers heat away quickly).

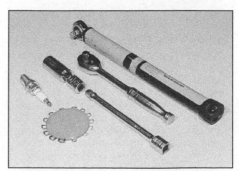

10.4 Tools required for spark plug removal, gap adjustment and refitting

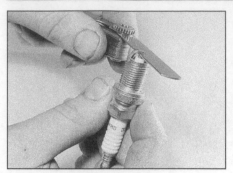

10.9a Measure the spark plug gap with a feeler blade . . .

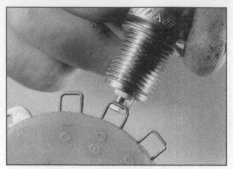

10.9b . . . or with a wire gauge

10.9c . . . and if necessary adjust the gap by bending the outer electrode

6 If the tip and insulator nose are covered with hard black-looking deposits, then this indicates that the mixture is too rich. Should the plug be black and oily, then it is likely that the engine is fairly worn, as well as the mixture being too rich.

7 If the insulator nose is covered with light tan to greyish-brown deposits, then the mixture is correct and it is likely that the engine is in good condition.

8 The spark plug electrode gap is of considerable importance as, if it is too large or too small, the size of the spark and its efficiency will be seriously impaired. The gap should be set to the value given in the Specifications at the beginning of this Chapter.

9 To set the gap, measure it with a feeler blade or wire gauge and then bend open, or close, the outer plug electrode until the correct gap is achieved **(see illustrations)**. The centre electrode should never be bent, as this will crack the insulator and cause plug failure, if nothing worse. If using feeler blades, the gap is correct when the appropriate-size blade is a firm sliding fit.

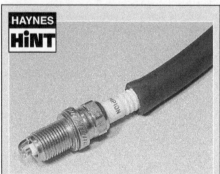

HAYNES HINT

It is very often difficult to insert spark plugs into their holes without cross-threading them. To avoid this possibility, fit a short length of 5/16 inch internal diameter rubber hose over the end of the spark plug. The flexible hose acts as a universal joint to help align the plug with the plug hole. Should the plug begin to cross-thread, the hose will slip on the spark plug, preventing thread damage to the aluminium cylinder head.

10 Special spark plug electrode gap adjusting tools are available from most motor accessory shops, or from some spark plug manufacturers.

11 Before fitting the spark plugs, check that the threaded connector sleeves are tight, and that the plug exterior surfaces and threads are clean.

12 Insert the plugs and carefully screw them in, making sure they are not cross-threaded. Use a rubber tube, if necessary, to prevent cross-threading **(see Haynes Hint)**. Then remove the hose, and tighten the plug to the specified torque using the spark plug socket and a torque wrench. Refit the remaining spark plugs in the same manner.

13 Connect the HT leads in their correct order, and refit any components removed for access.

11 Ignition system check

⚠️ *Warning: Voltages produced by an electronic ignition system are considerably higher than those produced by conventional ignition systems. Extreme care must be taken when working on the system with the ignition switched on. Persons with surgically-implanted cardiac pacemaker devices should keep well clear of the ignition circuits, components and test equipment.*

General component check

1 The spark plug (HT) leads should be checked whenever new spark plugs are fitted.

2 Ensure that the leads are numbered before removing them, to avoid confusion when refitting (see Section 10).

 HAYNES HINT *Pull the leads from the plugs by gripping the end fitting, not the lead, otherwise the lead connection may be fractured.*

3 Check inside the end fitting for signs of corrosion, which will look like a white crusty powder. Push the end fitting back onto the spark plug, ensuring that it is a tight fit on the plug. If not, remove the lead again and use

pliers to carefully crimp the metal connector inside the end fitting until it fits securely on the end of the spark plug.

4 Using a clean rag, wipe the entire length of the lead to remove any built-up dirt and grease. Once the lead is clean, check for burns, cracks and other damage. Do not bend the lead excessively, nor pull the lead length-wise - the conductor inside might break.

5 Disconnect the other end of the lead from the distributor cap. Again, pull only on the end fitting. Check for corrosion and a tight fit, in the same manner as the spark plug end. If an ohmmeter is available, check the resistance of the lead by connecting the meter between the spark plug end of the lead and the segment inside the distributor cap. If the resistance is not similar to that given in the Specifications, the lead should be renewed. Refit the lead securely on completion.

6 Check the remaining leads one at a time, in the same way.

7 If new spark plug (HT) leads are required, purchase a set for your specific vehicle and engine.

8 Unscrew its retaining screws and remove the distributor cap. Wipe it clean, and carefully inspect it inside and out for signs of cracks, black carbon tracks (tracking) and worn, burned or loose contacts. Check that the cap's carbon brush is unworn, free to move against spring pressure, and making good contact with the rotor arm. Also inspect the cap seal for signs of wear or damage, and renew it if necessary. Remove the rotor arm from the distributor shaft, and inspect the rotor arm **(see illustration)**. It is common practice to renew

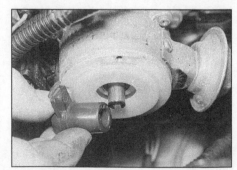

11.8 The rotor arm is a push fit on the distributor shaft

the cap and rotor arm whenever new spark plug (HT) leads are fitted. When fitting a new cap, remove the leads from the old cap one at a time, and fit them to the new cap in the exact same location. Do not simultaneously remove all the leads from the old cap, or the engine will misfire if they are refitted in the wrong order. When refitting, ensure that the rotor arm is securely pressed onto the shaft, and tighten the cap retaining screws securely.

9 Even with the ignition system in first-class condition, some engines may still occasionally experience poor starting, attributable to damp ignition components. To disperse moisture, a water-dispersant aerosol can be very effective.

Ignition timing - check and adjustment

10 Check the ignition timing as described in Chapter 5B.

12 Fuel filter renewal - carburettor models

Warning: Before carrying out the following operation, refer to the precautions given in "Safety first!" at the beginning of this manual, and follow them implicitly. Petrol is a highly-dangerous and volatile liquid, and the precautions necessary when handling it cannot be overstressed.

1 The fuel filter is situated in the fuel feed line between the fuel tank at the rear of the vehicle and the fuel pump mounted behind the cylinder head.

2 To remove the filter from the fuel line, first note its direction of fitting, then separate the filter from the hoses on each side. Be prepared for fuel leakage as the hoses are detached, and plug the hose ends to prevent a continuous loss of fuel if the new filter is not being fitted immediately.

3 When fitting the new filter, ensure that it is correctly orientated as noted during removal, and if necessary renew the hose clips. On completion, restart the engine and check for any signs of leakage from the filter hose connections.

13 Idle speed and mixture check and adjustment

1 Before checking the idle speed and mixture setting, always check the following first:
  a) Check the ignition timing (Chapter 5B).
  b) Check that the spark plugs are in good condition and correctly gapped (Section 10).
  c) Check that the accelerator cable and, on carburettor models, the choke cable, is correctly adjusted (see relevant Part of Chapter 4).

d) Check that the crankcase breather hoses are secure, with no leaks or kinks (Section 14).
  e) Check that the air cleaner filter element is clean (Section 20).
  f) Check that the exhaust system is in good condition (see relevant Part of Chapter 4).
  g) If the engine is running very roughly, check the compression pressures and valve clearances as described in Chapter 2A.
  h) On fuel-injection models, check that the fuel injection/ignition system warning light is not illuminated. If it comes on, take the vehicle to your Citroën dealer for a diagnostic check.

2 Take the vehicle on a journey of sufficient length to warm it up to normal operating temperature. Proceed as described under the relevant sub-heading. **Note:** *Adjustment should be completed within two minutes of return, without stopping the engine. If this cannot be achieved, or if the radiator electric cooling fan operates, first wait for the cooling fan to stop. Clear any excess fuel from the inlet manifold by racing the engine two or three times to between 2000 and 3000 rpm, then allow it to idle again.*

Carburettor models

Note: *The instructions in this sub-section apply to the Solex 32 PBISA carburettor fitted to 1124 cc engines. No specific information is available about the Solex 32 IBSN carburettor fitted to 954 cc engines, although it is thought to be similar.*

3 Ensure that all electrical loads are switched off and the choke lever is pushed fully in. Connect a tachometer (rev counter), to the engine, according to the manufacturer's instructions. The usual type of tachometer, which works from ignition system pulses, can be used. Alternatively, a specialised tachometer, if available, can be connected to the TDC diagnostic socket on the HT coil mounting bracket **(see illustration)**. Allow the engine to idle and measure the idle speed, and compare it with that specified.

4 The idle speed adjusting screw is on the throttle linkage on the carburettor. Using a suitable flat-bladed screwdriver, turn the idle

13.3 TDC diagnostic socket (arrowed) on the HT coil mounting bracket

screw in or out as necessary to obtain the specified speed **(see illustration)**.

5 The idle mixture (exhaust gas CO level) is set at the factory, and should require no further adjustment. If, due to a change in engine characteristics (carbon build-up, bore wear etc) or after a major carburettor overhaul, the mixture setting is lost, it can be reset. Note, however, that an exhaust gas analyser (CO meter) will be required to check the mixture, in order to set it with the necessary standard of accuracy; if this is not available, the vehicle must be taken to a Citroën dealer for the work to be carried out.

6 If an exhaust gas analyser is available, follow its manufacturer's instructions to check the exhaust gas CO level. If adjustment is required, it is made by turning the mixture adjustment screw. The screw may be covered with a tamperproof plug to prevent unnecessary adjustment. To gain access to the screw, use a sharp instrument to hook out the plug.

7 Using a suitable flat-bladed screwdriver, turn the mixture adjustment screw (in very small increments) until the CO level is correct. Turning the screw in (clockwise) weakens the mixture and reduces the CO level, turning it out will enrich the mixture and increase the CO level **(see illustration)**.

8 When adjustments are complete, disconnect any test equipment used, and fit a new tamperproof plug to the mixture adjustment screw. Recheck the idle speed and, if necessary, readjust it.

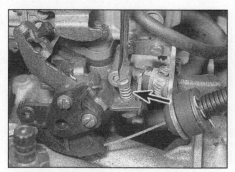

13.4 Idle speed adjustment screw (arrowed) - Solex 32 PBISA carburettor (1124 cc engines)

13.7 Mixture adjustment screw (arrowed) - Solex 32 PBISA carburettor (1124 cc engines)

Fuel-injection models

9 The idle speed and mixture are under full control of the Bosch Monopoint A2.2 fuel injection system ECU (see Chapter 4B). Experienced home mechanics with a considerable amount of skill and equipment (including a good-quality tachometer and a good-quality, carefully-calibrated exhaust gas analyser) may be able to *check* the idle speed and exhaust CO level, but no adjustments are possible. If the values are not according to the Specifications, there must be a fault in the fuel injection system, and the vehicle should be taken to a Citroën dealer for testing using special diagnostic equipment.

14 Emission control systems check

1 Details of the emission control system components are given in Chapter 4D.
2 Checking consists simply of a visual check for obvious signs of damaged or leaking hoses and joints.
3 Detailed checking and testing of the evaporative and/or exhaust emission systems (as applicable) should be entrusted to a Citroën dealer.

15 Auxiliary drivebelt check and renewal

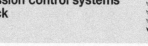

1 On all petrol models, only one auxiliary drivebelt is fitted, to drive the alternator.

Checking the auxiliary drivebelt condition

2 Apply the handbrake, then jack up the front of the vehicle and support it on axle stands (see "*Jacking and Vehicle Support*"). Remove the right-hand front roadwheel. If necessary, free the coolant hose from any retaining clips to improve access to the crankshaft sprocket bolt. Remove the belt guard, if fitted.
3 Using a suitable socket and extension bar fitted to the crankshaft sprocket bolt, rotate the crankshaft so that the entire length of the drivebelt can be examined. Examine the

drivebelt for cracks, splitting, fraying, or other damage. Check also for signs of glazing (shiny patches) and for separation of the belt plies. Renew the belt if worn or damaged.
4 If the condition of the belt is satisfactory, check the drivebelt tension as described below under the relevant sub-heading.

Removal

5 If not already done, carry out the operations described in paragraph 2.
6 Disconnect the battery negative lead.
7 Slacken both the alternator upper and lower mounting bolts, and the bolt securing the adjuster strap to the mounting bracket.
8 Back off the adjuster bolt to relieve the tension in the drivebelt, then slip the drivebelt from the pulleys (see illustrations).

Refitting

9 Fit the belt around the pulleys, ensuring that the belt is of the correct type if it is being renewed, and take up the slack in the belt by tightening the adjuster bolt.
10 Tension the drivebelt as described in the following paragraphs.

Tensioning

11 If not already done, carry out the operations described in paragraph 2.
12 Correct tensioning of the drivebelt will ensure that it has a long life. Beware, however, of overtightening, as this can cause wear in the alternator bearings.
13 The belt should be tensioned so that the free movement under firm thumb pressure, at the mid-point between the pulleys, is according to the Specifications.
14 To adjust, with the upper mounting bolt just holding the alternator firm, and the lower mounting bolt loosened, turn the adjuster bolt until the correct tension is achieved. Rotate the crankshaft through two complete turns, then recheck the tension. When the tension is correct, securely tighten both the alternator mounting bolts and, where necessary, the bolt securing the adjuster strap to its mounting bracket.
15 Reconnect the battery negative lead.
16 Secure the coolant hose to any clips that have been disconnected. Refit the roadwheel and lower the vehicle to the ground.

15.8a Loosen the alternator mounting bolts, then slacken the adjuster bolt (arrowed) . . .

15.8b . . . and slip the drivebelt off its pulleys

16 Clutch adjustment check and control mechanism lubrication

1 Check that the clutch pedal moves smoothly and easily through its full travel, and that the clutch itself functions correctly, with no trace of slip or drag.
2 Adjust the clutch cable as described in Chapter 6.
3 If excessive effort is required to operate the clutch, check first that the cable is correctly routed and undamaged, then remove the pedal and check that its pivot is properly greased. Also grease the cable ends where they connect with the operating levers. Refer to Chapter 6 for further information.

17 Valve clearance check and adjustment

1 The valve clearances must be correctly adjusted, as they vitally affect the performance of the engine. If the clearances are too big, the engine will be noisy (characteristic rattling or tapping noises) and engine efficiency will be reduced, as the valves open too late and close too early. A more serious problem arises if the clearances are too small. In this case, the valves may not close fully when the engine is hot, resulting in serious damage to the engine, for example burnt valve seats or cylinder head warping and cracking.
2 For details of how to remove the camshaft cover and check and adjust the valve clearances, see Chapter 2A.

18 Rear brake shoe and drum check

1 Remove the rear brake drums, and check the brake shoes for signs of wear or contamination. At the same time, also inspect the wheel cylinders for signs of leakage, and the brake drum for signs of wear. Refer to the relevant Sections of Chapter 9 for further information.

19 Road test

Instruments and electrical equipment

1 Check the operation of all instruments and electrical equipment.
2 Make sure that all instruments read correctly, and switch on all electrical equipment in turn to check that it functions properly.

Steering and suspension

3 Check for any abnormalities in the steering, suspension, handling or road "feel".
4 Drive the vehicle, and check that there are no unusual vibrations or noises.
5 Check that the steering feels positive, with no excessive "sloppiness", or roughness, and check for any suspension noises when cornering, or when driving over bumps.

Drivetrain

6 Check the performance of the engine, clutch, transmission and driveshafts.
7 Listen for any unusual noises from the engine, clutch and transmission.
8 Make sure that the engine runs smoothly when idling, and that there is no hesitation when accelerating.
9 Check that the clutch action is smooth and progressive, that the drive is taken up smoothly, and that the pedal travel is correct.

Also listen for any noises when the clutch pedal is depressed.
10 Check that all gears can be engaged smoothly, without noise, and that the gear lever action is not abnormally vague or "notchy".
11 Listen for a metallic clicking sound from the front of the vehicle, as the vehicle is driven slowly in a circle with the steering on full lock. Carry out this check in both directions. If a clicking noise is heard, this indicates wear in a driveshaft joint, in which case the joint will have to be renewed. This process usually requires renewal of the complete driveshaft (see Chapter 8).

Check the operation and performance of the braking system

12 Make sure that the vehicle does not pull to one side when braking, and that the wheels

do not lock prematurely when braking hard.
13 Check that there is no vibration through the steering when braking.
14 Check that the handbrake operates correctly, without excessive movement of the lever, and that it holds the vehicle stationary on a slope.
15 Test the operation of the brake servo unit as follows. With the engine off, depress the footbrake four or five times to exhaust the vacuum. Start the engine, holding the brake pedal depressed. As the engine starts, there should be a noticeable "give" in the brake pedal as vacuum builds up. Allow the engine to run for at least two minutes, and then switch it off. If the brake pedal is now depressed again, it should be possible to detect a hiss from the servo as the pedal is depressed. After about four or five applications, no further hissing should be heard.

Every 18 000 miles / 30 000 km or 18 months

20 Air filter renewal

1 Slacken the retaining clips (where fitted) and disconnect the vacuum and breather

20.1 Disconnect the vacuum and breather hoses (arrowed) from the front of the air cleaner duct

hoses from the front of the duct connecting the air cleaner housing to the carburettor or throttle body **(see illustration)**. Where crimped-type hose clips or ties are fitted, cut and discard them; replace them with standard worm-drive hose clips or new cable ties on refitting.
2 Slacken the retaining clips and remove the duct **(see illustration)**. Recover the rubber sealing ring(s) from the top of the carburettor/throttle body and/or the air cleaner housing (as applicable).
3 The air filter element is attached to the air cleaner housing cover. Release the retaining clips and remove the cover and filter element **(see illustrations)**.
4 Fit the new cover and element, and secure it in position with the retaining clips.
5 Refit the air cleaner duct assembly. Ensure that the duct is correctly seated on its sealing

rings, and securely tighten its retaining clips.
6 Reconnect the vacuum and breather hoses to the duct, and secure them in position with the retaining clips (where fitted).

21 Hinge and lock lubrication

1 Work around the vehicle, and lubricate the hinges of the bonnet and doors with a light machine oil.
2 Lightly lubricate the bonnet release mechanism and exposed section of the inner cable with a smear of grease.
3 Check carefully the security and operation of all hinges, latches and locks, adjusting them where required. Check the operation of the central locking system (if fitted).

20.2 Slacken the retaining clips (arrowed) and remove the duct

20.3a Release the retaining clips . . .

20.3b . . . and remove the cover and air filter element

Every 24 000 miles / 40 000 km or 2 years

22 Brake fluid renewal

⚠️ **Warning: Brake hydraulic fluid can harm your eyes and damage painted surfaces, so use extreme caution when handling and pouring it. Do not use fluid that has been standing open for some time, as it absorbs moisture from the air. Excess moisture can cause a dangerous loss of braking effectiveness.**

1 The procedure is similar to that for bleeding the hydraulic system as described in Chapter 9, except that allowance should be made for all the old fluid to be expelled when bleeding a section of the circuit.

2 The brake fluid reservoir should first be emptied of old fluid by siphoning, using a clean poultry baster, syringe, or similar, then refilled with fresh fluid. Do not operate the brakes while the reservoir is being emptied, or air will be drawn into the system.

3 Working as described in Chapter 9, open the first bleed screw in the sequence, and pump the brake pedal gently until nearly all the fluid has been emptied from the reservoir. Top-up to the "MAXI" level with new fluid, and continue pumping until only new fluid can be seen emerging from the bleed screw. Tighten the screw, and top the reservoir level up to the "MAXI" level line.

4 Work through all the remaining bleed screws in the sequence until new fluid can be seen at all of them. Be careful to keep the master cylinder reservoir topped-up to above the "DANGER" level at all times, or air may enter the system and greatly increase the length of the task.

HAYNES HiNT *Old hydraulic fluid is invariably much darker in colour than the new, making it easy to distinguish the two.*

5 When the operation is complete, check that all bleed screws are securely tightened, and that their dust caps are refitted. Wash off all traces of spilt fluid, and recheck the master cylinder reservoir fluid level.

6 Check the operation of the brakes before taking the vehicle on the road.

Every 36 000 miles / 60 000 km or 3 years

23 Manual transmission oil level check

Note: *A suitable square-section wrench may be required to undo the transmission filler/level plug on some models. These wrenches can be obtained from most motor factors, or from your Citroën dealer.*

Note: *A new sealing washer is required for the filler/level plug.*

1 Park the vehicle on a level surface, and wait at least 5 minutes after the engine has been switched off.

HAYNES HiNT *If the oil is checked immediately after driving the vehicle, some of the oil will remain distributed around the transmission components, giving an inaccurate level reading.*

2 To access the filler/level plug, firmly apply the handbrake, jack up the front left corner of the vehicle using a suitable trolley jack, and remove the roadwheel (see "Jacking and Vehicle Support"). Use jacks and axle stands to keep the vehicle perfectly level, to ensure accuracy, when refilling and checking the oil level. If necessary, support the vehicle at all four jacking points.

HAYNES HiNT *This job is most easily performed while working in a pit, so that the vehicle remains level without having to jack it up.*

3 Wipe clean the area around the filler/level plug, which is situated on the left-hand end of the transmission **(see illustration)**. Unscrew the plug and clean it. Discard the sealing washer.

4 The oil level should reach the lower edge of the filler/level hole. A certain amount of oil will have gathered behind the filler/level plug, and will trickle out when it is removed. This does not necessarily indicate that the level is correct.

5 Filling the transmission with oil is an extremely awkward operation; above all, allow plenty of time for the oil level to settle properly before checking it. If a large amount had to be added to the transmission, and a large amount flows out on checking the level, refit the filler/level plug and take the vehicle on a short journey so that the new oil is distributed fully around the transmission components, then re-check the level when it has settled again **(see Haynes Hint)**.

6 If the transmission has been overfilled so that oil flows out as soon as the filler/level plug is removed, check that the vehicle is completely level (front-to-rear and side-to-side), and allow the surplus to drain off into a suitable container.

7 When the level is correct, fit a new sealing washer to the filler/level plug. Refit the plug and tighten it to the specified torque. Wash off any spilt oil.

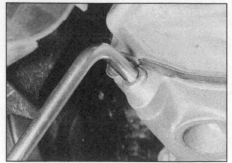

23.3 Using a square-section wrench to unscrew the transmission oil filler/level plug

HAYNES HiNT *To ensure that a true level is established, wait until the initial trickle has stopped, then add oil as necessary until a trickle of new oil can be seen emerging. The level will be correct when the flow ceases. Use only good-quality oil of the specified type.*

Every 48 000 miles / 80 000 km or 4 years

24 Fuel filter renewal - fuel-injection models

24.3 Note the direction of the arrow on the fuel filter body (arrowed) - fuel injection model

⚠ **Warning: Before carrying out the following operation, refer to the precautions given in "Safety first!" at the beginning of this manual, and follow them implicitly. Petrol is a highly-dangerous and volatile liquid, and the precautions necessary when handling it cannot be overstressed.**

1 The fuel filter is located underneath the rear of the vehicle. To gain access to the filter, chock the front wheels, then jack up the rear of the vehicle and support it on axle stands (see "Jacking and Vehicle Support").

2 Depressurise the fuel system, with reference to Chapter 4B.

3 Release the filter from its retaining clip on the underbody. Note the direction of the fuel flow arrow marked on the filter body **(see illustration)**. Release the retaining clips and disconnect the fuel hoses from the filter. Where the original Citroën crimped-type clips are still fitted, cut and discard them; replace them with standard worm-drive hose clips on installation.

4 Remove the filter from the vehicle. Dispose safely of the old filter; it will be highly inflammable, and may explode if thrown on a fire.

5 Refit the new filter, ensuring that the arrow on the filter body is pointing in the direction of the fuel flow, as noted when removing the old filter. The flow direction can otherwise be determined by tracing the fuel hoses back along their length. Connect the fuel hoses to the filter, and tighten their retaining clips. Secure the filter to the vehicle underbody using the retaining clip.

6 Start the engine, check the filter hose connections for leaks, then lower the vehicle to the ground.

Every 72 000 miles / 120 000 km or 6 years

25 Timing belt renewal

Refer to Chapter 2A.

26 Manual transmission oil renewal

Refer to Chapter 7.

Every 2 years, regardless of mileage

27 Coolant renewal

Cooling system draining

⚠ **Warning: Wait until the engine is cold before starting this procedure. Do not allow antifreeze to come into contact with your skin, or with the painted surfaces of the vehicle. Rinse off spills immediately with plenty of water. Never leave antifreeze lying around in an open container, or in a puddle in the driveway or on the garage floor. Children and pets are attracted by its sweet smell, but antifreeze can be fatal if ingested.**

HAYNES HiNT *If the cooling system is being drained for purposes other than renewal, it is not always necessary to drain the complete system. For example, if a component is to be disconnected in the upper part of the circuit, just drain the coolant until the level is below the component.*

1 With the engine completely cold, remove the expansion tank filler cap. Turn the cap anti-clockwise until it reaches the first stop. Wait until any pressure remaining in the system is released, then push the cap down, turn it anti-clockwise to the second stop, and lift it off.

2 Position a suitable container beneath the lower left-hand side of the radiator. Loosen the radiator drain plug and allow the coolant to drain into the container. **Note:** *If there is no drain plug fitted, disconnect the radiator bottom hose.*

3 To assist draining, open the cooling system

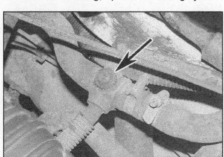

27.3 Bleed screw (arrowed) in the heater hose connector near the engine compartment bulkhead

bleed screws, located as follows.
a) At the top left-hand corner of the radiator.
b) On the thermostat housing at the left-hand end of the cylinder head.
c) In the heater hose connector near the engine compartment bulkhead **(see illustration)**.

4 When the flow of coolant stops, reposition the container below the cylinder block drain plug, located at the front of the cylinder block, near the flywheel. Remove the drain plug, and allow the coolant to drain into the container **(see illustration)**.

5 If the coolant has been drained for a reason

27.4 Cylinder block drain plug location (arrowed)

other than renewal, then provided it is clean and less than two years old, it can be re-used.

6 Refit the radiator drain plug and the cylinder block drain plug on completion of draining.

Cooling system flushing

7 If coolant renewal has been neglected, or if the antifreeze mixture has become diluted, then in time, the cooling system may gradually lose efficiency, as the coolant passages become restricted due to rust, scale deposits, and other sediment. The cooling system efficiency can be restored by flushing the system clean.

8 The radiator should be flushed independently of the engine, to avoid unnecessary contamination.

Radiator flushing

9 To flush the radiator, tighten the radiator drain plug and radiator bleed screw, and disconnect the top and bottom hoses.

10 Insert a garden hose into the radiator top inlet. Direct a flow of clean water through the radiator, and continue flushing until clean water emerges from the radiator bottom outlet.

11 If, after a reasonable period, the water still does not run clear, the radiator can be flushed with a good proprietary cleaning agent. It is important that their manufacturer's instructions are followed carefully. If the contamination is particularly bad, remove the radiator, then insert the hose in the radiator bottom outlet, and reverse-flush the radiator.

Engine flushing

12 To flush the engine, first refit the cylinder block drain plug, and tighten the cooling system bleed screws.

13 Remove the thermostat as described in Chapter 3, then temporarily refit the thermostat cover.

14 With the top and bottom hoses disconnected from the radiator, insert a garden hose into the radiator top hose. Direct a clean flow of water through the engine, and continue flushing until clean water emerges from the radiator bottom hose.

15 On completion of flushing, refit the thermostat and reconnect the hoses with reference to Chapter 3.

Cooling system filling

16 Before attempting to fill the cooling system, make sure that all hoses and clips are in good condition, and that the clips are tight. Note that an antifreeze mixture must be used all year round, to prevent corrosion of the engine components (see the following sub-Section). Also check that the radiator and cylinder block drain plugs are in place and tight.

17 Remove the expansion tank filler cap.

18 Open all the cooling system bleed screws (see paragraph 3).

19 Some of the cooling system hoses are positioned at a higher level than the top of the radiator expansion tank. It is therefore necessary to use a "header tank" when refilling the cooling system, to eliminate the possibility of air being trapped in the system.

 Although Citroën dealers use a special header tank, the same effect can be achieved by using a suitable bottle or plastic container sealed to the expansion tank filler neck.

20 Fit the "header tank" to the expansion tank, and slowly fill the system. Coolant will emerge from each of the bleed screws in turn, starting with the lowest screw. As soon as coolant free from air bubbles emerges from the lowest screw, tighten that screw, and watch the next bleed screw in the system. Repeat the procedure on the remaining bleed screws. Note that the bleed screws should be tightened in the following order:

a) Radiator bleed screw.
b) Thermostat housing bleed screw.
c) Heater hose bleed screw.

21 Ensure that the "header tank" is full (at least 0.5 litres of coolant). Start the engine, and run it at a fast idle speed (do not exceed 2000 rpm) until the cooling fan cuts in, and then cuts out. Stop the engine.

22 Remove the "header tank", taking great care not to scald yourself with the hot coolant, then fit the expansion tank cap.

23 Allow the engine to cool, then check the coolant level (see *"Weekly checks"*). Top-up the level if necessary.

Antifreeze mixture

24 The antifreeze should always be renewed at the specified intervals. This is necessary not only to maintain the antifreeze properties, but also to prevent corrosion which would otherwise occur as the corrosion inhibitors become progressively less effective.

25 Always use an ethylene-glycol based antifreeze which is suitable for use in mixed-metal cooling systems. The quantity of antifreeze and levels of protection are indicated in the Specifications.

26 Before adding antifreeze, the cooling system should be completely drained, preferably flushed, and all hoses checked for condition and security.

27 After filling with antifreeze, a label should be attached to the expansion tank filler neck, stating the type and concentration of antifreeze used, and the date installed. Any subsequent topping-up should be made with the same type and concentration of antifreeze.

28 Do not use engine antifreeze in the windscreen/rear window washer system, as it will cause damage to the vehicle paintwork. Instead, a screenwash additive should be added to the washer system in the quantities stated on the bottle.

Chapter 1 Part B:
Routine maintenance and servicing - diesel models

Contents

Degrees of difficulty

 Easy, suitable for novice with little experience

 Fairly easy, suitable for beginner with some experience

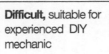 **Fairly difficult,** suitable for competent DIY mechanic

 Difficult, suitable for experienced DIY mechanic

 Very difficult, suitable for expert DIY or professional

Lubricants and fluids

Refer to "Weekly checks"

Capacities

Engine oil:
Excluding filter .... 4.0 litres
Including filter .... 4.6 litres
Difference between MAX and MIN dipstick marks .... 1.5 litres

Cooling system
Approximate .... 7.5 litres

Transmission
According to model* .... 1.8 to 2.2 litres
*Oil capacity varies according to model and production date. Fill the transmission until oil runs out of the filler/level hole, then check the level as described in Section 22.

Fuel tank
All models .... 47 litres

Braking system
Approximate .... 0.25 litres

Brake vacuum pump
Belt driven .... 40 cc

Washer reservoirs
Front .... 1.5 litres
Rear (if fitted) .... 1.5 litres

Engine

Oil filter type .... Champion F104
Air filter .... Champion W117
Auxiliary drivebelt deflection .... Approx 5.0 mm deflection midway between pulleys

Cooling system

Antifreeze mixture:
  28% antifreeze .... Protection down to -15°C
  50% antifreeze .... Protection down to -30°C
Note: Refer to antifreeze manufacturer for latest recommendations.

Fuel system

Fuel filter type:
  Roto-diesel .... Champion L132
  Bosch .... Champion L135
Idle speed:
  Roto-diesel injection pump (according to type, see Chapter 4C)
    From 1993 .... 800 ± 50 rpm
    Later models .... 775 ± 25 rpm
  Bosch injection pump .... 800 ± 50 rpm
Fast-idle speed .... 950 ± 50 rpm
Anti-stall speed:
  Roto-diesel injection pump (using 3mm peg and 3 mm feeler blade) .. 900 ± 100 rpm
  Bosch injection pump (using 1 mm feeler blade) .... Idle speed + 50 rpm

Brakes

Front brake pad friction material minimum thickness .... 2.0 mm
Rear brake shoe friction material minimum thickness .... 1.0 mm
Vacuum pump drivebelt tension (where applicable) .... Approx 5.0 mm deflection midway between pulleys

Steering

Power steering pump drivebelt tension (where applicable) .... Approx 5.0 mm deflection midway between pulleys

Torque wrench settings	Nm	lbf ft
Transmission filler/level plug	20	15
Transmission drain plug	30	22
Roadwheel bolts:		
3 bolts (pre-launch models before July 1988)	70	52
4 bolts	80	59

The maintenance intervals in this manual are provided with the assumption that you, not the dealer, will be carrying out the work. These are the minimum maintenance intervals recommended by us for vehicles driven daily. If you wish to keep your vehicle in peak condition at all times, you may wish to perform some of these procedures more often. We encourage frequent maintenance, because it enhances the efficiency, performance and resale value of your vehicle.

If the vehicle is driven in dusty areas, used to tow a trailer, or driven frequently at slow speeds (idling in traffic) or on short journeys, more frequent maintenance intervals are recommended.

When the vehicle is new, it should be serviced by a factory-authorised dealer service department, in order to preserve the factory warranty.

Every 250 miles (400 km) or weekly
☐ Refer to *"Weekly Checks"*

Every 6000 miles / 10 000 km or 6 months, whichever comes first
In addition to the "Weekly Checks", carry out the following:
☐ Renew the engine oil and filter (Section 3).
☐ Drain water from the fuel filter (Section 4)
☐ Check all underbonnet components and hoses for fluid leaks (Section 5).
☐ Check the suspension and steering components for condition and security (Section 6).
☐ Check the condition of the driveshaft rubber gaiters (Section 7).
☐ Examine the exhaust system for corrosion and leakage (Section 8).
☐ Check the condition of the front brake pads, and renew if necessary (Section 9).
☐ Check the operation of the handbrake (Section 10).

Every 12 000 miles / 20 000 km or 12 months, whichever comes first
In addition to the 6000 mile service, carry out the following:
☐ Check the idle speed and anti-stall speed (Section 11).
☐ Check the condition of the auxiliary drivebelt, and renew if necessary (Section 12).
☐ Check the condition of the brake vacuum pump drivebelt and power steering pump drivebelt, where applicable (Section 13).
☐ Check the brake vacuum pump oil level, on models with a belt-driven pump (Section 14).
☐ Check the clutch adjustment (Section 15).
☐ Lubricate the clutch control mechanism (Section 15).
☐ Check the condition of the rear brake shoes, and renew if necessary (Section 16).
☐ Carry out a road test (Section 17).
Note: *Change the coolant after the first 12 000 miles (Section 25)*

Every 18 000 miles / 30 000 km or 18 months, whichever comes first
In addition to the 6000 mile service, carry out the following:
☐ Renew the air filter (Section 18).
☐ Renew the fuel filter (Section 19).
☐ Lubricate all hinges and locks (Section 20).

Every 24 000 miles / 40 000 km or 2 years, whichever comes first
In addition to the 12 000 mile service, carry out the following:
☐ Renew the brake fluid (Section 21).

Every 36 000 miles / 60 000 km or 3 years, whichever comes first
In addition to the 12 000 mile service, carry out the following:
☐ Renew the air filter (Section 18).
☐ Renew the fuel filter (Section 19).
☐ Lubricate all hinges and locks (Section 20).
☐ Check the manual transmission oil level, and top-up if necessary (Section 22).

Every 48 000 miles / 80 000 km or 4 years, whichever comes first
In addition to the 24 000 mile service, carry out the following:
☐ Renew the timing belt (Section 23).

Every 72 000 miles / 120 000 km or 6 years, whichever comes first
☐ Renew the manual transmission oil (Section 24).

Every 12 months, regardless of mileage
☐ Renew the fuel filter, before winter (Section 19).

Every 2 years, regardless of mileage
☐ Renew the coolant (Section 25).
☐ Lubricate all hinges and locks (Section 20).

Underbonnet view of a C15 diesel model - up to February 1993

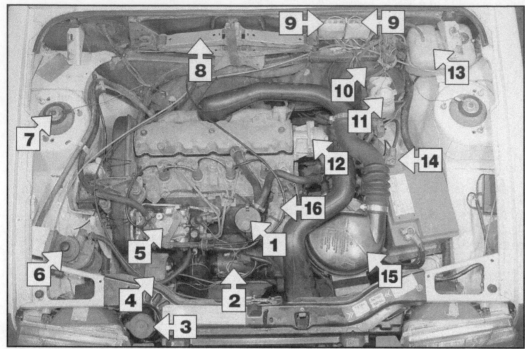

1 Engine oil dipstick and filler cap
2 Starter motor
3 Radiator filler cap
4 Alternator
5 Injection pump
6 Fuel filter/priming pump
7 Front suspension upper mounting
8 Wheel changing jack
9 Fuseboxes
10 Brake vacuum servo unit
11 Brake fluid reservoir
12 Brake vacuum pump
13 Washer fluid reservoir
14 Battery earth (negative) terminal
15 Air cleaner housing
16 Thermostat housing

Underbonnet view of a C15 diesel model - from January 1994

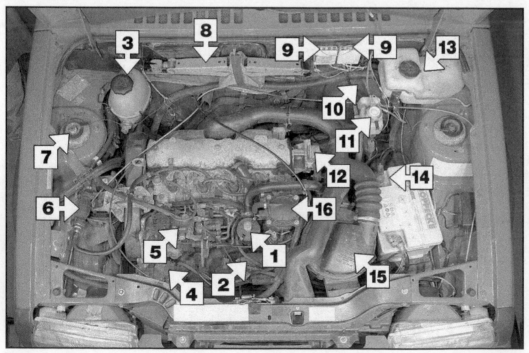

1 Engine oil dipstick and filler cap
2 Starter motor
3 Coolant filler cap and expansion tank
4 Alternator
5 Injection pump
6 Fuel priming bulb
7 Front suspension upper mounting
8 Wheel changing jack
9 Fuseboxes
10 Brake vacuum servo unit
11 Brake fluid reservoir
12 Brake vacuum pump
13 Washer fluid reservoir
14 Battery earth (negative) terminal
15 Air cleaner housing
16 Thermostat housing

Front underbody view of a C15 diesel model

1 Subframe
2 Anti-roll bar
3 Anti-roll guide bar
4 Brake hydraulic pipe
5 Exhaust pipe
6 Exhaust resonator
7 Fuel feed and return hoses
8 Steering track rod
9 Track control arm
10 Brake caliper
11 Coolant hose
12 Lower engine mounting
   bracket and driveshaft
   intermediate bearing
13 Driveshaft
14 Engine oil drain plug
15 Transmission
16 Final drive oil drain plug

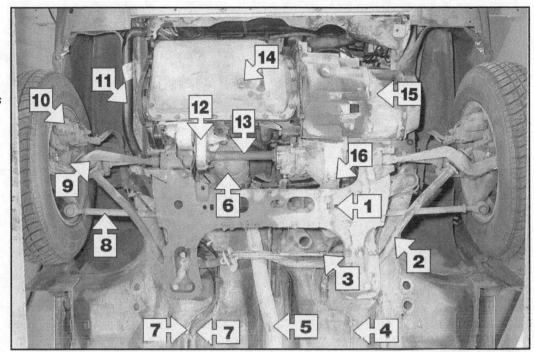

Rear underbody view of a C15 diesel model

1 Subframe
2 Anti-roll bar
3 Spare wheel harness
4 Spring
5 Shock absorber
6 Trailing arm
7 Brake drum
8 Brake pressure limiter
9 Handbrake cables
10 Brake hydraulic pipes
11 Fuel tank
12 Exhaust pipe
13 Silencer

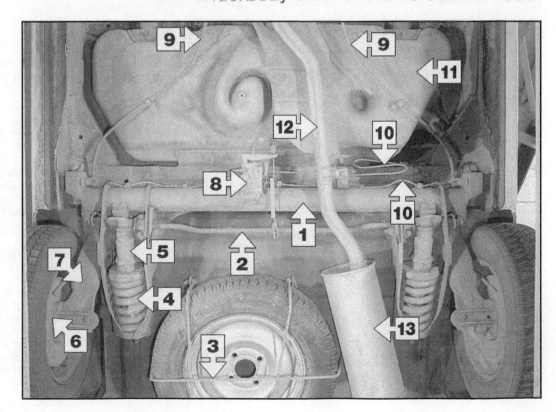

1 General information

This Chapter is designed to help the home mechanic maintain his/her vehicle for safety, economy, long life and peak performance.

The Chapter contains a master maintenance schedule, followed by Sections dealing specifically with each task in the schedule. Visual checks, adjustments, component renewal and other helpful items are included. Refer to the accompanying illustrations of the engine compartment and the underside of the vehicle for the locations of the various components.

Servicing your vehicle according to the mileage/time maintenance schedule and the following Sections will provide a planned maintenance programme, which should result in a long and reliable service life. This is a comprehensive plan, so maintaining some items but not others at the specified service intervals, will not produce the same results.

As you service your vehicle, you will discover that many of the procedures can - and should - be grouped together, because of the particular procedure being performed, or because of the proximity of two otherwise-unrelated components to one another. For example, if the vehicle is raised for any reason, the exhaust can be inspected at the same time as the suspension and steering components.

The first step in this maintenance programme is to prepare yourself before the actual work begins. Read through all the Sections relevant to the work to be carried out, then make a list and gather all the parts and tools required. If a problem is encountered, seek advice from a parts specialist, or a dealers service department.

2 Intensive maintenance

If, from the time the vehicle is new, the routine maintenance schedule is followed closely, and frequent checks are made of fluid levels and high-wear items, as suggested throughout this manual, the engine will be kept in relatively good running condition, and the need for additional work will be minimised.

It is possible that there will be times when the engine is running poorly due to the lack of regular maintenance. This is even more likely if a used vehicle, which has not received regular and frequent maintenance checks, is purchased. In such cases, additional work may need to be carried out, outside of the regular maintenance intervals.

If engine wear is suspected, a compression test or leakdown test (refer to Chapter 2B) will provide valuable information regarding the overall performance of the main internal components. Such a test can be used as a basis to decide on the extent of the work to be carried out. If, for example, a compression or leakdown test indicates serious internal engine wear, conventional maintenance as described in this Chapter will not greatly improve the performance of the engine, and may prove a waste of time and money, unless extensive overhaul work is carried out first.

The following series of operations are those most often required to improve the performance of a generally poor-running engine:

Primary operations

a) Clean, inspect and test the battery (see "Weekly checks").
b) Check all the engine related fluids (see "Weekly checks").
c) Check the condition and tension of the auxiliary drivebelt (Section 12).
d) Check the condition of the air filter, and renew if necessary (Section 18).
e) Check the fuel filter (Section 19).
f) Check the condition of all hoses, and check for fluid leaks (Section 5).
g) Check the idle speed and anti-stall speed (Section 11).

If the above operations do not prove fully effective, carry out the following secondary operations:

Secondary operations

All items listed under "Primary operations", plus the following:
a) Check the charging system (Chapter 5A)
b) Check the preheating system (Chapter 5C)
c) Check the fuel system (Chapter 4C)

Every 6000 miles / 10 000 km or 6 months

3 Engine oil and filter renewal

1 Frequent oil and filter changes are the most important preventative maintenance procedures which can be undertaken by the DIY owner. As engine oil ages, it becomes diluted and contaminated, which leads to premature engine wear.

3.4 Slacken the oil drain plug

2 Before starting this procedure, gather together all the necessary tools and materials. Also make sure that you have plenty of clean rags and newspapers handy, to mop up any spills. Ideally, the engine oil should be warm, as it will drain better, and more built-up sludge will be removed with it.
Caution: Take care not to touch the exhaust or any other hot parts of the engine when working under the vehicle. To avoid any possibility of scalding, and to protect yourself from possible skin irritants and other harmful contaminants in used engine oils, it is advisable to wear gloves when carrying out this work.
3 Access to the underside of the vehicle will be greatly improved if it can be raised on a lift, driven onto ramps, or jacked up and supported on axle stands (see *"Jacking, and Vehicle Support"*). Whichever method is chosen, make sure that the vehicle remains level, or if it is at an angle, that the drain plug is at the lowest point.
4 Slacken the drain plug about half a turn using a suitable square or hexagon key **(see illustration)**. Position the draining container under the drain plug, then remove the plug completely. Recover the sealing ring from the drain plug.

Try to keep the drain plug pressed into the sump while unscrewing it by hand the last couple of turns. As the plug releases from the threads, move it away sharply, so that the stream of oil issuing from the sump runs into the container, not up your sleeve!

5 Allow some time for the old oil to drain, noting that it may be necessary to reposition the container as the oil flow slows to a trickle.
6 After all the oil has drained, wipe off the drain plug with a clean rag, and fit a new sealing washer. Clean the area around the drain plug opening, and refit the plug. Tighten the plug securely.
7 Move the container into position under the oil filter, which is located on the front right-hand side of the cylinder block, next to the alternator.

3.8 Unscrewing the oil filter with a strap wrench

8 Using an oil filter removal tool, slacken the filter initially, then unscrew it by hand the rest of the way **(see illustration)**. Empty the oil in the old filter into the container.

9 Use a clean rag to remove all oil, dirt and sludge from the filter sealing area on the engine.

 HAYNES HINT *Check the old filter to make sure that the rubber sealing ring hasn't stuck to the engine. If it has, carefully remove it.*

10 Apply a light coating of clean engine oil to the sealing ring on the new filter, then screw it into position on the engine **(see illustration)**. Tighten the filter firmly by hand only - **do not use any tools**.

11 Remove the old oil and all tools from under the vehicle, then lower the vehicle to the ground (if applicable).

12 Remove the dipstick, then release the clips and remove the oil filler cap from the filler tube at the front of the engine. Fill the engine, using the correct grade and type of oil (refer to *"Weekly checks"* for details of topping up). A funnel, or an oil can with a spout may help to reduce spillage. Pour in half the specified quantity of oil first, then wait a few minutes for the oil to fall to the sump. Continue adding oil, a small quantity at a time, until the level is up to the lower mark on the dipstick. Then add a further quantity of oil, according to the Specifications, to bring the level up to the upper mark on the dipstick. Refit the filler cap.

13 Start the engine and run it for a few minutes. Check for leaks around the oil filter seal and the sump drain plug. Note that there may be a delay of a few seconds before the oil pressure warning light goes out when the engine is first started, as the oil circulates through the engine oil galleries before the pressure builds up.

14 Switch off the engine, and wait a few minutes for the oil to settle in the sump once more. With the new oil circulated and the filter completely full, recheck the level on the dipstick, and add more oil as necessary.

15 Dispose of the used engine oil safely, with reference to *"General repair procedures"*.

Note: *It is antisocial and illegal to dump oil down the drain. To find the location of your local oil recycling bank, call this number free.*

OIL CARE
FOLLOW THE CODE
OIL BANK LINE
0800 66 33 66

4 Fuel filter water draining

Up to February 1993

1 Diesel models up to February 1993 have either a Roto-diesel or Purflux fuel filter mounted on the right-hand wing behind the headlamp. The water drain screw is underneath the filter **(see illustration)**.

2 Place a small container under the filter and loosen the drain screw. Also loosen the air bleed screw, where fitted, on the filter head or inlet union bolt. Allow fuel and water to drain until fuel, free from water, emerges from the drain screw. Then tighten the drain screw, and tighten the air bleed screw where fitted.

From February 1993 onwards

3 Diesel models from February 1993 onwards have a combined fuel filter/heater unit mounted on the thermostat housing, with a water drain plug and tube at the base of the unit.

4 Cover the clutch bellhousing with a piece of plastic sheeting, to protect the clutch from fuel spillage, and place a suitable container beneath the drain tube. Open the drain plug by turning it anti-clockwise, and allow fuel and water to drain until fuel, free from water, emerges from the end of the tube **(see illustration)**. Close the drain plug and remove the plastic sheeting from above the clutch.

All models

5 Dispose of the drained fuel safely.

6 Start the engine. If difficulty is experienced, prime the fuel system as described in Chapter 4C.

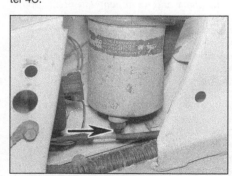

4.1 Water drain screw (arrowed) on Purflux fuel filter mounted on right wing - up to February 1993

3.10 Lubricate the oil filter sealing ring before fitting

5 Hose and fluid leak check

1 Visually inspect the engine joint faces, gaskets and seals for any signs of water or oil leaks. Pay particular attention to the areas around the camshaft cover, cylinder head, oil filter and sump joint faces. Bear in mind that, over a period of time, some very slight seepage from these areas is to be expected - what you are really looking for is any indication of a serious leak. Should a leak be found, renew the offending gasket or oil seal by referring to the appropriate Chapters in this manual.

2 Also check the security and condition of all the engine-related pipes and hoses. Ensure that all cable ties or securing clips are in place and in good condition. Clips which are broken or missing can lead to chafing of the hoses, pipes or wiring, which could cause more serious problems in the future.

3 Carefully check the radiator hoses and heater hoses along their entire length. Renew any hose which is cracked, swollen or deteriorated. Cracks will show up better if the hose is squeezed. Pay close attention to the hose clips that secure the hoses to the cooling system components. Hose clips can pinch and puncture hoses, resulting in cooling system leaks. If the original Citroën crimped-type hose clips are used, it may be a good idea to replace them with standard worm-drive clips.

4.4 Drain the water from the fuel filter mounted on the thermostat housing - February 1993 onwards

6.2 Checking a steering gear gaiter

4 Inspect all the cooling system components (hoses, joint faces etc.) for leaks. A leak in the cooling system will usually show up as white or rust-coloured deposits on the area adjoining the leak. Where any problems of this nature are found on system components, renew the component or gasket with reference to Chapter 3.

5 With the vehicle raised, inspect the fuel tank and filler neck for punctures, cracks and other damage. The connection between the filler neck and tank is especially critical. Sometimes, a rubber filler neck or connecting hose will leak due to loose retaining clamps or deteriorated rubber.

6 Carefully check all rubber hoses and metal fuel lines leading away from the fuel tank. Check for loose connections, deteriorated hoses, crimped lines, and other damage. Pay particular attention to the vent pipes and hoses, which often loop up around the filler neck and can become blocked or crimped. Follow the lines to the front of the vehicle, carefully inspecting them all the way. Renew damaged sections as necessary. Similarly, while the vehicle is raised, take the opportunity to inspect all underbody brake fluid pipes and hoses.

7 From within the engine compartment, check the security of all fuel hose attachments and pipe unions, and inspect the fuel hoses and vacuum hoses for kinks, chafing and deterioration.

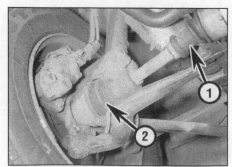

7.2 Right-hand driveshaft inner gaiter (1) and outer gaiter (2)

6 Suspension and steering check

Front suspension and steering

1 Apply the handbrake, jack up the front of the vehicle, and support it securely on axle stands (see "*Jacking and Vehicle Support*").

2 Visually inspect the balljoint dust covers and the steering rack-and-pinion gaiters for splits, chafing or deterioration **(see illustration)**. Any wear of these components will cause loss of lubricant, together with dirt and water entry, resulting in rapid deterioration of the balljoints or steering gear.

3 On vehicles equipped with power steering, check the fluid hoses for chafing and deterioration, and the pipe and hose unions for fluid leaks. Also check for signs of fluid leakage under pressure from the steering gear rubber gaiters, which would indicate failed fluid seals within the steering gear.

4 Grasp the roadwheel at the 12 o'clock and 6 o'clock positions, and try to rock it. Very slight free play may be felt, but if the movement is appreciable, further investigation is necessary to determine the source. Continue rocking the wheel while an assistant depresses the footbrake. If the movement is now eliminated or significantly reduced, it is likely that the hub bearings are at fault. If the free play is still evident with the footbrake depressed, then there is wear in the suspension joints or mountings.

5 Now grasp the wheel at the 9 o'clock and 3 o'clock positions, and try to rock it as before. Any movement felt now may again be caused by wear in the hub bearings or the steering track-rod balljoints. If the outer balljoint is worn, the visual movement will be obvious. If the inner joint is suspect, it can be felt by placing a hand over the rack-and-pinion rubber gaiter and gripping the track rod. If the wheel is now rocked, movement will be felt at the inner joint if wear has taken place.

6 Using a large screwdriver or flat bar, check for wear in the suspension mounting bushes by levering between the relevant suspension component and its attachment point. Some movement is to be expected, as the mountings are made of rubber, but excessive wear should be obvious. Also check the condition of any visible rubber bushes, looking for splits, cracks or contamination of the rubber.

7 With the vehicle standing on its wheels, have an assistant turn the steering wheel back and forth about an eighth of a turn each way. There should be very little, if any, lost movement between the steering wheel and roadwheels. If this is not the case, closely observe the joints and mountings previously described, but in addition, check the steering column universal joints for wear, and the rack-and-pinion steering gear itself.

Rear suspension check

8 Chock the front wheels, then jack up the rear of the vehicle and support it securely on axle stands (see "*Jacking and Vehicle Support*").

9 Working as described previously for the front suspension, check the rear hub bearings, and the suspension bushes and mountings for wear.

Shock absorber check

10 Check for any signs of fluid leakage around the front and rear shock absorbers. On the front shock absorber, check for fluid on the rubber gaiter around the piston rod. Should any fluid be noticed, the shock absorber is defective internally, and should be renewed. **Note:** *Shock absorbers should always be renewed in pairs on the same axle.*

11 The efficiency of the shock absorber may be checked by bouncing the vehicle at each corner. Generally speaking, the body will return to its normal position and stop after being depressed. If it rises and returns on a rebound, the shock absorber is probably suspect. Also examine the shock absorber mountings for signs of wear, and check that the rear shock absorber ball joints fit securely in their sockets.

7 Driveshaft and gaiter check

1 Apply the handbrake, jack up the front of the vehicle, and support it securely on axle stands (see "*Jacking and Vehicle Support*").

2 Turn the steering onto full lock, then slowly rotate the roadwheel. Inspect the condition of the inner and outer constant velocity (CV) joint rubber gaiters, squeezing the gaiters to open out the folds. Check for signs of cracking, splits or deterioration of the rubber, which may allow the grease to escape, and lead to water and grit entry into the joint. Also check the security and condition of the retaining clips **(see illustration)**. If any damage or deterioration is found, the gaiters should be renewed as described in Chapter 8.

3 At the same time, check the general condition of the CV joints themselves by first holding the driveshaft and attempting to rotate the wheel. Repeat this check by holding the inner joint and attempting to rotate the driveshaft. Any appreciable movement indicates wear in the joints, wear in the driveshaft splines, or a loose driveshaft retaining nut.

8 Exhaust system check

1 Chock the front wheels, then jack up the rear of the vehicle and support it securely on axle stands (see "*Jacking and Vehicle Support*").

2 With the engine cold (wait at least an hour after switching off the engine), check the complete exhaust system from the engine to the end of the tailpipe.

3 Check the exhaust pipes and connections for evidence of leaks, severe corrosion and damage. Make sure that all brackets and mountings are in good condition, and that all relevant nuts and bolts are tight. Leakage at any of the joints or in other parts of the system will usually show up as a black, sooty stain in the vicinity of the leak.

4 Proprietary pastes and bandages are available for repairing leaks. They work well in the short term, but eventually the exhaust section will need to be renewed as described in Chapter 4C.

5 Rattles and vibrations can often be traced to the exhaust system. Tap the silencer units with a soft mallet and listen for noises caused by corroded or displaced baffle material. *Caution: Do not strike the catalytic converter (where fitted), as this may damage the ceramic block inside.*
Note: *Catalytic converters are not normally fitted to diesel models, but may be fitted if required by some countries for emission control.*

6 Carefully rock the pipes and silencers from side to side on their mountings. If the components are able to come into contact with the body or suspension parts, look for broken or worn rubber mountings.

7 Extra clearance can be gained by slackening the clamps between adjacent sections of the exhaust pipe to loosen the joints where possible, and twisting the pipes as necessary to provide the additional clearance.

9 Front brake pad check

1 Firmly apply the handbrake, then jack up the front of the vehicle and support it securely on axle stands (see *"Jacking and Vehicle Support"*). Remove the front roadwheels.

2 For a quick check, the thickness of friction material remaining on each brake pad can be measured from the front of the caliper **(see illustration)**. If any pad's friction material is worn to the specified thickness or less, *all four pads must be renewed as a set.*

3 For a comprehensive check, the brake pads should be removed and cleaned. The operation of the caliper can then also be checked, and the condition of the brake disc itself can be fully examined on both sides. Refer to Chapter 9 for further information.

10 Handbrake check and adjustment

1 Chock the front wheels, then jack up the rear of the vehicle, and support securely on axle stands (see *"Jacking and Vehicle Support"*).

2 Apply the footbrake firmly several times to establish correct shoe-to-drum clearance, then apply and release the handbrake several times to ensure that the self-adjust mechanism has compensated fully for any wear in the linings.

3 Fully release the handbrake, and check that the rear wheels rotate freely, without binding. If not, check that all cables are routed correctly and that the cable components and levers move freely.

> **HAYNES HiNT** *If the handbrake mechanism fails to operate, or appears to be seized on one side of the vehicle only, remove the relevant brake drum (see Chapter 9) and check the handbrake lever pivot on the trailing brake shoe - it is possible for the lever to seize due to corrosion. If necessary, remove the lever, and clean the contact faces of the lever, brake shoe, and pivot.*

4 If all components are free to move, but the wheels still bind when rotated, adjustment is required as follows.

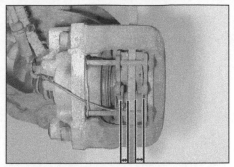

9.2 Brake pad friction material thickness (arrowed) can be checked with the pads in place

5 Pull up the boot from the handbrake to gain access to the equaliser and adjusting nut. The equaliser is a short metal bar at the rear of the lever, which lies transversely across the vehicle and pulls on both handbrake cables simultaneously.

6 Apply the handbrake and set the lever on the third notch on the ratchet mechanism. With the lever in this position, slacken the locknut and turn the adjusting nut until only a slight drag can be felt when the rear wheels are turned.

7 Make sure there is equal tension on the left and right-hand cables. If the tensions are unequal, the equaliser bar will be inclined at an angle. If necessary, adjust the cables individually until the equaliser becomes transverse. This can only be done if the cables are fitted with in-line adjusters (normally underneath the vehicle). If the cables cannot be equalised they will need to be renewed (see Chapter 9).

8 Fully release the handbrake lever and check that the wheels rotate freely. Then apply the handbrake lever one notch at a time until both rear wheels are locked. The adjustment is correct if the wheels become locked at 5 notches. Turn the central adjustment nut on the equaliser bar to achieve the correct adjustment, then tighten the locknut.

9 Refit the boot to the handbrake and lower the vehicle to the ground.

Every 12 000 miles / 20 000 km or 12 months

11 Idle speed and anti-stall speed check and adjustment

1 The usual type of tachometer (rev counter), which works from ignition system pulses, cannot be used on diesel engines. A TDC diagnostic socket is provided for use with Citroën test equipment which will not normally be available to the home mechanic. If it is felt that adjusting the idle speed "by ear" is not satisfactory, you will need to purchase or hire an appropriate tachometer, or else take the vehicle to a Citroën dealer or other specialist.

Idle speed - check and adjustment

2 Run the engine until it warms up to normal operating temperature.

3 Connect a tachometer, if available, to the TDC sensor diagnostic socket **(see illustration)**. Allow the engine to idle and check that the idling speed is according to the Specifications.

4 If adjustment is necessary on the Roto-diesel pump, slacken the locknut and turn the idle speed adjustment screw as required, then retighten the locknut **(see illustration)**.

11.3 Connect a tachometer to the TDC sensor diagnostic socket (arrowed)

11.4 Idle speed adjustment screw (arrowed) on the Roto-diesel injection pump

5 If adjustment is necessary on the Bosch pump, first slacken the locknut and unscrew the anti-stall adjustment screw until it is clear of the accelerator lever. Slacken the locknut and turn the idle speed adjustment screw as required, then retighten the locknut **(see illustration)**.

6 Adjust the anti-stall adjustment screw as described below.

Anti-stall speed - check and adjustment

Roto-diesel injection pump

7 Run the engine to normal operating temperature, then switch it off.

8 Insert a 3.0 mm shim or feeler blade between the accelerator lever and the anti-stall adjustment screw. Turn the stop lever clockwise until it is clear of the hole in the fast idle lever, then insert a 3.0 mm dowel rod or twist drill **(see illustration)**.

9 Start the engine and allow it to idle. Connect a tachometer, if available, to the TDC sensor diagnostic socket, and check that the engine runs at the anti-stall speed given in the Specifications.

10 If adjustment is necessary, loosen the locknut, turn the anti-stall adjustment screw as required, then tighten the locknut.

11 Remove the feeler blade and twist drill, then recheck the idling speed and adjust it if necessary as described above.

12 Turn the accelerator lever to increase the engine speed to 3000 rpm, then quickly release the lever. If the deceleration is too fast and the engine stalls, turn the anti-stall adjustment screw 1/4 turn anti-clockwise (viewed from flywheel end of engine). If the deceleration is too slow, resulting in poor engine braking, turn the screw 1/4 turn clockwise. Adjust as necessary, then securely retighten the locknut.

13 Recheck the idling speed and adjust it if necessary as described above.

14 With the engine idling, check the operation of the manual stop control by turning the stop lever clockwise. The engine must stop instantly.

15 Switch off the ignition.

Bosch injection pump

16 Run the engine to normal operating temperature. Connect a tachometer, if available, to the TDC sensor diagnostic socket and note the exact idling speed, then switch off the engine.

17 Insert a 1.0 mm feeler blade between the accelerator lever and the anti-stall adjustment screw.

18 Start the engine and allow it to idle. The engine speed should exceed the normal idling speed by 50 rpm.

19 If adjustment is necessary, slacken the locknut and turn the anti-stall adjustment screw as required, then retighten the locknut.

20 Remove the feeler blade and allow the engine to idle.

21 Move the fast idle lever fully towards the flywheel end of the engine so that it touches the stop, and check that the engine speed increases to the fast idle speed given in the Specifications. If necessary loosen the locknut and turn the stop adjusting screw as required, then retighten the locknut.

22 With the engine idling, check the operation of the manual stop control by turning the stop lever. The engine must stop instantly.

23 Switch off the ignition.

24 If the fast idle stop screw has been adjusted, wait for the engine to cool down and adjust the ferrule on the outer sleeve of the fast idle cable so that the fast idle lever just touches the stop screw. The ferrule is held by a locknut which has to be slackened and then re-tightened.

12 Auxiliary drivebelt check and renewal

1 This Section describes the auxiliary drivebelt, fitted to all diesel models, to drive the alternator from the crankshaft pulley at the timing end of the engine. Some models have additional belts at the transmission end of the engine, to drive the brake vacuum pump and power steering pump from a pulley fitted to the camshaft. For details see Section 13.

Checking the auxiliary drivebelt condition

2 Apply the handbrake, then jack up the front of the vehicle and support it securely on axle stands (see *"Jacking and Vehicle Support"*). Remove the right-hand front roadwheel. If necessary, free the coolant hose from any retaining clips to improve access to the crankshaft sprocket bolt.

3 Using a suitable socket and extension bar fitted to the crankshaft sprocket bolt, rotate the crankshaft so that the entire length of the drivebelt can be examined. Examine the drivebelt for cracks, splitting, fraying, or other damage. Check also for signs of glazing (shiny patches) and for separation of the belt plies. Renew the belt if worn or damaged.

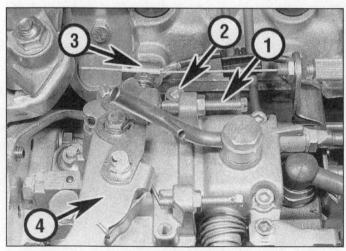

11.5 Bosch injection pump

1 Anti-stall adjustment screw   3 End fitting on fast idle lever
2 Idling adjustment screw   4 Accelerator lever

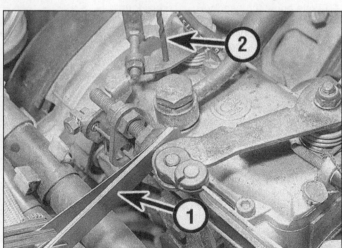

11.8 Anti-stall adjustment on the Roto-diesel injection pump using feeler blades (1) and twist drill (2)

4 If the condition of the belt is satisfactory, check the drivebelt tension as described below under the relevant sub-heading.

Removal

5 If not already done, carry out the operations described in paragraph 2.
6 Disconnect the battery negative lead.
7 Slacken the pivot bolt and adjuster locknut, then unscrew the adjustment bolt to release the tension. Remove the drivebelt from the pulleys **(see illustration)**.

Refitting

8 Fit the belt around the pulleys, ensuring that the belt is of the correct type if it is being renewed, and take up the slack in the belt by tightening the adjustment bolt.
9 Tension the drivebelt as described in the following paragraphs.

Tensioning

10 If not already done, carry out the operations described in paragraph 2.
11 Correct tensioning of the drivebelt will ensure that it has a long life. Beware, however, of overtightening, as this can cause wear in the alternator bearings.
12 The belt should be tensioned so that the free movement under firm thumb pressure, at the mid-point between the pulleys, is according to the Specifications.
13 To adjust, with the alternator mountings loose, tighten the adjustment bolt until the correct tension is achieved. Rotate the crankshaft through two complete turns, then recheck the tension. When the tension is correct, tighten the pivot bolt and adjuster locknut securely.
14 Reconnect the battery negative lead.
15 Secure the coolant hose to any clips that have been disconnected. Refit the roadwheel and lower the vehicle to the ground.

13 Brake servo vacuum pump and power steering pump drivebelt check and renewal

Note: *This Section applies only to models with the brake vacuum pump mounted on the transmission and driven by a belt from a camshaft-mounted pulley. On models equipped with power steering, the power steering pump and brake vacuum pump are driven by separate belts from a double pulley. Check and renew both belts at the same time.*

Brake servo vacuum pump drivebelt

Checking

1 Inspect the entire length of the drivebelt for signs of wear or damage. If the belt is found to be worn, frayed or cracked, it should be renewed as a precaution against breakage in service.

⚠️ *Warning: Breakage of the vacuum pump drivebelt in service will lead to a loss of servo assistance,*

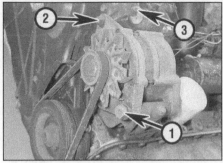

12.7 Alternator and drivebelt

1 Pivot bolt *3 Adjustment bolt*
2 Adjuster locknut

greatly reducing the efficiency of the braking system, and could lead to an accident.

Renewal

2 Slacken the pump adjuster bolt, pivot bolt and the adjuster strap lower mounting nut, then unhook the drivebelt from the pump pulley and remove it from the engine.
3 Locate the new drivebelt over the camshaft drive pulley and the pump pulley.
4 Tension the drivebelt by pivoting the pump around its mounting, until the deflection of the drivebelt under firm thumb pressure, midway between the pulleys, is according to the Specifications. Holding the pump in this position, securely tighten the adjuster strap nut, and the pump pivot and adjuster bolts. Re-check the drivebelt tension, and re-adjust it if necessary.

Power steering pump drivebelt

Checking

5 On models equipped with a power steering pump, inspect the entire length of the drivebelt for signs of wear or damage. If the belt is found to be worn, frayed or cracked, it should be renewed as a precaution against breakage in service.

⚠️ *Warning: Breakage of the power steering pump drivebelt in service will cause the steering to become heavier, although the vehicle can still be driven on manual steering. A sudden change in the steering characteristics while driving could lead to an accident.*

14.2 Filler/level plug (arrowed) on belt-driven brake servo vacuum pump

Renewal

6 Remove the brake vacuum pump drivebelt as described above.
7 Slacken the power steering pump adjuster bolt and pivot bolt, then unhook the drivebelt from the pump pulley and remove it from the engine.
8 Locate the new drivebelt over the camshaft drive pulley and the pump pulley.
9 Tension the drivebelt by pivoting the pump around its mounting, until the deflection of the drivebelt under firm thumb pressure, midway between the pulleys, is according to the Specifications. Holding the pump in this position, securely tighten the pump pivot and adjuster bolts.
10 Refit and tension a new brake vacuum pump drivebelt as described above.
11 Re-check the tension of both drivebelts, and re-adjust it if necessary.

14 Brake servo vacuum pump oil level check

Note: *This Section applies only to belt-driven brake vacuum pumps. On models with a vacuum pump mounted directly on the camshaft, the pump is lubricated from the engine oil.*
1 Remove the air cleaner and ducting, as required, to access the brake vacuum pump (see Chapter 4C).
2 With the vehicle on level ground, unscrew the filler/level plug and check that the oil level is up to the bottom of the hole **(see illustration)**. If not, top-up with the specified grade of oil (see *"Lubricants and fluids"*), then refit and tighten the plug.
3 Refit the air cleaner and ducting.

15 Clutch adjustment check and control mechanism lubrication

1 Check that the clutch pedal moves smoothly and easily through its full travel, and that the clutch itself functions correctly, with no trace of slip or drag.
2 Adjust the clutch cable as described in Chapter 6.
3 If excessive effort is required to operate the clutch, check first that the cable is correctly routed and undamaged, then remove the pedal and check that its pivot is properly greased. Also grease the cable ends where they connect with the operating levers. Refer to Chapter 6 for further information.

16 Rear brake shoe and drum check

1 Remove the rear brake drums, and check the brake shoes for signs of wear or contamination. At the same time, also inspect the wheel cylinders for signs of leakage, and the brake drum for signs of wear. Refer to the relevant Sections of Chapter 9 for further information.

17 Road test

Instruments and electrical equipment

1 Check the operation of all instruments and electrical equipment.

2 Make sure that all instruments read correctly, and switch on all electrical equipment in turn to check that it functions properly.

Steering and suspension

3 Check for any abnormalities in the steering, suspension, handling or road "feel".

4 Drive the vehicle, and check that there are no unusual vibrations or noises.

5 Check that the steering feels positive, with no excessive "sloppiness", or roughness, and check for any suspension noises when cornering, or when driving over bumps.

Drivetrain

6 Check the performance of the engine, clutch, transmission and driveshafts.

7 Listen for any unusual noises from the engine, clutch and transmission.

8 Make sure that the engine runs smoothly when idling, and that there is no hesitation when accelerating.

9 Check that the clutch action is smooth and progressive, that the drive is taken up smoothly, and that the pedal travel is correct. Also listen for any noises when the clutch pedal is depressed.

10 Check that all gears can be engaged smoothly, without noise, and that the gear lever action is not abnormally vague or "notchy".

11 Listen for a metallic clicking sound from the front of the vehicle, as the vehicle is driven slowly in a circle with the steering on full lock. Carry out this check in both directions. If a clicking noise is heard, this indicates wear in a driveshaft joint, in which case the joint will have to be renewed. This process usually requires renewal of the complete driveshaft (see Chapter 8).

Check the operation and performance of the braking system

12 Make sure that the vehicle does not pull to one side when braking, and that the wheels do not lock prematurely when braking hard.

13 Check that there is no vibration through the steering when braking.

14 Check that the handbrake operates correctly, without excessive movement of the lever, and that it holds the vehicle stationary on a slope.

15 Test the operation of the brake servo unit as follows. With the engine off, depress the footbrake four or five times to exhaust the vacuum. Start the engine, holding the brake pedal depressed. As the engine starts, there should be a noticeable "give" in the brake pedal as vacuum builds up. Allow the engine to run for at least two minutes, and then switch it off. If the brake pedal is now depressed again, it should be possible to detect a hiss from the servo as the pedal is depressed. After about four or five applications, no further hissing should be heard.

Every 18 000 miles / 30 000 km or 18 months

18 Air filter renewal

Removal

1 Remove the front ducting from the air cleaner assembly. Undo the clip connecting it to the air cleaner housing, then pull the ducting away from the front grille and withdraw it from the vehicle (see illustration).

2 Release the clips on the air cleaner cover and pull the cover upwards, then remove the air filter element (see illustrations).

Refitting

3 Refitting is the reverse of removal. Lightly oil the air filter element.

19 Fuel filter renewal

Note: Although not essential, it is always beneficial to change the fuel filter just before winter, regardless of mileage.

Up to February 1993

Roto-diesel filter

1 This job may be carried out leaving the filter head in situ. However due to limited access and the possibility of spilling fuel over the engine, it is recommended that the filter head is removed, together with the cartridge. To remove the assembly, unscrew the union bolts and disconnect the inlet and outlet fuel unions from the filter head. Recover the union washers. Unbolt the filter head from the bracket and withdraw it, together with the cartridge (see illustration).

2 Unscrew the through-bolt to release the end cap, then remove the cartridge and seals (see illustrations). If the assembly has been removed from the vehicle, place it in a container before opening it, to catch spilled fuel.

3 Clean the filter head and end cap. Locate the new seals in position, then fit the new cartridge using a reversal of the removal procedure. Refit the filter assembly to the vehicle, if removed.

Purflux filter

4 To renew the Purflux filter element, the complete assembly has to be removed from the vehicle. Unscrew the union bolts and disconnect the inlet and outlet fuel unions

18.1 Remove the front ducting from the air cleaner

18.2a Release the clips on the air cleaner cover . . .

18.2b . . . then open the cover and remove the air filter element

19.1 Removing the Roto-diesel filter assembly

19.2a Unscrew the through-bolt . . .

19.2b . . . and remove the Roto-diesel filter cartridge

19.4 Remove the Purflux filter assembly

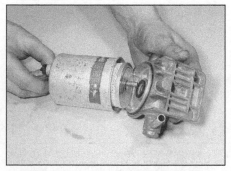

19.5a Unscrew the through-bolt and separate the chamber from the filter head . . .

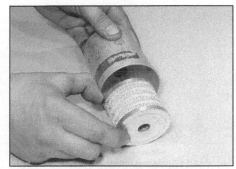

19.5b . . . remove the filter element . . .

from the filter head. Recover the union washers. Unbolt the filter head from the bracket and withdraw it, together with the lower chamber (see illustration).
5 Place the assembly in a container to catch any spilled fuel, then unscrew the through-bolt from underneath the chamber. Separate the chamber from the filter head and remove the filter element. Remove the inner and outer seals from the filter head, and the conical seal from the end of the spindle. Check that the spring is working correctly and if necessary remove it from the spindle (see illustrations).
6 Clean the filter head and chamber. Fit a new spring, if required. Locate the new seals in position, then fit the new cartridge using a reversal of the removal procedure. Refit the filter assembly to the vehicle,

February 1993 onwards

7 Diesel models from February 1993 onwards have a combined fuel filter/heater unit mounted on the thermostat housing, with a drain plug and tube at the base of the unit.
8 Cover the clutch bellhousing with a piece of plastic sheeting, to protect the clutch from fuel spillage, and place a suitable container beneath the drain tube. Open the drain plug by turning it anti-clockwise, and allow the fuel to drain completely (see illustration).

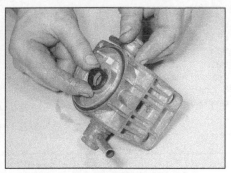

19.5c . . . remove the inner seal . . .

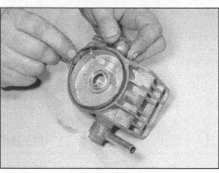

19.5d . . . and the outer seal from the filter head . . .

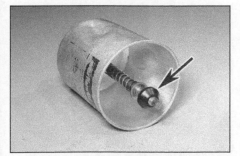

19.5e . . . and remove the conical seal (arrowed) from the end of the spindle - Purflux filter

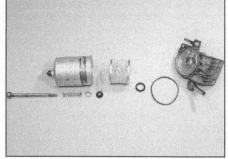

19.5f Complete disassembly of the Purflux filter

19.8 Drain the fuel from the filter mounted on the thermostat housing - February 1993 onwards

19.9a Undo the cover bolts . . .

9 Using a suitable Allen key or hexagon bit, remove the four filter housing cover bolts, then lift off the cover and remove the filter element from the housing, together with the sealing rubber (see illustrations).
10 The new filter element is supplied with a new sealing rubber. Make sure the sealing

rubber is correctly fitted to the upper edge of the filter element, and place the filter element in the housing.
11 Coat the threads of the filter cover securing bolts with thread-locking compound, then refit the cover and tighten the bolts securely.
12 Close the fuel filter drain plug and remove the plastic sheeting from above the clutch.

All models

13 Dispose of the drained fuel safely.
14 Prime the fuel system as described in Chapter 4C and start the engine.

20 Hinge and lock lubrication

1 Work around the vehicle, and lubricate the hinges of the bonnet and doors with a light machine oil.

19.9b . . . and remove the filter element from the housing, together with the sealing rubber (arrowed) - February 1993 onwards

2 Lightly lubricate the bonnet release mechanism and exposed section of inner cable with a smear of grease.
3 Check carefully the security and operation of all hinges, latches and locks, adjusting them where required. Check the operation of the central locking system (if fitted).

Every 24 000 miles / 40 000 km or 2 years

21 Brake fluid renewal

⚠️ *Warning: Brake hydraulic fluid can harm your eyes and damage painted surfaces, so use extreme caution when handling and pouring it. Do not use fluid that has been standing open for some time, as it absorbs moisture from the air. Excess moisture can cause a dangerous loss of braking effectiveness.*

1 The procedure is similar to that for bleeding the hydraulic system as described in Chapter 9, except that allowance should be made for all the old fluid to be expelled when bleeding a section of the circuit.

2 The brake fluid reservoir should first be emptied of old fluid by siphoning, using a clean poultry baster, syringe, or similar, then refilled with fresh fluid. Do not operate the brakes while the reservoir is being emptied, or air will be drawn into the system.
3 Working as described in Chapter 9, open the first bleed screw in the sequence, and pump the brake pedal gently until nearly all the fluid has been emptied from the reservoir. Top-up to the "MAXI" level with new fluid, and continue pumping until only new fluid can be seen emerging from the bleed screw. Tighten the screw, and top the reservoir level up to the "MAXI" level line.
4 Work through all the remaining bleed screws in the sequence until new fluid can be seen at all of them. Be careful to keep the

master cylinder reservoir topped-up to above the "DANGER" level at all times, or air may enter the system and greatly increase the length of the task.

 HAYNES HiNT *Old hydraulic fluid is invariably much darker in colour than the new, making it easy to distinguish the two.*

5 When the operation is complete, check that all bleed screws are securely tightened, and that their dust caps are refitted. Wash off all traces of spilt fluid, and recheck the master cylinder reservoir fluid level.
6 Check the operation of the brakes before taking the vehicle on the road.

Every 36 000 miles / 60 000 km or 3 years

22 Manual transmission oil level check

Note: *A new sealing washer is required for the filler/level plug.*

1 Park the vehicle on a level surface, and wait at least 5 minutes after the engine has been switched off.

HAYNES HiNT *If the oil is checked immediately after driving the vehicle, some of the oil will remain distributed around the transmission components, giving an inaccurate level reading.*

2 To access the filler/level plug, firmly apply the handbrake, jack up the front left corner of the vehicle using a suitable trolley jack, and remove the roadwheel (see *"Jacking and Vehicle Support"*). Use jacks and axle stands to keep the vehicle perfectly level, to ensure accuracy, when refilling and checking the oil level. If necessary, support the vehicle at all four jacking points.

 HAYNES HiNT *This job is most easily performed while working in a pit, so that the vehicle remains level without having to jack it up.*

3 Wipe clean the area around the filler/level plug, which is the largest bolt securing the end cover to the transmission (see illustration).

Unscrew the plug and clean it. Discard the sealing washer.

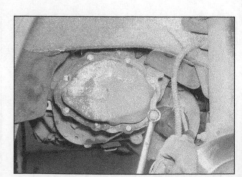

22.3 Remove the transmission oil filler/level plug

4 The oil level should reach the lower edge of the filler/level hole. A certain amount of oil will have gathered behind the filler/level plug, and will trickle out when it is removed. This does **not** necessarily indicate that the level is correct. To ensure that a true level has been established, wait until the initial trickle has stopped, then add oil as necessary until a trickle of new oil can be seen emerging. The level will be correct when the flow ceases. Use only good-quality oil of the type specified in *"Lubricants and fluids"* (**see illustrations**).

5 Filling the transmission with oil is an extremely awkward operation; above all, allow plenty of time for the oil level to settle properly before checking it. If a large amount had to be added to the transmission, and a large amount flows out on checking the level, refit the filler/level plug and take the vehicle on a short journey so that the new oil is distributed fully around the transmission components, then re-check the level when it has settled again.

6 If the transmission has been overfilled so

22.4a Topping up the manual transmission oil level

22.4b The oil level is correct when the oil stops flowing out of the filler/level hole

that oil flows out as soon as the filler/level plug is removed, check that the vehicle is completely level (front-to-rear and side-to-side), and allow the surplus to drain off into a suitable container.

7 When the level is correct, fit a new sealing washer to the filler/level plug. Refit the plug and tighten it to the specified torque. Wash off any spilt oil.

Note: *From January 1991 onwards, there is a web at the lower edge of the filler/level hole, to increase the oil capacity. Make sure the oil delivery nozzle goes over the web and into the casing when filling the transmission.*

Every 48 000 miles / 80 000 km or 4 years

23 Timing belt renewal

Caution: If the timing belt breaks in service, extensive damage may be caused to the engine. Renewal at or before the specified interval is strongly recommended.

Refer to Chapter 2B.

Every 72 000 miles / 120 000 km or 6 years

24 Manual transmission oil renewal

Refer to Chapter 7.

Every 2 years, regardless of mileage

25 Coolant renewal

Note: *The coolant should be renewed at the first 12 000 miles (20 000 km) from new, or from when the engine is renewed. This is necessary to flush out corrosive elements that can build up to high levels during the early life of the engine. Thereafter renew as described in the maintenance schedule.*

Cooling system draining

 Warning: Wait until the engine is cold before starting this procedure. Do not allow antifreeze to come

into contact with your skin, or with the painted surfaces of the vehicle. Rinse off spills immediately with plenty of water. Never leave antifreeze lying around in an open container, or in a puddle in the driveway or on the garage floor. Children and pets are attracted by its sweet smell, but antifreeze can be fatal if ingested.

 If the cooling system is being drained for purposes other than renewal, it is not always necessary to drain the complete system. For example, if a component is to be disconnected in the upper part of the circuit, just drain the coolant until the level is below the component.

1 With the engine completely cold, remove the expansion tank filler cap. Turn the cap anti-clockwise until it reaches the first stop. Wait until any pressure remaining in the system is released, then push the cap down, turn it anti-clockwise to the second stop, and lift it off.

2 Position a suitable container beneath the lower left-hand side of the radiator. Loosen the radiator drain plug and allow the coolant to drain into the container. **Note:** *If there is no drain plug fitted, disconnect the radiator bottom hose.*

3 To assist draining, open the cooling system bleed screws, located as follows.

a) On the thermostat housing, up to February 1993 (**see illustration**).

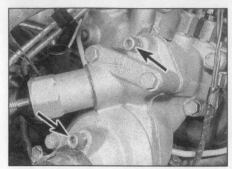

25.3a Coolant bleed screws (arrowed) on the thermostat housing - up to February 1993

25.3b Coolant bleed screw in hose running to top of fuel filter/thermostat housing - February 1993 to January 1994

25.3c Coolant bleed screw (arrowed) in pipework to header tank - January 1994 onwards

b) *In the coolant hose running to the top of the fuel filter/thermostat housing, from February 1993 to January 1994 (see illustration).*
c) *In the four-way union in the pipework connecting the fuel filter/thermostat housing and radiator to the header tank, from January 1994 onwards (see illustration).*
d) *At the top left-hand corner of the radiator.*
e) *In the heater hose connector near the engine compartment bulkhead, on some models.*

4 When the flow of coolant stops, reposition the container below the cylinder block drain plug, located at the rear of the cylinder block, near the flywheel. Remove the drain plug, and allow the coolant to drain into the container.

5 If the coolant has been drained for a reason other than renewal, then provided it is clean and less than two years old, it can be re-used.

6 Refit the radiator drain plug and the cylinder block drain plug on completion of draining.

Cooling system flushing

7 If coolant renewal has been neglected, or if the antifreeze mixture has become diluted, then in time, the cooling system may gradually lose efficiency, as the coolant passages become restricted due to rust, scale deposits, and other sediment. The cooling system efficiency can be restored by flushing the system clean.

8 The radiator should be flushed independently of the engine, to avoid unnecessary contamination.

Radiator flushing

9 To flush the radiator, tighten the radiator drain plug and bleed screw, and disconnect the top and bottom hoses. On models from January 1994 onwards with a remote expansion tank, also disconnect the expansion tank hose from the right-hand side of the radiator.

10 Insert a garden hose into the radiator top inlet. Direct a flow of clean water through the radiator, and continue flushing until clean water emerges from the radiator bottom outlet.

11 If, after a reasonable period, the water still does not run clear, the radiator can be flushed with a good proprietary cleaning agent. It is

important that their manufacturer's instructions are followed carefully. If the contamination is particularly bad, remove the radiator, then insert the hose in the radiator bottom outlet, and reverse-flush the radiator.

Engine flushing

12 To flush the engine, first refit the cylinder block drain plug, and tighten the cooling system bleed screws.

13 Remove the thermostat as described in Chapter 3, then temporarily refit the thermostat cover.

14 With the top and bottom hoses disconnected from the radiator, insert a garden hose into the radiator top hose. Direct a clean flow of water through the engine, and continue flushing until clean water emerges from the radiator bottom hose.

15 On completion of flushing, refit the thermostat and reconnect the hoses with reference to Chapter 3.

Cooling system filling

16 Before attempting to fill the cooling system, make sure that all hoses and clips are in good condition, and that the clips are tight. Note that an antifreeze mixture must be used all year round, to prevent corrosion of the engine components (see the following sub-Section). Also check that the radiator and cylinder block drain plugs are in place and tight.

17 Remove the expansion tank filler cap.

18 Open all the cooling system bleed screws (see paragraph 3).

Models with filler cap on radiator

19 On models with the expansion tank and filler cap integral with the radiator, some of the cooling system hoses are positioned at a higher level than the top of the expansion tank. It is therefore necessary to use a "header tank" when refilling the cooling system, to eliminate the possibility of air being trapped in the system.

 HAYNES HiNT *Although Citroën dealers use a special header tank, the same effect can be achieved by using a suitable bottle or plastic container sealed to the expansion tank filler neck.*

20 Fit the "header tank" to the expansion tank, and slowly fill the system. Coolant will emerge from each of the bleed screws in turn, starting with the lowest screw. As soon as coolant free from air bubbles emerges from the lowest screw, tighten that screw, and watch the next bleed screw in the system. Repeat the procedure on the remaining bleed screws.

21 Ensure that the "header tank" is full (at least 0.5 litres of coolant). Start the engine, and run it at a fast idle speed (do not exceed 2000 rpm) until the cooling fans cut in, and then cut out. Stop the engine.

22 Remove the "header tank", taking great care not to scald yourself with the hot coolant, then fit the expansion tank cap.

All models

23 Allow the engine to cool, then check the coolant level (see *"Weekly checks"*). Top-up the level if necessary.

Antifreeze mixture

24 The antifreeze should always be renewed at the specified intervals. This is necessary not only to maintain the antifreeze properties, but also to prevent corrosion which would otherwise occur as the corrosion inhibitors become progressively less effective.

25 Always use an ethylene-glycol based antifreeze which is suitable for use in mixed-metal cooling systems. The quantity of antifreeze and levels of protection are indicated in the Specifications.

26 Before adding antifreeze, the cooling system should be completely drained, preferably flushed, and all hoses checked for condition and security.

27 After filling with antifreeze, a label should be attached to the expansion tank filler neck, stating the type and concentration of antifreeze used, and the date installed. Any subsequent topping-up should be made with the same type and concentration of antifreeze.

28 Do not use engine antifreeze in the windscreen/rear window washer system, as it will cause damage to the vehicle paintwork. Instead, a screenwash additive should be added to the washer system in the quantities stated on the bottle.

Chapter 2 Part A:
Petrol engine in-vehicle repair procedures

Contents

Degrees of difficulty

Easy, suitable for novice with little experience	**Fairly easy,** suitable for beginner with some experience	**Fairly difficult,** suitable for competent DIY mechanic 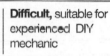	**Difficult,** suitable for experienced DIY mechanic	**Very difficult,** suitable for expert DIY or professional

Specifications

General

Designation:
954 cc engine	TU9
1124 cc engine	TU1

Engine codes*:
954 cc carburettor engine	C1A
1124 cc carburettor engine	H1A
1124 cc fuel-injected engine	HDZ
No 1 cylinder location	Flywheel/transmission end of block
Direction of crankshaft rotation	Clockwise (viewed from right-hand side of vehicle)
Firing order	1-3-4-2

Bore:
954 cc engine	70.00 mm
1124 cc engine	72.00 mm

Stroke:
954 cc engine	62.00 mm
1124 cc engine	69.00 mm

Compression ratio:
954 cc engine	9.4 : 1
1124 cc engine	9.4 : 1

Maximum power (DIN):
954 cc engine	50 bhp (37.3 kW) @ 6000 rpm
1124 cc engine	60 bhp (44.1 kW) @ 6200 rpm

Maximum torque (DIN):
954 cc engine	74 Nm (54.6 lbf ft) @ 3700 rpm
1124 cc engine	90 Nm (66.4 lbf ft) @ 3800 rpm

*The engine code is the first three letters of the engine number, stamped on a plate which is riveted to the left-hand end of the front face of the cylinder block.

Camshaft

Drive method	Toothed belt
Number of bearings	5

Camshaft bearing journal diameter (outside diameter):

No 1	36.950 to 36.925 mm
No 2	40.650 to 40.625 mm
No 3	1.250 to 41.225 mm
No 4	1.850 to 41.825 mm
No 5	2.450 to 42.425 mm

Cylinder head bearing journal diameter (inside diameter):

No 1	7.000 to 37.039 mm
No 2	0.700 to 47.739 mm
No 3	1.300 to 41.339 mm
No 4	1.900 to 41.939 mm
No 5	2.500 to 42.539 mm

Valve clearances (engine cold)

Inlet valves	0.20 mm
Exhaust valves	0.40 mm

Lubrication system

Oil pump type	Gear-type, chain-driven off the crankshaft
Minimum oil pressure at 90°C	4 bars at 4000 rpm
Oil pressure warning switch operating pressure	0.5 bars

Torque wrench settings

	Nm	lbf ft
Big-end bearing cap nuts	40	30
Camshaft cover nuts	16	12
Camshaft sprocket retaining bolt	80	59
Camshaft thrust fork retaining bolt	16	12
Crankshaft pulley retaining bolts	8	6
Crankshaft sprocket retaining bolt	100	74
Cylinder head bolts:		
Stage 1	20	15
Stage 2	Angle-tighten through 240°	
Engine/transmission mountings*:		
Right-hand mounting-to-body bracket:		
To engine	50	37
To mounting rubber	35	26
Left-hand stud bracket-to-transmission nuts	18	13
Flywheel retaining bolts	65	48
Main bearing ladder casting:		
11 mm bolts:		
Stage 1	20	15
Stage 2	Angle-tighten through 45°	
6 mm bolts	8	6
Oil pump retaining bolts	8	6
Sump drain plug	30	22
Sump retaining nuts and bolts	8	6
Timing belt cover bolts	8	6
Timing belt tensioner pulley nut	23	17

See Chapter 2B Specifications for additional torque wrench settings for left-hand and lower mountings.

1 General information

How to use this Chapter

This Part of Chapter 2 describes those repair procedures that can reasonably be carried out on the TU series petrol engine while it remains in the vehicle. If the engine has been removed from the vehicle and is being dismantled as described in Part C, any preliminary dismantling procedures can be ignored.

Note that, while it may be possible physically to overhaul items such as the piston/connecting rod assemblies while the engine is in the vehicle, such tasks are not normally carried out as separate operations. Usually, several additional procedures (not to mention the cleaning of components and of oilways) have to be carried out. For this reason, all such tasks are classed as major overhaul procedures, and are described in Part C of this Chapter.

Part C describes the removal of the engine/transmission unit from the vehicle, and the full overhaul procedures that can then be carried out.

Engine description

The TU series engine is a well-proven engine which has been fitted to many previous Citroën and Peugeot vehicles. The engine is of the in-line four-cylinder, overhead camshaft (OHC) type, mounted transversely at the front of the vehicle. The clutch and transmission are attached to its left-hand end. The engine sizes are 954 cc with carburettor, and 1124 cc with either carburettor or monopoint fuel injection.

The crankshaft runs in five main bearings. Thrustwashers are fitted to the No 2 main bearing (upper half) to control the crankshaft endfloat.

The connecting rods rotate on horizontally-split bearing shells at their big-ends. The pistons are attached to the connecting rods by gudgeon pins, which are an interference fit in the connecting rod small-end eyes. The aluminium-alloy pistons are fitted with three piston rings - two compression rings and an oil control ring.

The cylinder bores have replaceable wet liners. Sealing O-rings are fitted at the base of each liner, to prevent the escape of coolant into the sump.

The inlet and exhaust valves are each closed by coil springs, and operate in guides pressed into the cylinder head. The valve seat inserts are also pressed into the cylinder head, and can be renewed separately if worn.

The camshaft is driven by a toothed timing belt, and operates the eight valves via rocker arms. Valve clearances are adjusted by a screw-and-locknut arrangement. The camshaft rotates directly in the cylinder head. The timing belt also drives the coolant pump.

Lubrication is by means of an oil pump, which is driven (via a chain and sprocket) off the right-hand end of the crankshaft. It draws oil through a strainer located in the sump, and then forces it through an externally-mounted filter into galleries in the cylinder block/crankcase. From there, the oil is distributed to the crankshaft (main bearings) and camshaft. The big-end bearings are supplied with oil via internal drillings in the crankshaft, while the camshaft bearings also receive a pressurised supply. The camshaft lobes and valves are lubricated by splash, as are all other engine components.

Throughout this manual, it is often necessary to identify the engines not only by their capacity, but also by their engine code. On petrol engines, the engine number can be found on a plate riveted to the left-hand end of the front face of the cylinder block. The first part of the engine number gives the engine code - e.g. "H1A" (see illustration).

Repair operations possible with the engine in the vehicle

The following work can be carried out with the engine in the vehicle:

a) Compression pressure - testing.
b) Cylinder head cover - removal and refitting.
c) Timing belt covers - removal and refitting.
d) Timing belt - removal, refitting and adjustment.
e) Timing belt tensioner and sprockets - removal and refitting.
f) Camshaft oil seal - renewal.
g) Camshaft and rocker arms - removal, inspection and refitting.*
h) Cylinder head - removal and refitting.
i) Cylinder head and pistons - decarbonising (refer to Part C of this Chapter).
j) Sump - removal and refitting.
k) Oil pump - removal, inspection and refitting.
l) Crankshaft oil seals - renewal.

m) Engine/transmission mountings - inspection and renewal.
n) Flywheel - removal, inspection and refitting.
*The cylinder head must be removed for the successful completion of this work. Refer to Section 12 for details.

2 Compression test - description and interpretation

1 When engine performance is down, or if misfiring occurs which cannot be attributed to the ignition or fuel systems, a compression test can provide diagnostic clues as to the engine's condition. If the test is performed regularly, it can give warning of trouble before any other symptoms become apparent.
2 The engine must be fully warmed-up to normal operating temperature, the battery must be fully charged, and all the spark plugs must be removed (see Chapter 1A). The aid of an assistant will also be required.
3 Disable the ignition system by disconnecting the ignition HT coil lead from the distributor cap and earthing it on the cylinder block. Use a jumper lead or similar wire to make a good connection.
4 On fuel injection models, disconnect the fuel injection system relay unit (see Chapters 4B and 12) to disable the fuel pump and prevent contamination of the catalytic converter with unburnt fuel.
5 Fit a compression tester to the No 1 cylinder spark plug hole at the transmission end of the block. The type of tester which screws into the plug thread is to be preferred.
6 Have the assistant hold the throttle wide open, and crank the engine on the starter motor. After one or two revolutions, the compression pressure should build up to a maximum figure, and then stabilise. Record the highest reading obtained.
7 Repeat the test on the remaining cylinders, recording the pressure in each.
8 All cylinders should produce very similar pressures. A difference of more than 2 bars between any two cylinders indicates a fault. The compression should build up quickly in a healthy engine. Low compression on the first stroke, followed by gradually-increasing pressure on successive strokes, indicates worn piston rings. A low compression reading on the first stroke, which does not build up during successive strokes, indicates leaking valves or a blown head gasket (a cracked head could also be the cause). Deposits on the undersides of the valve heads can also cause low compression.
9 Although Citroën do not specify exact compression pressures, as a guide, any cylinder pressure of below 10 bars can be considered as less than healthy. Refer to a Citroën dealer or other specialist if in doubt as to whether a particular pressure reading is acceptable.

1.11 Engine code is stamped on a plate (arrowed) attached to the front of the cylinder block

10 If the pressure in any cylinder is low, carry out the following test to isolate the cause. Introduce a teaspoonful of clean oil into that cylinder through its spark plug hole, and repeat the test.
11 If the addition of oil temporarily improves the compression pressure, this indicates that bore or piston wear is responsible for the pressure loss. No improvement suggests that leaking or burnt valves, or a blown head gasket, may be the cause.
12 A low reading from two adjacent cylinders is almost certainly due to the head gasket having blown between them. The presence of coolant in the engine oil, or vice-versa, will confirm this.
13 If one cylinder is about 20 percent lower than the others and the engine has a slightly rough idle, a worn camshaft lobe could be the cause.
14 If the compression reading is unusually high, the combustion chambers are probably coated with carbon deposits. If this is the case, the cylinder head should be removed and decarbonised.
15 On completion of the test, refit the spark plugs and reconnect the ignition system.

3 Top dead centre (TDC) for No 1 piston - locating

Caution: Do not attempt to rotate the engine whilst the crankshaft/camshaft are locked in position. If the engine is to be left in this state for a long period of time, it is a good idea to place warning notices inside the vehicle, and in the engine compartment. This will reduce the possibility of the engine being accidentally cranked on the starter motor, which is likely to cause damage with the locking tools in place.

1 On all models, timing holes are drilled in the camshaft sprocket and in the flywheel. The holes are used to ensure that the crankshaft and camshaft are correctly positioned when assembling the engine (to prevent the possibility of the valves contacting the pistons when refitting the cylinder head), or when refitting the timing belt. When the timing holes are aligned with access holes in the cylinder

3.4 Insert a 6 mm bolt (arrowed) through hole in cylinder block flange and into timing hole in the flywheel . . .

3.5 . . . then insert a 10 mm bolt through the camshaft sprocket timing hole, and locate it in the cylinder head

head and the front of the cylinder block, suitable diameter tools can be inserted to lock both the camshaft and crankshaft in position, preventing them from rotating. Proceed as follows. **Note:** *With the timing holes aligned, No 1 cylinder at the transmission end of the cylinder block is at TDC on its compression stroke.*

2 Remove the timing belt upper cover, as described in Section 6.

3 The crankshaft must now be turned until the timing hole in the camshaft sprocket is aligned with the corresponding hole in the cylinder head. The holes are aligned when the

4.3 Disconnect the breather hose from the cylinder head cover . . .

camshaft sprocket hole is in the 2 o'clock position, when viewed from the right-hand end of the engine. The crankshaft can be turned by using a spanner on the crankshaft sprocket bolt, noting that it should always be rotated in a clockwise direction (viewed from the right-hand end of the engine).

 Turning the crankshaft will be much easier if the spark plugs are removed first (see Chapter 1A).

4 With the camshaft sprocket hole correctly positioned, insert a 6 mm diameter bolt, or drill bit, through the hole in the front, left-hand flange of the cylinder block, and locate it in the timing hole in the flywheel **(see illustration)**. Note that it may be necessary to rotate the crankshaft slightly, to get the holes to align.

5 With the flywheel correctly positioned, insert a 10 mm diameter bolt, or drill bit, through the timing hole in the camshaft sprocket, and locate it in the hole in the cylinder head **(see illustration)**.

6 The crankshaft and camshaft are now locked in position, preventing unnecessary rotation.

4 Camshaft cover - removal and refitting

Removal

1 Disconnect the battery negative lead.

2 Where necessary, undo the bolts securing the HT lead retaining clips to the rear of the cylinder head cover, and position the clips clear of the cover.

3 Slacken the retaining clip, and disconnect the breather hose from the left-hand end of the cylinder head cover **(see illustration)**. Where the original crimped-type Citroën hose clip is fitted, cut it off and discard it. Use a standard worm-drive clip on refitting.

4 Undo the two retaining nuts, and remove the washer from each of the cylinder head cover studs **(see illustration)**.

5 Lift off the cylinder head cover, and remove it along with its rubber seal **(see illustration)**. Examine the seal for signs of damage and deterioration, and if necessary, renew it.

6 Remove the spacer from each stud, and lift off the oil baffle plate **(see illustrations)**.

4.4 . . . then slacken and remove the cover retaining nuts and washers . . .

4.5 . . . and lift off the cylinder head cover

4.6a Lift off the spacers . . .

Refitting

7 Carefully clean the cylinder head and cover mating surfaces, and remove all traces of oil.

8 Fit the rubber seal over the edge of the cylinder head cover, ensuring that it is correctly located along its entire length **(see illustration)**.

9 Refit the oil baffle plate to the engine, and locate the spacers in their recesses in the baffle plate.

10 Carefully refit the cylinder head cover to the engine, taking great care not to displace the rubber seal.

11 Check that the seal is correctly located, then refit the washers and retaining nuts to the cover, and tighten them to the specified torque.

12 Where necessary, refit the HT lead clips to the rear of the head cover, and securely tighten their retaining bolts.

13 Reconnect the breather hose to the cylinder head cover, securely tightening its retaining clip, and reconnect the battery negative lead.

5 Crankshaft pulley - removal and refitting

1 The crankshaft pulley is fitted over the lower timing belt cover. Follow the procedures given in Section 6, but leave the lower cover in place.

4.6b . . . and remove the oil baffle plate

6 Timing belt covers - removal and refitting

Removal

Upper cover

1 Slacken and remove the two retaining bolts (one at the front and one at the rear), and remove the upper timing cover from the cylinder head **(see illustrations)**.

Centre cover

2 Remove the upper cover as described in paragraph 1, then free the wiring from its

4.8 On refitting, ensure that the rubber seal is correctly located on the cylinder head cover

retaining clips on the centre cover **(see illustration)**.

3 Slacken and remove the retaining bolts, and manoeuvre the centre cover out from the engine compartment **(see illustration)**.

Lower cover

4 Remove the auxiliary drivebelt as described in Chapter 1A.

5 Remove the upper and centre covers as described in paragraphs 1 to 3.

6 Undo the three crankshaft pulley retaining bolts and remove the pulley, noting which way round it is fitted **(see illustrations)**.

7 Slacken and remove the single retaining bolt, and slide the lower cover off the end of the crankshaft **(see illustration)**.

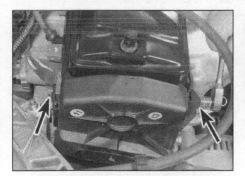

6.1a Undo the two retaining bolts (arrowed) . . .

6.1b . . . and remove the upper timing belt cover

6.2 Free the wiring loom from its retaining clips . . .

6.3 . . . then undo the retaining bolts (locations arrowed) and remove the centre timing belt cover

6.6a Undo the three retaining bolts (arrowed) . . .

6.6b . . . and remove the crankshaft pulley

6.7 Undo the retaining bolt and remove the lower timing belt cover

Refitting

Upper cover

8 Refit the cover, ensuring that it is correctly located with the centre cover, and tighten its retaining bolts.

Centre cover

9 Manoeuvre the centre cover back into position, ensuring it is correctly located with the lower cover, and tighten its retaining bolts.
10 Clip the wiring loom into its retaining clips on the front of the centre cover, then refit the upper cover as described in paragraph 8.

Lower cover

11 Locate the lower cover over the timing belt sprocket, and tighten its retaining bolt.
12 Fit the pulley to the end of the crankshaft, ensuring that it is fitted the correct way round, and tighten its retaining bolts to the specified torque.
13 Refit the centre and upper covers as described above, then refit and tension the auxiliary drivebelt as described in Chapter 1A.

---

7 Timing belt - general, removal, refitting and tensioning

General

1 The timing belt drives the camshaft and coolant pump from a toothed sprocket on the front of the crankshaft. If the belt breaks or slips in service, the pistons are likely to hit the valve heads, resulting in extensive (and

7.8 Mark the direction of rotation on the belt, if it is to be re-used

expensive) damage.
2 The timing belt should be renewed at the specified intervals (see Chapter 1A), or earlier if it is contaminated with oil, or if it is at all noisy in operation (a "scraping" noise due to uneven wear).
3 If the timing belt is being removed, it is a wise precaution to check the condition of the coolant pump at the same time (check for signs of coolant leakage). This may avoid the need to remove the timing belt again at a later stage, should the coolant pump fail.

Removal

4 Disconnect the battery negative terminal.
5 Align the engine assembly/valve timing holes as described in Section 3, and lock both the camshaft sprocket and the flywheel in position. *Do not* attempt to rotate the engine whilst the locking tools are in position.
6 Remove the timing belt centre and lower covers as described in Section 6.
7 Loosen the timing belt tensioner pulley retaining nut. Pivot the pulley in a clockwise direction, using a square-section key fitted to the hole in the pulley hub, then retighten the retaining nut.
8 If the timing belt is to be re-used, use white paint or similar to mark the direction of rotation on the belt (if markings do not already exist) **(see illustration)**. Slip the belt off the sprockets.
9 Check the timing belt carefully for any signs of uneven wear, splitting, or oil contamination. Pay particular attention to the roots of the teeth. If signs of oil contamination are found, trace the source of the oil leak, and rectify it. Wash down the engine timing belt area and all related components, to remove all traces of oil.

> **HAYNES HINT** *Renew the belt if there is the slightest doubt about its condition. If the engine is undergoing an overhaul, and has covered more than 36 000 miles (60 000 km) with the existing belt fitted, renew the belt as a matter of course, regardless of its apparent condition. The cost of a new belt is nothing when compared to the cost of repairs, should the belt break in service.*

Refitting and tensioning

10 Before refitting, thoroughly clean the timing belt sprockets. Check that the tensioner pulley rotates freely, without any sign of roughness. If necessary, renew the tensioner pulley as described in Section 8. Make sure that the locking tools are still in place, as described in Section 3.
11 Manoeuvre the timing belt into position, ensuring that the arrows on the belt are pointing in the direction of rotation (clockwise when viewed from the right-hand end of the engine).

12 Do not twist the timing belt sharply while refitting it. Fit the belt over the crankshaft and camshaft sprockets. Make sure that the "front run" of the belt is taut - i.e., ensure that any slack is on the tensioner pulley side of the belt. Fit the belt over the water pump sprocket and tensioner pulley. Ensure that the belt teeth are seated centrally in the sprockets.
13 Loosen the tensioner pulley retaining nut. Pivot the pulley anti-clockwise to remove all free play from the timing belt, then retighten the nut.
14 Citroën tool No 4507-T.J. is used for tensioning the timing belt. This is a rod which fits into the hole in the tensioner pulley, and extends horizontally to the left with a weight hanging on the end. If the tool is not available, it can be fabricated using a suitable square-section bar attached to an arm. A hole should be drilled in the arm at a distance of 80 mm from the centre of the square-section bar. Fit the tool to the hole in the tensioner pulley, keeping the tool arm as close to the horizontal as possible, and hang a 1.5 kg (3.3 lb) weight from the hole in the tool. In the absence of an object of the specified weight, a spring balance can be used to exert the required force, ensuring that the spring balance is held at 90° to the tool arm. Slacken the pulley retaining nut, allowing the weight or force exerted (as applicable) to push the tensioner pulley against the belt, then retighten the pulley nut.
15 If a tool is not available, an approximate setting may be achieved as follows. Slacken the pulley retaining nut, and pivot the tensioner pulley anti-clockwise until it is just possible to twist the timing belt through 90° by finger and thumb, midway between the crankshaft and camshaft sprockets. The square hole in the tensioner pulley hub should be directly below the retaining nut, and the deflection of the belt at the mid-point between the sprockets should be approximately 6.0 mm. If this method is used, the belt tension should be checked by a Citroën dealer at the earliest possible opportunity.
16 Remove the locking tools from the camshaft sprocket and flywheel.
17 Using a socket and extension bar on the crankshaft sprocket bolt, rotate the crankshaft through four complete rotations in a clockwise direction (viewed from the right-hand end of the engine). *Do not* at any time rotate the crankshaft anti-clockwise.
18 Slacken the tensioner pulley nut, re-tension the belt using one of the methods just described, then tighten the tensioner pulley nut to the specified torque.
19 Rotate the crankshaft through a further two turns clockwise, and check that both the camshaft sprocket and flywheel timing holes are still correctly aligned.
20 If all is well, refit the timing belt covers as described in Section 6, and reconnect the battery negative terminal.

8.10 Use the fabricated tool shown to lock flywheel ring gear and prevent the crankshaft rotating

8.11a Remove the crankshaft sprocket retaining bolt . . .

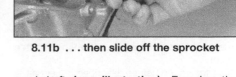

8.11b . . . then slide off the sprocket

8 Timing belt tensioner and sprockets - removal, inspection and refitting

Removal

Note: *This Section describes the removal and refitting of the components concerned as individual operations. If more than one of them is to be removed at the same time, start by removing the timing belt as described in Section 7. Remove the actual component as described below, ignoring the preliminary dismantling steps.*

1 Disconnect the battery negative terminal.
2 Position the engine assembly/valve timing holes as described in Section 3, and lock both the camshaft sprocket and flywheel in position. *Do not* attempt to rotate the engine whilst the locking tools are in position.

Camshaft sprocket

3 Remove the upper and centre timing belt covers as described in Section 6.
4 Loosen the timing belt tensioner pulley retaining nut. Rotate the pulley in a clockwise direction, using a square-section key fitted to the hole in the pulley hub, then retighten the retaining nut.
5 Disengage the timing belt from the sprocket, and move the belt clear, taking care not to bend or twist it sharply. Remove the locking tool from the camshaft sprocket.
6 Slacken the camshaft sprocket retaining bolt and remove it, along with its washer. To prevent the camshaft rotating as the bolt is slackened, a sprocket-holding tool will be required. *Do not* attempt to use the locking tool to prevent the sprocket from rotating whilst the bolt is slackened.

 HAYNES HINT *In the absence of the special Citroën tool to prevent the camshaft rotating, an acceptable substitute can be fabricated as follows. Use two lengths of steel strip (one long, the other short), and three nuts and bolts; one nut and bolt forms the pivot of a forked tool, with the remaining two nuts and bolts at the tips of the "forks" to engage with the sprocket spokes as shown (see illustration 8.18).*

7 With the retaining bolt removed, slide the sprocket off the end of the camshaft. If the locating peg is a loose fit in the rear of the sprocket, remove it for safe-keeping. Examine the camshaft oil seal for signs of oil leakage and, if necessary, renew it as described in Section 9.

Crankshaft sprocket

8 Remove the centre and lower timing belt covers as described in Section 6.
9 Loosen the timing belt tensioner pulley retaining nut. Rotate the pulley in a clockwise direction, using a square-section key fitted to the hole in the pulley hub, then retighten the retaining nut.
10 To prevent crankshaft rotation whilst the sprocket retaining bolt is slackened, select top gear, and have an assistant apply the brakes firmly. If the engine has been removed from the vehicle, lock the flywheel ring gear, using an arrangement similar to that shown **(see illustration)**. *Do not* be tempted to use the flywheel locking tool to prevent the crankshaft from rotating; temporarily remove the locking tool from the rear of the flywheel prior to slackening the pulley bolt, then refit it once the bolt has been slackened.
11 Unscrew the retaining bolt and washer, then slide the sprocket off the end of crankshaft **(see illustrations)**. Refit the locking tool through the timing hole into the rear of the flywheel.
12 If the Woodruff key is a loose fit in the crankshaft, remove it and store it with the sprocket for safe-keeping. If necessary, also slide the flanged spacer off the end of the

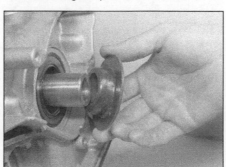

8.12 Remove the flanged spacer if necessary

crankshaft **(see illustration)**. Examine the crankshaft oil seal for signs of oil leakage and, if necessary, renew it as described in Section 15.

Tensioner pulley

13 Remove the centre timing belt cover as described in Section 6.
14 Slacken and remove the timing belt tensioner pulley retaining nut, and slide the pulley off its mounting stud. Examine the mounting stud for signs of damage and, if necessary, renew it.

Inspection

15 Clean the sprockets thoroughly, and renew any that show signs of wear, damage or cracks.
16 Clean the tensioner assembly, but do not use any strong solvent which may enter the pulley bearing. Check that the pulley rotates freely about its hub, with no sign of stiffness or of free play. Renew the tensioner pulley if there is any doubt about its condition, or if there are any obvious signs of wear or damage.

Refitting

Camshaft sprocket

17 Refit the locating peg (where removed) to the rear of the sprocket, then locate the sprocket on the end of the camshaft. Ensure that the locating peg is correctly engaged with the cut-out in the camshaft end.
18 Refit the sprocket retaining bolt and washer. Tighten the bolt to the specified torque, whilst retaining the sprocket with the tool used on removal **(see illustration)**.

8.18 Using a home-made tool to hold the camshaft sprocket stationary whilst the retaining bolt is tightened

19 Realign the timing hole in the camshaft sprocket (see Section 3) with the corresponding hole in the cylinder head, and refit the locking tool.

20 Refit the timing belt to the camshaft sprocket. Ensure that the "front run" of the belt is taut - i.e., ensure that any slack is on the tensioner pulley side of the belt. Do not twist the belt sharply while refitting it, and ensure that the belt teeth are seated centrally in the sprockets.

21 Loosen the tensioner pulley retaining nut. Rotate the pulley anti-clockwise to remove all free play from the timing belt, then retighten the nut.

22 Tension the belt as described in Section 7.

23 Refit the timing belt covers as described in Section 6, and reconnect the battery negative terminal.

Crankshaft sprocket

24 Where removed, locate the Woodruff key in the crankshaft end, then slide on the flanged spacer, aligning its slot with the Woodruff key.

25 Align the crankshaft sprocket slot with the Woodruff key, and slide it onto the end of the crankshaft.

26 Temporarily remove the locking tool from the rear of the flywheel, then refit the crankshaft sprocket retaining bolt and washer. Tighten the bolt to the specified torque, whilst preventing crankshaft rotation using the method employed on removal. Refit the locking tool to the rear of the flywheel.

27 Relocate the timing belt on the crankshaft sprocket. Ensure that the "front run" of the belt is taut - i.e., ensure that any slack is on the tensioner pulley side of the belt. Do not twist the belt sharply while refitting it, and ensure that the belt teeth are seated centrally in the sprockets.

28 Loosen the tensioner pulley retaining nut. Rotate the pulley anti-clockwise to remove all free play from the timing belt, then retighten the nut.

29 Tension the belt as described in Section 7.

30 Refit the timing belt covers as described in Section 6, and reconnect the battery negative terminal.

Tensioner pulley

31 Refit the tensioner pulley to its mounting stud, and fit the retaining nut.

10.5 Adjusting a valve clearance

32 Ensure that the "front run" of the belt is taut - i.e., ensure that any slack is on the pulley side of the belt. Check that the belt is centrally located on all its sprockets. Rotate the pulley anti-clockwise to remove all free play from the timing belt, then tighten the pulley retaining nut securely.

33 Tension the belt as described in Section 7.

34 Refit the timing belt covers as described in Section 6, and reconnect the battery negative terminal.

9 Camshaft oil seal - renewal

Note: *If the camshaft oil seal is to be renewed with the timing belt still in place, check first that the belt is free from oil contamination. (Renew the belt as a matter of course if signs of oil contamination are found; see Section 7.) Cover the belt to protect it from oil contamination while work is in progress. Ensure that all traces of oil are removed from the area before the belt is refitted.*

1 Remove the camshaft sprocket as described in Section 8.

2 Punch or drill two small holes opposite each other in the oil seal. Screw a self-tapping screw into each, and pull on the screws with pliers to extract the seal.

3 Clean the seal housing, and polish off any burrs or raised edges, which may have caused the seal to fail in the first place.

4 Lubricate the lips of the new seal with clean engine oil, and drive it into position until it seats on its locating shoulder.

> *Use a tubular drift, such as a socket, which bears only on the hard outer edge of the seal. Take care not to damage the seal lips during fitting. Note that the seal lips should face inwards.*

5 Refit the camshaft sprocket as described in Section 8.

10 Valve clearances - checking and adjustment

Note: *The valve clearances must be checked and adjusted only when the engine is cold.*

1 The importance of having the valve clearances correctly adjusted cannot be overstressed, as they vitally affect the performance of the engine. If the clearances are too big, the engine will be noisy (characteristic rattling or tapping noises) and engine efficiency will be reduced, as the valves open too late and close too early. A more serious problem arises if the clearances are too small. If this is the case, the valves may not close fully when the engine is hot, resulting in serious damage to the engine (e.g.

burnt valve seats and/or cylinder head warping/cracking). The clearances are checked and adjusted as follows.

2 Remove the camshaft cover as described in Section 4.

3 Firmly apply the handbrake, then jack up the front right corner of the vehicle and support it on an axle stand (see "*Jacking and vehicle support*"). Remove the front roadwheel. The engine can now be turned using a socket and extension bar fitted to the crankshaft sprocket/pulley bolt. It will be easier if the spark plugs are removed first - see Chapter 1A.

4 It is important that the clearance of each valve is checked and adjusted only when the valve is fully closed, with the rocker arm resting on the heel of the cam (directly opposite the peak). This can be ensured by carrying out the adjustments in the following sequence, noting that No 1 cylinder is at the transmission end of the engine. The correct valve clearances are given in the Specifications at the start of this Chapter. The valve locations can be determined from the position of the manifolds.

Valve fully open	Adjust valves
No 1 ex.	*No 3 in. and No 4 ex.*
No 3 ex.	*No 4 in. and No 2 ex.*
No 4 ex.	*No 2 in. and No 1 ex.*
No 2 ex.	*No 1 in. and No 3 ex.*

5 With the relevant valve fully open, check the clearances of the two valves specified. Clearances are checked by inserting a feeler gauge of the correct thickness between the valve stem and the rocker arm adjusting screw. The feeler gauge should be a light, sliding fit. If adjustment is necessary, slacken the adjusting screw locknut, and turn the screw as necessary. Once the correct clearance is obtained, hold the adjusting screw and securely tighten the locknut **(see illustration)**. Recheck the valve clearance, and adjust again if necessary.

6 Rotate the crankshaft until the next valve in the sequence is fully open, and check the clearances of the next two specified valves.

7 Repeat the procedure until all eight valve clearances have been checked (and if necessary, adjusted), then refit the camshaft cover as described in Section 4, and lower the vehicle to the ground.

11 Camshaft and rocker arms - removal, inspection and refitting

General information

1 The rocker arm assembly is secured to the top of the cylinder head by the cylinder head bolts. Although in theory it is possible to undo the head bolts and remove the rocker arm assembly without removing the head, in practice, this is not recommended. Once the bolts have been removed, the head gasket will be disturbed, and the gasket will almost certainly leak or blow after refitting. For this reason, removal of the rocker arm assembly

11.4 Remove the circlip and slide the components off the end of the rocker arm

11.5a To remove the left-hand pedestal, lock two nuts together and unscrew the stud . . .

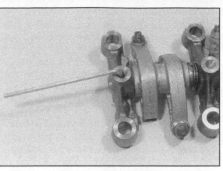

11.5b . . . then remove the grub screw

cannot be done without removing the cylinder head and renewing the head gasket.

2 The camshaft is slid out of the right-hand end of the cylinder head, and therefore it cannot be removed without first removing the cylinder head, due to a lack of clearance.

Removal

Rocker arm assembly

3 Remove the cylinder head as described in Section 12.

4 To dismantle the rocker arm assembly, carefully prise off the circlip from the right-hand end of the rocker shaft. Retain the rocker pedestal, to prevent it being sprung off the end of the shaft. Slide the various components off the end of the shaft, keeping all components in their correct fitted order. Make a note of each component's correct fitted position/orientation as it is removed, to ensure that it is fitted correctly on reassembly **(see illustration)**.

5 To separate the left-hand pedestal and shaft, first unscrew the cylinder head cover retaining stud from the top of the pedestal. This can be achieved using a stud extractor, or two nuts locked together. With the stud removed, unscrew the grub screw from the top of the pedestal, and withdraw the rocker shaft **(see illustrations)**.

Camshaft

6 Remove the cylinder head as described in Section 12.

7 With the head on a bench, remove the locking tool, then remove the camshaft sprocket as described in Section 8.

8 Undo the retaining bolt, and remove the camshaft thrust fork from the cylinder head **(see illustration)**.

9 Using a large flat-bladed screwdriver, carefully prise the oil seal out of the right-hand end of the cylinder head, then slide out the camshaft **(see illustrations)**. Discard the seal - a new one must be used on refitting.

Inspection

Rocker arm assembly

10 Examine the rocker arm bearing surfaces which contact the camshaft lobes for wear ridges and scoring. Renew any rocker arms on which these conditions are apparent. Renew worn components as necessary. The rocker arm assembly can be dismantled as described in paragraphs 4 and 5.

 If a rocker arm bearing surface is badly scored, also examine the corresponding lobe on the camshaft for wear, as it is likely that both will be worn.

11 Inspect the ends of the (valve clearance) adjusting screws for signs of wear or damage, and renew as required.

12 If the rocker arm assembly has been dismantled, examine the rocker arm and shaft bearing surfaces for wear ridges and scoring. If there are obvious signs of wear, the rocker arm(s) and/or the shaft must be renewed.

Camshaft

13 Examine the camshaft bearing surfaces and cam lobes for signs of wear ridges and

scoring. Renew the camshaft if any of these conditions are apparent. Examine the condition of the bearing surfaces, both on the camshaft journals and in the cylinder head. If the head bearing surfaces are worn excessively, the cylinder head will need to be renewed. If the necessary measuring equipment is available, camshaft bearing journal wear can be checked by direct measurement, noting that No 1 journal is at the transmission end of the head.

14 Examine the thrust fork for signs of wear or scoring, and renew as necessary.

Refitting

Rocker arm assembly

15 If the rocker arm assembly was dismantled, refit the rocker shaft to the left-hand pedestal, aligning its locating hole with the pedestal threaded hole. Refit the grub screw, and tighten it securely. With the grub screw in position, refit the cylinder head cover mounting stud to the pedestal, and tighten it securely. Apply a smear of clean engine oil to the shaft, then slide on all removed components, ensuring each is correctly fitted in its original position. Once all components are in position on the shaft, compress the right-hand pedestal and refit the circlip. Ensure that the circlip is correctly located in its groove on the shaft.

16 Refit the cylinder head and rocker arm assembly as described in Section 12.

Camshaft

17 Ensure that the cylinder head and camshaft bearing surfaces are clean, then liberally oil the camshaft bearings and lobes.

11.8 Undo the retaining bolt, and remove the camshaft thrust fork (arrowed) . . .

11.9a . . . prise out the oil seal . . .

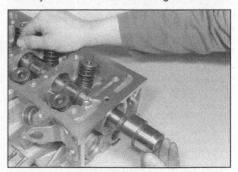

11.9b . . . and slide out the camshaft

Slide the camshaft back into position in the cylinder head.

18 Locate the thrust fork with the left-hand end of the camshaft. Refit the fork retaining bolt, tightening it to the specified torque.

 On carburettor engines, take care that the fuel pump operating lever is not trapped by the camshaft as it is slid into position. To prevent this, remove the fuel pump before refitting the camshaft, then refit it afterwards.

19 Lubricate the lips of the new seal with clean engine oil, then drive it into position until it seats on its locating shoulder. Use a tubular drift, such as a socket, which bears only on the hard outer edge of the seal. Take care not to damage the seal lips during fitting. Note that the seal lips should face inwards.

20 Refit the camshaft sprocket as described in Section 8.

21 Refit the cylinder head as described in Section 12.

12 Cylinder head - removal and refitting

Removal

1 Disconnect the battery negative lead.

2 Drain the cooling system as described in Chapter 1A.

3 Remove the camshaft cover as described in Section 4.

4 Align the engine assembly/valve timing holes as described in Section 3, and lock both the camshaft sprocket and flywheel in position. *Do not* attempt to rotate the engine whilst the locking tools are in position.

5 Note that the following text assumes that the cylinder head will be removed with both inlet and exhaust manifolds attached; this is easier, but makes it a bulky and heavy assembly to handle. If you prefer to remove the manifolds first, proceed as described in the relevant Part of Chapter 4.

6 Working as described in the relevant Part of Chapter 4, disconnect the exhaust system front pipe from the manifold. On fuel injection models, disconnect or release the lambda sensor wiring, so that it is not strained by the weight of the exhaust.

7 Remove the air cleaner housing and intake duct assembly as described in the relevant Part of Chapter 4.

8 On carburettor engines, disconnect the following from the carburettor and inlet manifold as described in Chapter 4A:
a) *Fuel feed hose from the pump (plug the openings, to prevent loss of fuel and the entry of dirt into the system).*
b) *Accelerator cable.*
c) *Choke cable.*
d) *Carburettor coolant hoses.*

e) *Vacuum servo unit vacuum hose, coolant hose and all other relevant breather/vacuum hoses from the manifold.*

9 On fuel injection engines, carry out the following operations as described in Chapter 4B:
a) *Depressurise the fuel system, and disconnect the fuel feed and return hoses from the throttle body (plug all openings, to prevent loss of fuel and the entry of dirt into the system).*
b) *Disconnect the accelerator cable.*
c) *Disconnect the relevant electrical connectors from the throttle body.*
d) *Disconnect the vacuum servo unit vacuum hose, coolant hose(s) and all the other relevant vacuum/breather hoses from the inlet manifold.*

10 Remove the upper and centre timing belt covers as described in Section 6.

11 Loosen the timing belt tensioner pulley retaining nut. Pivot the pulley in a clockwise direction, using a square-section key fitted to the hole in the pulley hub, then retighten the retaining nut.

12 Disengage the timing belt from the camshaft sprocket, and position the belt clear of the sprocket. Ensure that the belt is not bent or twisted sharply, if it is to be re-used.

13 Slacken the retaining clips, and disconnect the coolant hoses from the thermostat housing (on the left-hand end of the cylinder head).

14 Depress the retaining clip(s), and disconnect the wiring connector(s) from the electrical switch and/or sensor(s) which are screwed into the thermostat housing (as appropriate).

15 Disconnect the LT wiring connectors from the distributor and HT coil. Release the TDC sensor wiring connector from the side of the coil mounting bracket, and disconnect the vacuum pipe from the distributor vacuum diaphragm unit. If the cylinder head is to be dismantled for overhaul, remove the distributor and ignition HT coil as described in Chapter 5B, disconnect the HT leads from the spark plugs, and remove the distributor cap and lead assembly.

 If the cylinder numbers are not already marked on the HT leads, number each lead, to avoid the possibility of the leads being incorrectly connected on refitting.

16 Slacken and remove the bolt securing the engine oil dipstick tube to the cylinder head, and withdraw the tube from the cylinder block.

17 Working in the reverse of the sequence shown in illustration 12.36a, progressively slacken the ten cylinder head bolts by half a turn at a time, until all bolts can be unscrewed by hand.

18 With all the cylinder head bolts removed, lift the rocker arm assembly off the cylinder head. Note the locating pins which are fitted to the base of each rocker arm pedestal. If any pin is a loose fit in the head or pedestal, remove it for safe-keeping.

19 The joint between the cylinder head and gasket and the cylinder block/crankcase must now be broken without disturbing the wet liners. To break the joint, obtain two L-shaped metal bars which fit into the cylinder head bolt holes. Gently "rock" the cylinder head free towards the front of the vehicle. Do not try to swivel the head on the cylinder block/crankcase; it is located by dowels, as well as by the tops of the liners. **Note:** *If care is not taken and the liners are moved, there is also a possibility of the bottom seals being disturbed, causing leakage after refitting the head.* When the joint is broken, lift the cylinder head away; and seek assistance if possible as it is a heavy assembly, especially if it is removed complete with the manifolds.

20 Remove the gasket from the top of the block, noting the two locating dowels. If the locating dowels are a loose fit, remove them and store them with the head for safe-keeping. Do not discard the old gasket yet - on some models, it will be needed to ensure that the correct new gasket is obtained (see paragraphs 26 and 27).

21 Note: *Do not attempt to rotate the crankshaft with the cylinder head removed, otherwise the wet liners may be displaced. Operations that require the rotation of the crankshaft (e.g. cleaning the piston crowns), should only be carried out once the cylinder liners are firmly clamped in position.*

 In the absence of the special Citroën liner clamps, the liners can be clamped in position using large flat washers positioned underneath suitable-length bolts. Alternatively, the original head bolts could be temporarily refitted, with spacers fitted to their shanks.

22 If the cylinder head is to be dismantled for overhaul, remove the camshaft as described in Section 11, then refer to Part C of this Chapter.

Preparation for refitting

23 The mating faces of the cylinder head and cylinder block/crankcase must be perfectly clean before refitting the head. Use a hard plastic or wood scraper to remove all traces of gasket and carbon; also clean the piston crowns. Refer to paragraph 21 before turning the crankshaft. Take particular care during the cleaning operations, as the soft aluminium alloy is damaged easily. Also, make sure that the carbon is not allowed to enter the oil and water passages - this is particularly important for the lubrication system, as carbon could block the oil supply to the engine's components. Using adhesive tape and paper, seal the water, oil and bolt holes in the cylinder block/crankcase. To prevent carbon entering the gap between the pistons and bores, smear a little grease in the gap. After cleaning each piston, use a small brush to remove all traces of grease and carbon from the gap, then wipe away the remainder with a clean rag. Clean all the pistons in the same way.

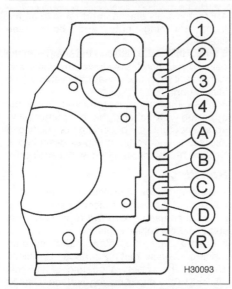

12.27 TU engine gasket markings

1, 2, 3, 4   *Engine type identification cut-outs*
A, B, C, D *Gasket manufacturer and asbestos content identification cut-outs*
R   *Gasket thickness identification cut-out*

24 Check the mating surfaces of the cylinder block/crankcase and the cylinder head for nicks, deep scratches and other damage. If slight, they may be removed carefully with a file, but if excessive, machining may be the only alternative to renewal.

25 If warping of the cylinder head gasket surface is suspected, use a straight-edge to check it for distortion. Refer to Part C of this Chapter if necessary.

26 When purchasing a new cylinder head gasket, it is essential that a gasket of the correct thickness is obtained. There are two different thicknesses available – the standard gasket which is fitted at the factory, and a slightly thicker gasket (+ 0.2 mm) for use once the head gasket face has been machined. If the cylinder head has been machined, it should have the letter "R" stamped adjacent to the No 3 exhaust port, and the gasket should also have the letter "R" stamped adjacent to No 3 cylinder on its front upper face. The gaskets can also be identified as described in the following paragraph, using the cut-outs on the left-hand end of the gasket.

27 With the gasket fitted the correct way up on the cylinder block, there is a single cut-out at the rear of the left-hand side (position 1) on all engines except the C1A which has no cut-out at all **(see illustration)**. In the centre of the gasket, there is another series of up to four cut-outs identifying the manufacturer of the gasket, and whether or not it contains asbestos (these cut-outs are of little importance). The important cut-out location is at the front of the gasket. On the standard-thickness gasket, there will be no cut-out in this position. On the thicker, "repair" gasket, there will be a single cut-out. Identify the

12.31 Locate the cylinder head gasket on the block . . .

gasket type, and ensure that the new gasket obtained is of the correct thickness. If there is any doubt as to which gasket is fitted, take the old gasket along to your Citroën dealer, and have the dealer confirm the gasket type.

28 Check the condition of the cylinder head bolts, and particularly their threads, whenever they are removed. Wash the bolts in suitable solvent, and wipe them dry. Check each for any sign of visible wear or damage, renewing any bolt if necessary. Measure the length of each bolt, to check for stretching (although this is not a conclusive test, in the event that all ten bolts have stretched by the same amount). Although Citroën do not actually specify that the bolts must be renewed, it is strongly recommended that the bolts should be renewed as a complete set whenever they are disturbed.

29 Before refitting the cylinder head, check the cylinder liner protrusion as described in Part C of this Chapter.

Refitting

30 Wipe clean the mating surfaces of the cylinder head and cylinder block/crankcase. Check that the two locating dowels are in position at each end of the cylinder block/crankcase surface and, if necessary, remove the cylinder liner clamps.

31 Position a new gasket on the cylinder block/crankcase surface **(see illustration)**, ensuring that its identification cut-outs are at the left-hand end of the gasket.

32 Check that the flywheel and camshaft sprocket are still correctly locked in position with their respective locking tools. With the aid of an assistant, carefully refit the cylinder

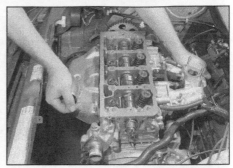

12.32 . . . then lower the cylinder head into position . . .

12.33 . . . and refit the rocker arm assembly

head assembly to the block, aligning it with the locating dowels **(see illustration)**.

33 Ensure that the locating pins are in position in the base of each rocker pedestal, then refit the rocker arm assembly to the cylinder head **(see illustration)**.

34 Apply a smear of grease to the threads and contact faces of the cylinder head bolts. Citroën recommend the use of Molykote G Rapid Plus grease, and a sachet is supplied with the head gasket set. In the absence of the specified grease, a good-quality high-melting-point grease may be used.

35 Carefully enter each bolt into its relevant hole (*do not drop them in*). Screw them in, by hand only, until finger-tight.

36 Working progressively and in the sequence shown, tighten the cylinder head bolts to their specified Stage 1 torque setting, using a torque wrench and socket **(see illustrations)**.

37 Once all the bolts have been tightened to their Stage 1 setting, working again in the

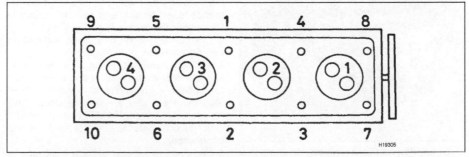

12.36a Cylinder head bolt tightening sequence (cylinder numbering also shown)

12.36b Working in the sequence shown, tighten the head bolts first to the Stage 1 torque setting . . .

12.37 . . . then through the angle specified for Stage 2

given sequence, angle-tighten the bolts through the specified Stage 2 angle, using a socket and extension bar. It is recommended that an angle-measuring gauge is used during this stage of tightening, to ensure accuracy **(see illustration)**. If a gauge is not available, use white paint to make alignment marks between the bolt head and cylinder head before tightening. The marks can then be used to check that the bolt has been rotated through the correct angle during tightening.

38 With the cylinder head bolts correctly tightened, refit the dipstick tube to the engine, and securely tighten its retaining bolt.

39 Refit the timing belt to the camshaft sprocket. Ensure that the "front run" of the belt is taut - i.e., ensure that any slack is on the tensioner pulley side of the belt. Do not twist the belt sharply while refitting it, and ensure that the belt teeth are seated centrally in the sprockets.

40 Loosen the tensioner pulley retaining nut. Pivot the pulley anti-clockwise to remove all free play from the timing belt, then retighten the nut.

41 Tension the belt as described in Section 7, then refit the centre and upper timing belt covers as described in Section 6.

42 If the head was stripped for overhaul, refit the distributor and HT coil as described in Chapter 5B, ensuring that the HT leads are correctly reconnected. If the head was not stripped, reconnect the wiring connector and vacuum pipe to the distributor, and the HT lead to the coil. Clip the TDC sensor wiring connector onto the coil bracket.

43 Reconnect the wiring connector(s) to the coolant switch/sensor(s) on the left-hand end of the head (as appropriate).

44 Reconnect the coolant hoses to the thermostat housing, securely tightening their retaining clips.

45 Working as described in the relevant Part of Chapter 4, carry out the following tasks:
a) *Refit all disturbed wiring, hoses and control cable(s) to the inlet manifold and fuel system components.*
b) *On carburettor models, reconnect and adjust the choke and accelerator cables.*
c) *On fuel injection models, reconnect and adjust the accelerator cable.*
d) *Reconnect the exhaust system front pipe to the manifold. Where applicable, reconnect the lambda sensor wiring connector.*
e) *Refit the air cleaner housing and intake duct.*

46 Check and, if necessary, adjust the valve clearances as described in Section 10, then refit the camshaft cover as described in Section 4.

47 On completion, reconnect the battery (see Chapter 5A), and refill the cooling system (see Chapter 1A).

13 Sump - removal and refitting

Removal

1 Firmly apply the handbrake, then jack up the front of the vehicle and support it on axle stands (see *Jacking and vehicle support*). Disconnect the battery negative lead.

2 Drain the engine oil, then clean and refit the engine oil drain plug, tightening it to the specified torque. If the engine is nearing its service interval when the oil and filter are due for renewal, it is recommended that the filter is also removed, and a new one fitted. After reassembly, the engine can then be refilled with fresh oil. Refer to Chapter 1A for further information.

3 Remove the exhaust system front pipe as described in the relevant Part of Chapter 4.

4 Progressively slacken and remove all the sump retaining nuts and bolts **(see illustration)**.

5 Break the joint by striking the sump with the palm of your hand, then lower the sump and withdraw it from underneath the vehicle **(see illustration)**.

6 While the sump is removed, take the opportunity to check the oil pump pick-up/strainer for signs of clogging or splitting. If necessary, remove the pump as described in Section 14, and clean or renew the strainer.

Refitting

7 Clean all traces of sealant from the mating surfaces of the cylinder block/crankcase and sump, then use a clean rag to wipe out the sump and the engine's interior.

8 Ensure that the sump and cylinder block/crankcase mating surfaces are clean and dry, then apply a coating of suitable sealant to the sump mating surface. Citroën recommend the use of Auto-Joint E10 sealant (available from your Citroën dealer). In the absence of the specified sealant, any good-quality sealant may be used.

9 Offer up the sump, locating it on its retaining studs, and refit its retaining nuts and bolts. Tighten the nuts and bolts evenly and progressively to the specified torque.

10 Refit the exhaust front pipe as described in the relevant Part of Chapter 4.

11 Fit a new oil filter (if removed) and replenish the engine oil as described in Chapter 1A.

14 Oil pump - removal, inspection and refitting

Removal

1 Remove the sump as described in Section 13.

2 Slacken and remove the three bolts securing the oil pump to the base of the main bearing ladder **(see illustration)**. Disengage the pump sprocket from the chain, and remove the oil pump. If the pump locating dowel is a loose fit, remove and store it with the retaining bolts for safe-keeping.

3 If it is required to remove the oil pump drive

13.4 Slacken and remove the sump retaining nuts and bolts . . .

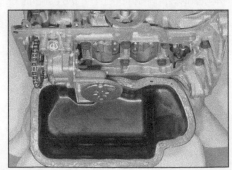

13.5 . . . then remove the sump from the engine

14.2 Oil pump is retained by three bolts

chain, the crankshaft right-hand oil seal must be removed as described in Section 15, then the spacer removed from the front of the crankshaft. The oil pump drive sprocket must then be removed from the key on the crankshaft after lifting the chain over its teeth. With the sprocket removed, the chain can be withdrawn upwards through the aperture in the front of the cylinder block, and over the nose of the crankshaft.

Inspection

4 Examine the oil pump sprocket for signs of damage and wear, such as chipped or missing teeth. If the sprocket is worn, the pump assembly must be renewed, since the sprocket is not available separately. It is also recommended that the chain and drive sprocket, fitted to the crankshaft, be renewed at the same time.

5 Slacken and remove the five bolts securing the strainer cover to the pump body, then lift off the strainer cover. Remove the relief valve piston and spring, noting which way round they are fitted.

6 Examine the pump rotors and body for signs of wear ridges and scoring. If worn, the complete pump assembly must be renewed.

7 Examine the relief valve piston for signs of wear or damage, and renew if necessary. The condition of the relief valve spring can only be measured by comparing it with a new one. If there is any doubt about its condition, it should also be renewed. Both the piston and spring are available individually.

8 Thoroughly clean the oil pump strainer with a suitable solvent, and check it for signs of clogging or splitting. If the strainer is damaged, the strainer and cover assembly must be renewed.

9 Locate the relief valve spring and piston in the strainer cover, then refit the cover to the pump body. Align the relief valve piston with its bore in the pump. Refit the five cover retaining bolts, tightening them securely.

Refitting

10 Where necessary, fit the oil pump drive chain and sprocket to the right-hand end of the crankshaft, using a reversal of the removal procedure, then fit a new right-hand oil seal with reference to Section 15.

11 Ensure that the locating dowel is in position, then engage the pump sprocket with its drive chain. Seat the pump on the main bearing ladder. Refit the pump retaining bolts, and tighten them to the specified torque.

12 Refit the sump as described in Section 13.

13 Before running the engine, disable the ignition system by disconnecting the ignition HT coil lead from the distributor cap and earthing it on the cylinder block. Use a jumper lead or similar wire to make a good connection. On fuel injection models, also disconnect the fuel injection system relay unit (see Chapters 4B and 12) to disable the fuel pump and prevent contamination of the catalytic converter with unburnt fuel. Turn the engine on the starter motor until the oil pressure is restored and the oil pressure warning light goes out. Then reconnect the HT lead (and the injection relay if disconnected) and start the engine to check for oil leaks.

15 Crankshaft oil seals - renewal

Right-hand oil seal

1 Remove the crankshaft sprocket and flanged spacer as described in Section 8. Secure the timing belt clear of the working area, so that it cannot be contaminated with oil. Make a note of the correct fitted depth of the seal in its housing.

2 Punch or drill two small holes opposite each other in the seal. Screw a self-tapping screw into each, and pull on the screws with pliers to extract the seal. Alternatively, the seal can be levered out of position using a flat-bladed screwdriver, taking great care not to damage the crankshaft shoulder or seal housing (see illustration).

3 Clean the seal housing, and polish off any burrs or raised edges, which may have caused the seal to fail in the first place.

4 Lubricate the lips of the new seal with clean engine oil, and carefully locate the seal on the end of crankshaft. Note that its sealing lip must face inwards. Take care not to damage the seal lips during fitting.

5 Using a tubular drift (such as a socket) which bears only on the hard outer edge of the seal, tap the seal into position, to the same depth in the housing as the original was prior to removal. The inner face of the seal must end up flush with the inner wall of the crankcase.

6 Wash off any traces of oil, then refit the crankshaft sprocket as described in Section 8.

Left-hand oil seal

7 Remove the flywheel as described in Section 16.

8 Make a note of the correct fitted depth of the seal in its housing. Punch or drill two small holes opposite each other in the seal. Screw a self-tapping screw into each, and pull on the

15.2 Using a screwdriver to lever out the crankshaft right-hand oil seal

screws with pliers to extract the seal.

9 Clean the seal housing, and polish off any burrs or raised edges, which may have caused the seal to fail in the first place.

10 Lubricate the lips of the new seal with clean engine oil, and carefully locate the seal on the end of the crankshaft.

11 Using a tubular drift (such as a socket), which bears only on the hard outer edge of the seal, drive the seal into position, to the same depth in the housing as the original was prior to removal.

12 Wash off any traces of oil, then refit the flywheel as described in Section 16.

16 Flywheel - removal, inspection and refitting

Removal

1 Remove the transmission as described in Chapter 7, then remove the clutch assembly as described in Chapter 6.

2 Prevent the flywheel from turning by locking the ring gear teeth with a similar arrangement to that shown (see illustration). Alternatively, bolt a strap between the flywheel and the cylinder block/crankcase. *Do not* attempt to lock the flywheel in position using the locking tool described in Section 3.

3 Slacken and remove the flywheel retaining bolts, and discard them; they must be renewed whenever they are disturbed.

4 Remove the flywheel. Do not drop it, as it is

16.2 Use the fabricated tool shown to lock flywheel ring gear and prevent the crankshaft rotating

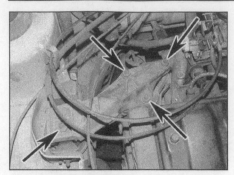

17.7 Undo the nuts (arrowed) and remove the right-hand engine mounting-to-body bracket

very heavy. If the locating dowel is a loose fit in the crankshaft end, remove and store it with the flywheel for safe-keeping.

Inspection

5 If the flywheel's clutch mating surface is deeply scored, cracked or otherwise damaged, the flywheel must be renewed. However, it may be possible to have it surface-ground; seek the advice of a Citroën dealer or engine reconditioning specialist.

6 If the ring gear is badly worn or has missing teeth, it must be renewed. This job is best left to a Citroën dealer or engine reconditioning specialist. The temperature to which the new ring gear must be heated for installation is critical and, if not done accurately, the hardness of the teeth will be destroyed.

Refitting

7 Clean the mating surfaces of the flywheel and crankshaft. Remove any remaining locking compound from the threads of the crankshaft holes, using the correct-size tap, if available.

> **HAYNES HINT** *If a tap is not available, cut two slots into the threads of an old flywheel bolt and use the bolt to remove the locking compound from the threads.*

8 If the new flywheel retaining bolts are not supplied with their threads already pre-coated, apply a suitable thread-locking compound to the threads of each bolt. Citroën recommend the use of Frenetanch E3. If this is not available, ensure that a good-quality locking compound is used.

9 Ensure that the locating dowel is in position. Offer up the flywheel, locating it on the dowel, and fit the new retaining bolts.

10 Lock the flywheel using the method employed on dismantling, and tighten the retaining bolts to the specified torque.

11 Refit the clutch as described in Chapter 6. Remove the locking tool, and refit the transmission as described in Chapter 7.

---

17 Engine/transmission mountings - inspection, removal and refitting

Inspection

1 If improved access is required, raise the front of the vehicle and support it securely on axle stands (see "*Jacking and vehicle support*").

2 Check the mounting rubber to see if it is cracked, hardened or separated from the metal at any point. Renew the mounting if any such damage or deterioration is evident.

3 Check that all the mounting's fasteners are securely tightened. Use a torque wrench to check if possible.

4 Using a large screwdriver or a crowbar, check for wear in the mounting by carefully levering against it to check for free play. Where this is not possible, enlist the aid of an assistant to move the engine/transmission unit back and forth, or from side to side, while you watch the mounting. While some free play is to be expected even from new components, excessive wear should be obvious. If excessive free play is found, check first that the fasteners are correctly secured, then renew any worn components as described below.

Removal and refitting

Right-hand mounting-to-body bracket

5 Place a jack beneath the engine, with a block of wood on the jack head. Raise the jack until it is supporting the weight of the engine.

6 Release the accelerator cable from its clip on the bodywork. On carburettor models, also unclip the choke cable. Move the cables to one side, away from the engine mounting.

7 Undo the nut securing the bracket to the mounting rubber. Undo the three nuts securing the bracket to the water pump housing at the rear of the cylinder block. Remove the bracket, noting the location of any shims **(see illustration)**.

8 Refitting is the reverse of removal. Tighten the bracket securing nuts to the specified torque.

Right-hand mounting rubber

9 The right-hand engine mounting rubber is the same on both petrol and diesel models. Removal and refitting is described in Chapter 2B.

Left-hand mounting

10 The left-hand engine mounting bracket and rubber is the same on both petrol and diesel models. Removal and refitting is described in Chapter 2B. Remove the air inlet ducting as required, to access the engine mounting components, but leave the air cleaner in place.

11 The transmission stud is fitted to a bracket, secured to the transmission by three nuts. To remove it, first remove the mounting bracket and rubber as described in Chapter 2B, then undo the three nuts on the transmission stud bracket. Refitting is the reverse of removal, tightening the nuts to the specified torque.

Lower mounting

12 The lower mounting supports both the engine and the right-hand driveshaft intermediate bearing. The rubber bush is not available as a separate component, and if damaged, the complete mounting bracket must be renewed. To remove the bracket, it is necessary to disconnect the driveshaft from the transmission. The mounting components are essentially the same on both petrol and diesel models, although the part numbers are different. Removal and refitting is described in Chapter 2B.

Chapter 2 Part B:
Diesel engine in-vehicle repair procedures

Contents

Degrees of difficulty

Easy, suitable for novice with little experience	Fairly easy, suitable for beginner with some experience	Fairly difficult, suitable for competent DIY mechanic	Difficult, suitable for experienced DIY mechanic	Very difficult, suitable for expert DIY or professional

Specifications

General

Type	Four-cylinder, in-line, four-stroke, overhead camshaft, compression-ignition, mounted transversely and inclined 30° to rear. Transmission mounted on left-hand end of engine.
Code	XUD 7 - 161A
Engine size	1769 cc
Number of cylinders	4
Bore and stroke	80.0 x 88.0 mm
Compression ratio	23.0 : 1
Compression pressures (engine hot, cranking speed):	
Minimum	18 bar
Normal	25 to 30 bar
Maximum difference between any two cylinders	5 bar
Maximum torque (ISO)	110 Nm at 2200 rpm
Maximum power (ISO)	45 kW at 4600 rpm
Maximum speed:	
No load	5100 rpm
Laden	4600 rpm
Firing order	1-3-4-2 (No 1 at flywheel end)

Valves

Valve clearances (cold):	
Inlet	0.10 to 0.25 mm
Exhaust	0.25 to 0.40 mm

Camshaft

Endfloat	0.07 to 0.16 mm

Lubrication system

Oil pressure (at engine temperature of 90°C):	
Minimum	2.5 bar at 800 rpm
Maximum	3.5 to 5.0 bar at 4000 rpm
Oil pressure switch operating pressures:	
On	0.58 to 0.44 bar
Off	0.8 bar maximum
Oil pump:	
Type	Two gear
Pressure relief valve opens	4.0 bar
Gear endfloat	0.12 mm
Clearance between gear lobes and housing	0.064 mm

Torque wrench settings

	Nm	lbf ft
Big-end bearing cap nuts:		
Stage 1	20	15
Stage 2	Angle tighten a further 70°	
Brake vacuum pump pulley bolt	35	26
Camshaft bearing cap nuts	18	13
Camshaft cover bolts	20	15
Camshaft sprocket bolt	35	26
Crankshaft pulley bolt:**		
Stage 1	40	30
Stage 2	Angle tighten a further 50°	
Crankshaft right-hand oil seal housing bolts	11	8
Cylinder head bolts:+		
Up to 1991:		
Stage 1	30	22
Stage 2	70	52
Stage 3	Angle-tighten a further 120°	
1992:		
Stage 1	70	52
Stage 2	Angle tighten a further 140°	
1993-on:++		
Stage 1	20	15
Stage 2	60	44
Stage 3	Angle tighten a further 180°	
Engine mounting, left hand:		
Centre nut*	35	26
Small nuts	18	13
Centre stud to transmission**	50	37
Engine mounting, lower:		
Mounting rubber central nut and bolt*	35	26
Torque link to subframe nut and bolt	50	37
Mounting bracket to engine block bolts	18	13
Engine mounting-to-body bracket, right-hand:		
Mounting nuts to engine	35	26
Mounting rubber central nut	28	21
Flywheel bolts**	50	37
Injection pump bracket bolts	20	15
Injection pump sprocket nut	50	37
Injection pump sprocket puller bolts	10	7
Main bearing cap bolts	70	52
Oil gallery plug	28	21
Oil pressure switch	30	22
Oil pump cover bolts	9	7
Oil pump mounting bolts	13	10
Sump bolts	19	14
Timing belt tensioner adjustment bolt	18	13
Timing belt tensioner pivot nut	18	13
Timing cover bolts, lower	12	9

*   *Use new self-locking nut.*
**   *Use locking fluid.*
+   *Use Molykote G Rapide Plus or other suitable high-temperature grease.*
++   *Measure the bolts to see if they are suitable for re-use (see Section 14).*

1 General information

How to use this Chapter

This Part of Chapter 2 describes the repair procedures that can reasonably be carried out on the XUD7 diesel engine while it remains in the vehicle. If the engine has been removed from the vehicle and is being dismantled as described in Part C, any preliminary dismantling procedures can be ignored.

Note that, while it may be possible physically to overhaul items such as the piston/connecting rod assemblies while the engine is in the vehicle, such tasks are not usually carried out as separate operations. Usually, several additional procedures are required (not to mention the cleaning of components and of oilways); for this reason, all such tasks are classed as major overhaul procedures, and are described in Part C of this Chapter.

Part C describes the removal of the engine/transmission unit from the vehicle, and the full overhaul procedures that can then be carried out.

Engine description

The XUD7 engine is the 1769 cc version of the XUD range of engines which are well-proven modern diesel units and have appeared in many Citroën, Peugeot and Talbot passenger cars and light commercial vehicles. The engine is the four-cylinder overhead camshaft design, mounted transversely and inclined 30° to the rear, with the transmission mounted on the left-hand side.

A toothed timing belt drives the camshaft, injection pump and water pump. Bucket tappets are fitted between the camshaft and valves, and valve clearance adjustment is by means of selective shims.

The camshaft is supported by three bearings machined directly in the cylinder head.

The crankshaft runs in five main bearings of the usual shell type. Endfloat is controlled by thrustwashers either side of No 2 main bearing.

The pistons are selected to be of matching weight, and incorporate fully floating gudgeon pins retained by circlips.

The oil pump is chain driven from the right-hand end of the crankshaft.

Repair operations possible with the engine in the vehicle

The following operations can be carried out without having to remove the engine from the vehicle:

a) Timing belt - removal and refitting
b) Camshaft - removal and refitting
c) Cylinder head - removal and refitting
d) Camshaft oil seals - renewal
e) Crankshaft oil seals - renewal
f) Sump - removal and refitting
g) Oil pump - removal and refitting
h) Pistons and connecting rods - removal and refitting
i) Flywheel - removal and refitting

2 Compression and leakdown test - description and interpretation

Note: *A compression tester specifically designed for diesel engines must be used for this test.*

Compression test

Description

1 When engine performance is down, or if misfiring occurs which cannot be attributed to the ignition or fuel systems, a compression test can provide diagnostic clues as to the engine's condition. If the test is performed regularly, it can give warning of trouble before any other symptoms become apparent.

2 A compression tester specifically intended for diesel engines must be used, because of the higher pressures this type of engine produces. The tester is connected to an adapter that screws into the glow plug or injector hole. It is unlikely to be worthwhile buying such a tester for occasional use, but it may be possible to borrow or hire one. If not, have the test performed by a garage, or dealer.

3 Unless specific instructions to the contrary are supplied with the tester, observe the following points:

a) The battery must be in a good state of charge, the air filter must be clean, and the engine must be at normal operating temperature
b) All the injectors or glow plugs should be removed before starting the test. If removing the injectors, also remove the fire shield washers, otherwise they may be blown out
c) The stop solenoid must be disconnected, to prevent the engine from running or fuel from being discharged

4 There is no need to hold the accelerator pedal down during the test, because the diesel engine air inlet is not throttled.

5 The actual compression pressures measured are not so important as the balance between cylinders. Values are given in the Specifications.

6 The cause of poor compression is less easy to establish on a diesel engine than a petrol engine. The effect of introducing oil into the cylinders ('wet testing') is not conclusive, because there is a risk that oil will sit in the swirl chamber or in the recess on the piston crown instead of passing to the rings. However, the following can be used as a rough guide to diagnosis.

Interpretation

7 All cylinders should produce very similar pressures. Any difference greater than that specified indicates the existence of a fault. Note that the compression should build up quickly in a healthy engine. Low compression on the first stroke, followed by gradually increasing pressure on successive strokes, indicates worn piston rings. A low compression reading on the first stroke, which does not build up during successive strokes, indicates leaking valves or a blown head gasket (a cracked head could also be the cause). Deposits on the undersides of the valve heads can also cause low compression.

8 A low reading from two adjacent cylinders is almost certainly due to the head gasket having blown between them. The presence of coolant in the engine oil will confirm this.

9 If the compression reading is unusually high, the cylinder head surfaces, valves and pistons are probably coated with carbon deposits. If this is the case, the cylinder head should be removed and decarbonised.

Leakdown test

Description

10 A leakdown test measures the rate at which compressed air fed into the cylinder is lost. It is an alternative to a compression test, and in many ways is better, since the escaping air provides easy identification of where pressure loss is occurring (piston rings, valves or head gasket).

11 The equipment needed for leakdown testing is unlikely to be available to the home mechanic. If poor compression is suspected, have the test performed by a suitably equipped garage.

3 Top dead centre (TDC) for No 4 piston - locating

Note: *Three 8 mm diameter bolts and one 8 mm diameter rod or drill will be required as locking tools for this procedure.*
Note: *Improved access to the timing belt system can be obtained by moving the engine to the left, as described in Section 19.*

1 Top dead centre (TDC) is the highest point in the cylinder that each piston reaches as the crankshaft turns. Each piston reaches TDC at the end of the compression stroke, and again at the end of the exhaust stroke. For the purpose of timing the engine, TDC refers to the position of No 4 piston at the end of its compression stroke. On all engines in this manual, No 1 piston is at the flywheel end of the engine.

2 Remove the left- and right-hand timing belt covers as described in Section 6.

3 Apply the handbrake, then jack up the front right-hand corner of the vehicle until the wheel is just clear of the ground (see "Jacking and vehicle support"). Support the vehicle on an axle stand and engage 4th or 5th gear.

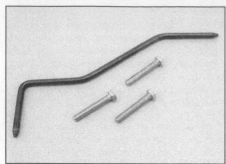

3.5a TDC setting and locking tools for Citroën diesel engines

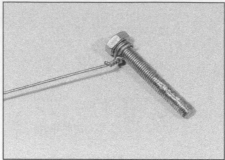

3.5b Home-made TDC setting tool

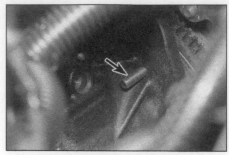

3.5c Rod (arrowed) inserted through cylinder block into TDC hole in flywheel with starter motor removed

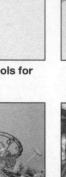

3.6 TDC locking bolts (arrowed) in camshaft and fuel injection pump sprockets (engine removed for clarity)

4 Rotate the engine by turning the right-hand wheel until the three bolt holes in the camshaft and injection pump sprockets (one hole in the camshaft sprocket, two holes in

3.8 Warning notice in engine compartment

the injection pump sprocket) are aligned with the corresponding holes in the engine right-hand plate.
5 Insert an 8.0 mm diameter rod or drill

through the hole in the left-hand flange of the cylinder block next to the starter motor, so that it enters the TDC setting hole in the flywheel. The hole in the cylinder block is awkwardly placed but can be accessed using locking tools, available from your Citroën dealer. You can make up an improvised tool from an M8 bolt with the threads filed away, attached to a piece of welding rod. Alternatively, remove the starter motor and insert a rod or twist drill **(see illustrations)**. Turn the engine back and forward slightly until the tool enters the hole in the flywheel, then leave the tool in position.
6 Insert an M8 bolt through the hole in the camshaft sprocket, and two M8 bolts through the holes in the fuel injection pump sprocket. Screw all three bolts into the engine finger-tight **(see illustration)**.
7 The crankshaft, camshaft and injection pump are now "locked" in position with No 4 piston at TDC.
8 If the engine is to be left in this state for a long period of time, it is a good idea to place warning notices inside the vehicle, and in the engine compartment. This will reduce the possibility of the engine being accidentally cranked on the starter motor, which is likely to cause damage with the locking tools in place **(see illustration)**.

4 Camshaft cover - removal and refitting

Note: *A new camshaft cover gasket must be used on refitting.*

Removal

1 On models from February 1993 onwards with the fuel filter/heater on top of the thermostat housing, the fuel hose is attached to a bracket on the camshaft cover. Undo the securing bolt and remove the bracket **(see illustration)**.
2 Disconnect the oil filler breather hose from the front of the camshaft cover **(see illustration)**.
3 Undo the three camshaft cover securing bolts and recover the metal and fibre washers under each bolt. Carefully move aside any cables and remove the camshaft cover **(see illustrations)**.

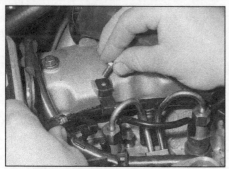

4.1 Remove the fuel hose bracket from the camshaft cover (1993 onwards)

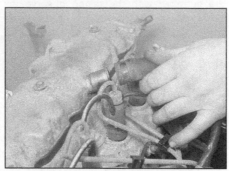

4.2 Disconnect the breather hose from the camshaft cover

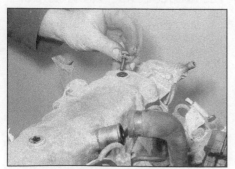

4.3a Undo the three camshaft cover securing bolts . . .

4.3b . . . then remove the cover

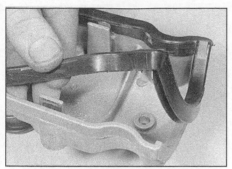

4.4 Remove the rubber gasket from the camshaft cover

5.2 Remove the crankshaft pulley bolt and thrustwasher . . .

5.3 . . . then remove the pulley

4 Remove the rubber gasket from the cover **(see illustration)**.

Refitting

5 Refitting is the reverse of removal, using a new camshaft cover gasket.

5 Crankshaft pulley - removal and refitting

Note: *A new crankshaft pulley bolt will be required on refitting.*

Removal

1 Remove the auxiliary drivebelt, as described in Chapter 1B.
2 The crankshaft pulley bolt is very tight. To prevent the crankshaft from turning as the pulley bolt is unscrewed, remove the starter

motor (see Chapter 5A) and lock the flywheel using a suitable notched tool engaged in the ring gear teeth. Alternatively, have an assistant hold a screwdriver, inserted between the ring gear teeth and the transmission location dowel. Unscrew the pulley bolt and recover the thrustwasher **(see illustration)**. **Note:** *Do not attempt to use the TDC locking tools as a counterhold while slackening the bolt.*
3 Remove the pulley from the end of the crankshaft, and recover the Woodruff key if it is loose **(see illustration)**. If the pulley is tight on the crankshaft, it can be removed using a puller as follows.
4 Refit the pulley bolt without the thrustwasher, but do not screw it fully home.
5 Improvise a suitable puller, using a short length of metal bar, two M6 bolts, and a large nut and bolt. Pass the two M6 bolts through the bar, and screw them into the tapped holes in the pulley. Tighten the large bolt, forcing it against the head of the pulley bolt while counterholding the nut, to force the pulley from the crankshaft.

Refitting

6 Refit the Woodruff key to the end of the crankshaft if it has come out, then refit the pulley.
7 Prevent the crankshaft from turning, as during removal, then fit a new pulley securing bolt with three drops of locking fluid on the threads. Ensure that the thrust washer is in place on the bolt head.
8 Tighten the bolt to the specified torque, then through the specified angle.

9 Refit and tension the auxiliary drivebelt as described in Chapter 1B.

6 Timing belt covers - removal and refitting

Note: *Improved access to the timing belt system can be obtained by moving the engine to the left, as described in Section 19.*
Note: *Some later XUD engines have an alternative timing belt cover design with upper, middle and lower covers, but there is no information about whether or not these have been fitted to C15 vans.*

Right-hand cover

Removal

1 Remove the lower clip from the right-hand timing cover, if fitted **(see illustration)**.
2 Release the upper spring clip from the cover, then pull the cover upwards so that it comes off the two studs and remove it from the engine **(see illustrations)**.

Refitting

3 Refitting is the reverse of removal.

Left-hand cover

Removal

4 Remove the right-hand cover as described above.
5 Release the two spring clips and manipulate the left-hand cover off the two studs, then withdraw the cover upwards **(see illustrations)**.

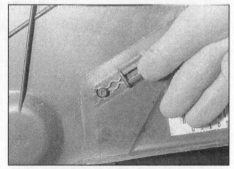

6.1 Remove the lower clip from the right-hand timing cover

6.2a Release the spring clip . . .

6.2b . . . and pull the right-hand cover upwards off the two studs

6.5a Release the two spring clips . . .

Refitting

6 Refitting is the reverse of removal.

Lower cover

Removal

7 Remove the auxiliary drivebelt (see Chapter 1B), then remove the crankshaft pulley (see Section 5).

8 Undo the two securing bolts and remove the lower cover **(see illustration)**.

Refitting

9 Refitting is the reverse of removal. Tighten the bolts to the specified torque.

---

7 Timing belt - general, removal, inspection, refitting and tensioning

General

1 The timing belt drives the camshaft, injection pump, and water pump from a toothed sprocket on the right-hand end of the crankshaft. The brake vacuum pump and power steering pump (if fitted) are also dependent on the timing belt because they are driven from the left-hand end of the camshaft. If the belt breaks or slips in service the pistons are likely to hit the valve heads resulting in expensive damage.

2 The timing belt should be renewed at the intervals specified in Chapter 1B. However, if it is contaminated with oil or if it is at all noisy in operation (a "scraping" noise due to uneven

7.8a Loosen the timing belt tensioner pivot nut . . .

7.8b . . . and the adjustment bolt . . .

6.5b . . . and manipulate the left-hand timing cover off the two studs (arrowed)

wear) it should be renewed earlier. Where a Bosch injection pump is fitted, excessive play in the right-hand bearing can wear the sides of the timing belt.

3 If the timing belt is being removed, it is a wise precaution to check the coolant pump for signs of leakage. This may avoid the need to remove the timing belt again at a later stage, should the coolant pump fail.

Removal

4 Remove the auxiliary drivebelt (see Chapter 1B), then remove the crankshaft pulley (see Section 5).

5 Support the weight of the engine using a hoist or trolley jack (see "*Jacking and vehicle support*"), then remove the right-hand engine mounting-to-body bracket. Alternatively, for improved access to the timing belt system, use an overhead hoist on both the left and right lifting lugs, and move the engine to the left (see Section 19).

6 Remove all three timing belt covers (see Section 6).

7 Set No 4 piston at TDC, then lock the crankshaft, camshaft and fuel pump in position as described in Section 3. If the engine is to be left in this state for some time, place suitable notices in the vehicle to avoid the possibility of cranking the engine on the starter motor, causing damage.

8 Loosen the timing belt tensioner pivot nut and adjustment bolt, then turn the tensioner bracket anti-clockwise against the spring to release the tension on the drivebelt. Use a 10 mm square socket wrench extension in the hole provided to turn the bracket against the

7.8c . . . then turn the tensioner bracket anti-clockwise with a 10 mm square socket wrench extension

6.8 Undo the two securing bolts and remove the lower cover

spring tension **(see illustrations)**. Then retighten the adjustment bolt to hold the tensioner in the released position.

9 Mark the timing belt with an arrow to indicate its normal direction of turning, then remove it from the camshaft, injection pump, water pump and crankshaft sprockets **(see illustration)**.

Inspection

10 Inspect the belt for cracks, fraying, and damage to the teeth. Pay particular attention to the roots of the teeth. If any damage is evident, or if the belt is contaminated with oil, fuel or coolant, it must be renewed and any leak rectified.

Refitting and tensioning

11 Begin refitting by ensuring that the M8 bolts are still fitted to the camshaft and fuel injection pump sprockets, and the rod or drill is positioned in the TDC hole in the flywheel.

12 Locate the timing belt on the crankshaft sprocket. If the original timing belt is being refitted, make sure the rotation arrow is facing the correct way.

13 Engage the timing belt with the crankshaft sprocket, hold it in position, then feed it over the remaining sprockets in the following order:

a) *idler roller*
b) *Fuel injection pump*
c) *Camshaft*
d) *Tensioner roller*
e) *Coolant pump.*

14 Be careful not to kink or twist the belt. To ensure correct engagement, locate only a half width on the injection pump sprocket before

7.9 Mark the timing belt with an arrow, then remove it from the sprockets

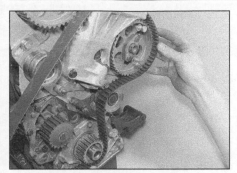

7.14a Fitting the timing belt over the injection pump sprocket . . .

7.14b . . . the camshaft sprocket . . .

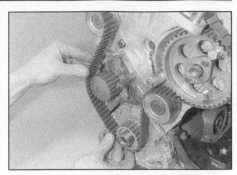

7.14c . . . and the water pump sprocket

feeding the timing belt onto the camshaft sprocket, keeping the belt taut and fully engaged with the crankshaft sprocket. Then locate the timing belt fully onto the sprockets **(see illustrations)**.

15 With the pivot nut loose, slacken the tensioner adjustment bolt while holding the bracket against the spring tension. Slowly release the bracket until the roller presses against the timing belt. Retighten the adjustment bolt.

16 Remove the bolts from the camshaft and injection pump sprockets. Remove the metal dowel rod from the cylinder block.

17 Rotate the engine two complete turns in its normal direction. Do not rotate the engine backwards as the timing belt must be kept tight between the crankshaft, injection pump and camshaft sprockets.

18 Loosen the tensioner adjustment bolt to allow the tensioner spring to push the roller against the timing belt, then tighten both the adjustment bolt and pivot nut to the specified torque.

19 Recheck the engine timing by inserting the bolts in the camshaft and fuel pump sprockets, and inserting the rod or drill in the flywheel TDC hole as described in Section 3. If the tools cannot be inserted, the timing belt has been incorrectly fitted (possibly one tooth out on one of the sprockets) and the refitting and tensioning procedure will have to be repeated.

20 Refit the three timing cover sections as described in Section 6.

21 Refit the right-hand engine mounting-to-body bracket and tighten the nuts, then remove the trolley jack or hoist.

22 Refit the crankshaft pulley as described in Section 5, using a new bolt and locking fluid.

23 Refit the auxiliary drivebelt and tension it as described in Chapter 1B.

8 Timing belt sprockets - removal and refitting

Note: *Improved access to the timing belt system can be obtained by moving the engine to the left, as described in Section 19.*

Camshaft sprocket

Removal

1 Remove the left- and right-hand timing belt covers as described in Section 6.

2 Slacken the camshaft sprocket bolt, but do not remove it at this stage. Use a counterhold to prevent the camshaft from turning, using either of the following methods:

a) Make up an improvised tool which can be inserted into the sprocket holes **(see illustration)**.

b) Remove the camshaft cover (see Section 4), then use a spanner on the lug between Nos 3 and 4 camshaft lobes **(see illustration)**.

3 Set No 4 piston at TDC, then lock the crankshaft, camshaft and fuel injection pump sprockets in position (see Section 3). **Note:** *Do not attempt to use the locking tools as a counterhold while slackening the camshaft sprocket bolt.*

4 Loosen the timing belt tensioner pivot nut and adjustment bolt, then turn the tensioner

bracket anti-clockwise against the spring to release the tension on the drivebelt. Use a 10 mm square socket wrench extension in the hole provided to turn the bracket against the spring tension. Then retighten the adjustment bolt to hold the tensioner in the released position.

5 Unscrew the camshaft sprocket bolt and recover the thrustwasher.

6 Withdraw the TDC locking bolt from the camshaft sprocket.

7 Withdraw the sprocket, manipulating the timing belt from it as it is withdrawn **(see illustration)**. Recover the Woodruff key from the end of the camshaft if it is loose. **Do not** allow the camshaft to rotate, otherwise the valves will strike the pistons of Nos 1 and 4 cylinders. If necessary, release the timing belt from the injection pump sprocket, remove the rod or drill from the TDC hole in the flywheel, then turn the crankshaft one quarter turn to position all the pistons half way down the cylinders.

Refitting

8 Fit the Woodruff key (if removed) and the camshaft sprocket to the camshaft. Fit the sprocket securing bolt and tighten it while holding the camshaft stationary. Do not allow the camshaft to rotate, unless the crankshaft has been turned as described in paragraph 7.

9 Refit the camshaft cover, if it has been removed (see Section 4).

10 Align the holes in the camshaft sprocket and the engine right-hand plate, and refit the M8 timing bolt to lock the camshaft in the TDC position.

8.2a Using an improvised tool to prevent the camshaft sprocket from turning

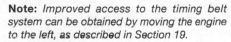

8.2b Lug (arrowed) for holding the camshaft

8.7 Remove the camshaft sprocket

8.16 Remove the crankshaft sprocket

8.25 Slacken the fuel injection pump sprocket nut using an improvised tool as a counterhold

11 If the crankshaft was turned a quarter turn from TDC, as in paragraph 7, turn it back by the same amount so that pistons 1 and 4 are at TDC. Do not turn the crankshaft more than a quarter turn, otherwise pistons 2 and 3 will pass their TDC positions and may strike the valves. Refit the rod or drill to the TDC hole in the flywheel.

12 Fit the timing belt around the fuel injection pump sprocket (where applicable), and the camshaft sprocket, and tension the timing belt (see Section 7).

13 Refit the left- and right-hand timing belt covers (see Section 6).

Crankshaft sprocket

Removal

14 Remove the crankshaft pulley (see Section 5).

15 Proceed as described in paragraphs 1, 3 and 4.

16 Withdraw the sprocket, manipulating the timing belt from it as it is removed. Recover the Woodruff key from the end of the crankshaft if it is loose **(see illustration)**. **Do not** allow the camshaft to rotate, otherwise the valves will strike the pistons of Nos 1 and 4 cylinders. If necessary, release the timing belt from the injection pump sprocket, remove the rod or drill from the TDC hole in the flywheel, then turn the crankshaft one quarter turn to position all the pistons half way down the cylinders.

Refitting

17 Refit the Woodruff key, if removed, then

8.26a Unscrew the nut . . .

refit the crankshaft sprocket with the flange nearest the cylinder block.

18 If the crankshaft was turned a quarter turn from TDC, as in paragraph 16, turn it back by the same amount so that pistons 1 and 4 are at TDC. Do not turn the crankshaft more than a quarter turn, otherwise pistons 2 and 3 will pass their TDC positions and may strike the valves. Refit the rod or drill to the TDC hole in the flywheel.

19 Fit the timing belt around the crankshaft sprocket, and tension the timing belt as described in Section 7.

20 Refit the crankshaft pulley (see Section 5).

21 Refit the left- and right-hand timing belt covers (see Section 6).

Fuel injection pump sprocket
Removal

22 Proceed as described in paragraphs 1, 3 and 4.

23 Make alignment marks on the fuel injection pump sprocket and the timing belt, to ensure that the sprocket and timing belt are correctly aligned on refitting.

24 On some models, the sprocket may be fitted with a built-in puller, which consists of a plate bolted to the sprocket. The plate contains a captive nut (the sprocket securing nut) which is screwed onto the fuel injection pump shaft. On models not fitted with a built-in puller, a suitable puller can be made up from a short length of bar, and two M7 bolts screwed into the holes provided in the sprocket.

25 Slacken the sprocket nut, using an

8.26b . . . and remove the fuel injection pump sprocket

improvised tool as a counterhold, fitted to the sprocket holes **(see illustration)**. **Note:** *Do not attempt to use the locking tools as a counterhold while slackening the fuel pump sprocket nut.*

26 On models with a built-in puller, unscrew the sprocket nut until the sprocket is freed from the taper on the pump shaft, then withdraw the sprocket. Recover the Woodruff key from the end of the pump shaft if it is loose **(see illustrations)**. If required, the puller assembly can be removed from the sprocket by removing the two bolts and washers **(see Haynes Hint)**.

27 On models not fitted with a built-in puller, partially unscrew the sprocket securing nut, then fit the improvised puller and tighten the two bolts, forcing the bar against the sprocket nut until the sprocket is freed from the taper on the pump shaft. Withdraw the sprocket and recover the Woodruff key from the end of the pump shaft if it is loose. Remove the puller from the sprocket.

Refitting

28 Refit the Woodruff key to the pump shaft, if it has been removed, making sure it is correctly located in its groove.

29 Where applicable, if the built-in puller assembly has been removed from the sprocket, refit it and tighten the bolts to the specified torque, ensuring that the washers are in place.

30 Refit the sprocket, then tighten the securing nut to the specified torque, using a counterhold as during removal.

31 Make sure that the M8 locking bolts are fitted to the camshaft and fuel injection pump sprockets, and the rod or drill is in position in the TDC hole in the flywheel.

32 Fit the timing belt around the fuel injection pump sprocket, ensuring that the marks made before removal are aligned.

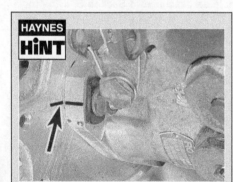

If extra space is required to remove the fuel injection pump sprocket, mark the pump in relation to the mounting bracket (arrowed), then slacken the three front mounting nuts and the rear mounting bolt, and pull the pump backwards slightly. Make sure the engine runs correctly after refitting, and if necessary check the static timing with reference to Chapter 4C.

9.1a Right-hand engine mounting bracket securing bolts (1) and timing belt tensioner plunger (2)

9.1b One additional bolt (arrowed) secures the mounting bracket from behind the engine right-hand plate

33 Tension the timing belt (see Section 7).
34 Refit the left- and right-hand timing belt covers (see Section 6).

Coolant pump sprocket

35 The coolant pump sprocket is integral with the pump, and cannot be removed. See Chapter 3 for details of how to remove the coolant pump.

9 Timing belt tensioner removal and refitting

Note: *Citroën tool No 7009-T1 or an equivalent improvised tool is required for this job.*
Note: *Improved access to the timing belt system can be obtained by moving the engine to the left, as described in Section 19.*

Removal

1 The timing belt tensioner is operated by a spring and plunger housed in the right-hand engine mounting bracket, which is bolted to the engine right-hand plate. There is one additional bolt, accessible from behind the plate (see illustrations).
2 Remove the left- and right-hand timing covers (see Section 6)

3 Turn the engine using the front right-hand wheel so that the No 4 piston is at TDC and insert the locking tools in the camshaft sprocket, fuel injection pump sprocket and flywheel (see Section 3).
4 Loosen the timing belt tensioner pivot nut and adjustment bolt, then turn the bracket anti-clockwise until the adjustment bolt is in the middle of the slot and retighten the bolt. Use a 10 mm square socket wrench extension in the hole provided to turn the bracket against the spring tension.
5 A tool is now required, to hold the tensioner plunger in the mounting bracket. Use Citroën tool No 7009-T1 which slides into the two lower bolt holes of the mounting bracket. Alternatively, fabricate a similar tool using sheet metal and two long M10 bolts and nuts (see illustration).
6 Undo the two lower engine mounting bracket bolts, then fit the special tool. Grease the inner surface of the tool to prevent any damage to the end of the tensioner plunger (see illustration). Unscrew the pivot nut and adjustment bolt and withdraw the tensioner bracket, complete with roller.
7 Undo the remaining bolts from the engine mounting bracket, including the one behind the engine right-hand plate, then withdraw the bracket (see illustration).

8 Compress the tensioner plunger into the mounting bracket, remove the special tool, then withdraw the plunger and spring.

Refitting

9 Refitting is the reverse of removal. Adjust the timing belt as described in Section 7.

10 Timing belt idler roller - removal and refitting

Note: *Improved access to the timing belt system can be obtained by moving the engine to the left, as described in Section 19.*

Removal

1 Remove the left- and right-hand timing covers (see Section 6).
2 Turn the engine using the front right-hand wheel so that the No 4 piston is at TDC and insert the locking tools in the camshaft sprocket, fuel injection pump sprocket and flywheel (see Section 3).
3 Loosen the timing belt tensioner pivot nut and adjustment bolt, then turn the bracket anti-clockwise until the adjustment bolt is in the middle of the slot and retighten the bolt.

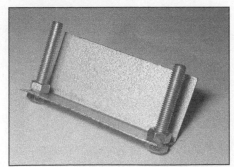

9.5 Improvised tool for holding tensioner plunger in engine mounting bracket

9.6 Tool in place to hold tensioner plunger in engine mounting bracket - timing belt removed for clarity

9.7 Remove the right-hand engine mounting bracket

10.6 Remove the timing cover spacer and stud (arrowed)

Use a 10 mm square socket wrench extension in the hole provided to turn the bracket against the spring tension.

4 Remove the auxiliary drivebelt (see Chapter 1B), then remove the crankshaft pulley (see Section 5).

5 Remove the lower timing belt cover (see Section 6).

6 Remove the spacer from the stud for the upper timing cover sections. Note the position of the stud, then unscrew and remove it **(see illustration)**.

7 Undo the remaining bolts securing the idler roller bracket to the cylinder block, noting that the upper bolt also secures the engine mounting bracket.

8 Slightly loosen the remaining engine mounting bracket bolts, then slide out the idler roller and bracket.

11.2 Remove the camshaft right-hand oil seal

11.3 Use an M10 bolt, washers and socket to fit the camshaft right-hand oil seal

Refitting

9 Refitting is the reverse of removal, but note the following additional points:
- a) Use a new crankshaft pulley bolt and apply three drops of locking fluid to the threads.
- b) Tighten all bolts to torque, where specified.
- c) Adjust the timing belt (see Section 7).
- d) Tension the alternator drivebelt (see Chapter 1B).

11 Camshaft oil seals - renewal

Right-hand oil seal

Note: *Improved access to the timing belt system can be obtained by moving the engine to the left, as described in Section 19.*

1 Remove the camshaft sprocket, and recover the Woodruff key if it is loose (see Section 8). **Do not** rotate the camshaft with the engine at the TDC position, otherwise the valves will strike the pistons of Nos 1 and 4 cylinders.

2 Pull out the oil seal using a hooked instrument **(see illustration)**.

3 Clean the oil seal seating. Smear the lip of the new oil seal with oil then fit it over the end of the camshaft, open end first, and press it in until flush with the end face of the cylinder head. Use an M10 bolt, washers and a socket to press it in **(see illustration)**.

4 Fit the Woodruff key (if removed) and the camshaft sprocket to the camshaft, insert the bolt and tighten it while holding the camshaft stationary.

5 Refit the M8 timing bolt to the camshaft sprocket.

6 Refit and adjust the timing belt, referring to Section 7. The remaining procedure is the reverse of removal.

Left-hand oil seal - models with belt-driven brake vacuum pump

7 Remove the air cleaner and ducting as required (see Chapter 4C).

8 Remove the drivebelt(s) from the camshaft pulley as described in Chapter 1B.

9 Unscrew the centre bolt and remove the camshaft pulley. If the centre bolt is very tight, remove the camshaft cover (see Section 4), and use a spanner as a counterhold on the lug between Nos 3 and 4 camshaft lobes, to prevent damage to the timing belt. If the pulley does not come off with the bolt removed, use a suitable puller. Recover the Woodruff key if it is loose.

10 Pull out the oil seal using a hooked instrument.

11 Clean the oil seal seating. Smear the lip of the new oil seal with oil then fit it over the end of the camshaft, open end first, and press it in until flush with the end face of the cylinder head **(see illustration)**. Use a bolt, washers and a socket to press it in.

12 Refit the Woodruff key (if removed) and the pump pulley to the camshaft and tighten the centre bolt. Refit and tension the drivebelt(s) as described in Chapter 1B.

13 Refit the air cleaner and inlet ducting.

Left-hand oil seal - models with direct-drive brake vacuum pump

14 No oil seal is fitted to the flywheel end of the camshaft. Sealing is provided by an O-ring fitted to the vacuum pump flange. The O-ring can be renewed after unbolting the pump from the cylinder head (see Chapter 9). Note the smaller O-ring which seals the oil feed gallery to the pump. This may also cause leakage from the pump/cylinder head mating faces if it deteriorates or fails **(see illustration)**.

12 Camshaft - removal and refitting

Note: *Improved access to the timing belt system can be obtained by moving the engine to the left, as described in Section 19.*

Removal

1 Remove the camshaft cover (see Section 4).

2 Remove the camshaft sprocket, taking care not to allow the valves to hit pistons 1 and 4 which are at TDC. If necessary, rotate the engine one quarter turn away from TDC so that all the pistons are half-way down their cylinders (see Section 8).

11.14 Camshaft left-hand oil seal (1) and oil feed gallery O-ring (2) on direct-drive brake vacuum pump

11.11 Camshaft left-hand oil seal flush with end face of cylinder head - belt-driven brake vacuum pump models

12.7 Lifting the camshaft from the cylinder head - model without a pump pulley

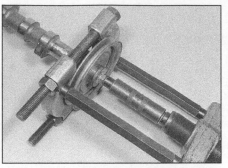

12.8 Using a puller to remove the pump pulley from the camshaft. This will be a double pulley on models with power steering.

12.12 The DIST marking must be at the timing belt end

3 On models with a brake vacuum pump driven directly from the camshaft, unbolt the pump and move it to one side, with reference to Chapter 9.

4 On models with a belt-driven brake vacuum pump (and optionally a power steering pump), remove the air cleaner and ducting as required to improve access (see Chapter 4C), then remove the drivebelt(s) as described in Chapter 1B.

5 Mark the position of the camshaft bearing caps, numbering them from the flywheel end and making the marks on the manifold side.

6 Progressively unscrew the nuts, then remove the bearing caps.

7 Lift the camshaft and withdraw it through

12.14a Fitting a camshaft end bearing cap

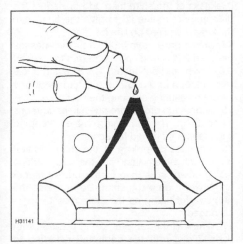

12.14b Areas on camshaft end bearing caps to apply sealing compound

the engine right-hand plate, complete with the brake vacuum pump drivebelt pulley if fitted **(see illustration)**. Remove the oil seal from the timing end of the camshaft.

8 On models with a pump pulley, hold the camshaft stationary with a spanner on the special lug between the 3rd and 4th cams, then unscrew the bolt from the end of the camshaft and remove the pump pulley. Use a puller if it is tight **(see illustration)**. Recover the Woodruff key if it is loose.

9 Remove the oil seal from the flywheel end of the camshaft.

10 Clean all the components including the bearing surfaces in the cylinder head. Examine the components carefully for wear and damage, and in particular check the surface of the cams for scoring and pitting. Renew components as necessary and obtain new oil seals.

Refitting

11 Begin reassembly by lubricating the cams and bearing journals with engine oil.

12 Locate the camshaft on the cylinder head, passing it through the engine right-hand plate and with the tips of cams 4 and 6 facing downwards and resting on the bucket tappets. The cast DIST marking on the camshaft should be at the timing belt end of the cylinder head **(see illustration)** and the key slot for the camshaft sprocket should be facing upwards.

13 Fit the centre bearing cap the correct way round as previously noted, then screw on the nuts and tighten them two or three turns.

12.15a Tightening the camshaft bearing cap nuts

12.15b Checking the camshaft endfloat

14 Apply sealing compound to the end bearing caps on the areas as shown. Fit them in the correct positions and tighten the nuts two or three turns **(see illustrations)**.

15 Tighten all the nuts progressively to the specified torque, making sure that cams 4 and 6 remain facing downwards **(see illustration)**. Check that the camshaft endfloat is as given in the Specifications using feeler blades **(see illustration)**. The only answer if it is not correct is to renew the cylinder head.

16 If the original camshaft is being refitted and it is known that the valve clearances are correct, go to the next paragraph, otherwise check and adjust the valve clearances as described in Section 13. Note that as the timing belt is disconnected at this stage, the crankshaft must be turned one quarter turn either way from the TDC position so that all the pistons are halfway down the cylinders. This will prevent the valves striking the pistons when the camshaft is rotated. Release the timing belt from the injection pump sprocket while turning the engine as the timing bolts are still in position.

17 Smear the lips of the oil seals with oil, then fit them over each end of the camshaft, open end first, and press them in until flush with the end faces of the end caps. Use an M10 bolt, washers and a socket to press in the oil seals (see Section 11).

18 On models with a belt-driven brake vacuum pump, fit the Woodruff key and pump pulley to the flywheel end of the camshaft, insert the bolt and tighten it while holding the camshaft stationary.

13.6 Checking the valve clearances with feeler blades

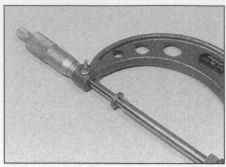

13.11 Checking the shim thickness with a micrometer

19 Fit the Woodruff key and camshaft sprocket to the timing end of the camshaft. Insert the bolt and tighten it to the specified torque while holding the camshaft stationary.
20 Refit the camshaft cover, together with a new gasket, and tighten the bolts. Refit the oil filler cap/breather hose.
21 On models with a brake vacuum pump driven directly from the camshaft, fit new O-rings to the pump recesses, then refit the pump and tighten the securing bolts with reference to Chapter 9.
22 On models with a belt-driven brake vacuum pump (and optionally a power steering pump), refit and tension the drivebelt(s) as described in Chapter 1B.
23 Align the holes and refit the M8 timing bolt to the camshaft sprocket.
24 If the crankshaft was turned a quarter turn from TDC, turn the crankshaft back the quarter turn so that pistons 1 and 4 are again at TDC, and fit the TDC locking tool to the flywheel. Do not turn the engine more than a quarter turn, otherwise pistons 2 and 3 will pass their TDC positions and will strike valves 4 and 6.
25 Refit and adjust the timing belt, referring to Section 7. The remaining procedure is the reverse of removal.

13 Valve clearances - checking and adjustment

Note: *This is not a routine operation. It should only be necessary at high mileage, after overhaul, or when investigating noise or power loss which may be attributable to the valve gear.*

Checking

1 Apply the handbrake, then jack up the front right-hand corner of the vehicle until the wheel is just clear of the ground, and support the vehicle on an axle stand (see "*Jacking and vehicle support*"). Engage 4th or 5th gear so that the engine may be rotated by turning the right-hand wheel.
2 Disconnect the battery negative lead.
3 Remove the camshaft cover (see Section 4).
4 On a piece of paper draw the outline of the engine with the cylinders numbered from the flywheel end and also showing the position of each valve, together with the specified valve clearance. Above each valve draw two lines for noting (1) the actual clearance and (2) the amount of adjustment required.

5 Turn the engine until the inlet valve of No 1 cylinder (nearest the flywheel) is fully closed and the apex of the cam is facing directly away from the bucket tappet.
6 Using feeler blades measure the clearance between the base of the cam and the bucket tappet **(see illustration)**. Record the clearance on line (1).
7 Repeat the measurement for the other seven valves, turning the engine as necessary so that the cam lobe in question is always facing directly away from the particular bucket tappet.
8 Calculate the difference between each measured clearance and the desired value and record it on line (2). Since the clearance is different for inlet and exhaust valves, make sure that you are aware which valve you are dealing with. The valve sequence from either end of the engine is:

Inlet - Exhaust - Exhaust - Inlet - Inlet - Exhaust - Exhaust - Inlet

9 If all the clearances are within tolerance, refit the valve cover using a new gasket if necessary. If any clearance measured is outside the specified tolerance, adjustment must be carried out as described below.

Adjustment

10 Remove the camshaft as described in Section 12.
11 Withdraw the first bucket tappet and its shim. Be careful that the shim does not fall out of the tappet. Clean the shim and measure its thickness with a micrometer **(see illustration)**.
12 Refer to the clearance recorded for the valve concerned. If the clearance was more than the amount required the shim thickness must be increased by the difference recorded (2); if too small the thickness must be decreased.
13 Draw three more lines beneath each valve on the calculation paper as shown **(see illustration)**. On line (4) note the measured thickness of the shim then add or deduct the difference from line (2) to give the final shim thickness required on line (5).
14 Shims are available in thicknesses between 2.225 mm and 3.025 mm in steps of 0.025 mm, and between 3.100 mm and 3.550 mm in steps of 0.075 mm. Clean new shims before measuring or fitting them.
15 Repeat the procedure given in paragraphs 11 to 13 on the remaining valves, keeping each tappet identified for position.
16 When reassembling, oil the shim and fit it on the valve stem first with the size marking facing downwards then oil the bucket tappet and lower it onto the shim. Do not raise the tappet after fitting as the shim may become dislodged.
17 When all the tappets are in position with their shims, refit the camshaft referring to Section 12, but recheck the clearances to make sure they are correct.

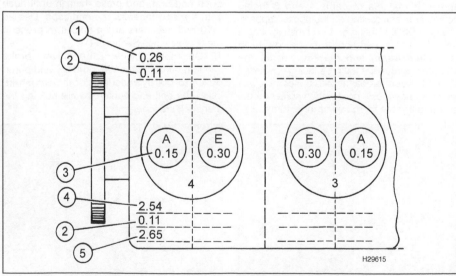

13.13 Example of valve shim thickness calculation

A Inlet   *E Exhaust*

14.13 Leak off pipe connected between two injectors

14.15 Disconnect the injector pipes

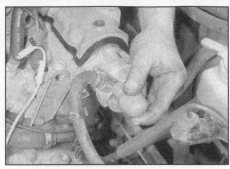

14.17 Disconnect the brake vacuum hose

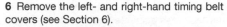

14 Cylinder head -
removal and refitting

Note: *This is an involved procedure, and it is suggested that the Section is read thoroughly before starting work. To aid refitting, make notes on the locations of all relevant brackets and the routing of hoses and cables before removal. A new cylinder head gasket must be used on refitting, and new cylinder head bolts may be required.*

⚠️ **Warning: Exercise extreme caution when working on the high-pressure pipework from the pump to the fuel injectors. The fuel may be under pressure for some time after the engine has been switched off. Never expose the hands or any part of the body to fuel under pressure because it can penetrate the skin, with possibly fatal results. Place rags around the pipework before you undo the unions.**

Caution: Take great care not to allow dirt into the injectors or fuel pipes during this procedure. Cover all open unions to keep dirt out, using suitable plugs, or small plastic bags secured with rubber bands.

Removal

1 Apply the handbrake, then jack up the front right-hand corner of the vehicle until the wheel is just clear of the ground, and support the vehicle on an axle stand (see "*Jacking and vehicle support*"). Engage 4th or 5th gear so that the engine may be rotated by turning the right-hand wheel.
2 Drain the cooling system (see Chapter 1B).
3 Disconnect the battery negative lead.
4 Remove the air cleaner and ducting (see Chapter 4C). Remove the duct from the inlet manifold and plug the orifice with a rag to prevent the entry of dirt.
5 Support the weight of the engine using a hoist or trolley jack (see "*Jacking and vehicle support*"), then remove the right-hand engine mounting-to-body bracket. Alternatively, for improved access to the timing belt system, use an overhead hoist on both the left and right lifting lugs, and move the engine to the left (see Section 19).

6 Remove the left- and right-hand timing belt covers (see Section 6).
7 Set No 4 piston at TDC, then lock the crankshaft, camshaft and fuel injection pump sprockets in position (see Section 3).
8 Loosen the timing belt tensioner pivot nut and adjustment bolt, then turn the bracket anti-clockwise to release the tension on the drivebelt. Use a 10 mm square socket wrench extension in the hole provided to turn the bracket against the spring tension. Then retighten the pivot nut to hold the tensioner in the released position.
9 Remove the timing belt from the camshaft sprocket and tie it to one side without bending it excessively.
10 Unscrew the M8 bolt holding the camshaft sprocket in the timing position. Also unscrew the tensioner adjustment bolt, and the two upper bolts securing the engine mounting bracket to the engine right-hand plate.
11 At this stage the right-hand engine mounting-to-body bracket may be temporarily refitted and the hoist or trolley jack removed.
12 Disconnect the heater hose from the rear of the cylinder head, at the flywheel end.
13 Disconnect the leak-off return hose from the appropriate fuel injector and plug the end. If required, also remove the leak-off hoses from between the fuel injectors **(see illustration)**. On models up to February 1993, the leak-off return hose connects the No 4 injector (at the timing belt end) to the fuel pump return hose. From February 1993 onwards, it connects to the No 1 injector (at the flywheel end) to the fuel system automatic bleed valve.
14 On models from February 1993 onwards, detach the fuel system automatic bleed valve from the cylinder head and place it to one side.
15 Unscrew the union nuts securing the injection pipes to the injectors and fuel injection pump, and remove the pipes as two assemblies **(see illustration)**. Plug the ends of all open unions, including the injectors.
16 Unbolt the thermostat housing from the cylinder head and place it to one side. Disconnect any hoses, as required, to facilitate removal. On models from February 1993 onwards, the fuel filter/heater unit is mounted

on the thermostat housing. Plug the ends of any fuel hoses that have to be disconnected.
17 On models with a brake vacuum pump driven directly from the camshaft, disconnect the brake vacuum hose **(see illustration)**.
18 On models with a belt-driven brake vacuum pump (and optionally a power steering pump), remove the drivebelt(s) as described in Chapter 1B.
19 Remove the camshaft cover (see Section 4).
20 Unbolt the left-hand engine lifting bracket.
21 Disconnect the electrical cable from the No 1 glow plug at the transmission end of the engine.
22 Hold the camshaft stationary with a spanner on the special lug between the 3rd and 4th cams or by using a lever in the sprocket holes, then unscrew the camshaft sprocket bolt and withdraw the sprocket. Recover the Woodruff key if it is loose. Do not rotate the camshaft otherwise the valves will strike the pistons of Nos 1 and 4 cylinders. If necessary release the timing belt from the injection pump sprocket and turn the engine one quarter turn in either direction to position all the pistons halfway down the cylinders to prevent any damage.

> **HAYNES HiNT** *If there is insufficient room to remove the camshaft sprocket, leave it loosely fitted for the time being, then remove it when the cylinder head is lifted off.*

23 Unscrew and remove the exhaust manifold-to-downpipe bolts, together with their springs and collars, referring to Chapter 4C. Recover the sealing ring from the downpipe or manifold, as applicable.
24 Progressively unscrew the cylinder head bolts **(see illustration)** in the reverse order to that shown for tightening (refer to paragraph 39). Remove the washers.
25 Make sure all hoses and cables that would obstruct removal of the cylinder head have been disconnected or moved to one side.
26 Release the cylinder head from the cylinder block and location dowel by rocking it. The Citroën tool for doing this consists of two metal dowel rods with 90° angled ends.

14.24 Unscrew the cylinder head bolts

14.30 Checking the piston protrusion

Do not prise between the mating faces of the cylinder head and block, as this may damage the gasket faces.

27 Lift the cylinder head from the block and remove the gasket. At the same time, remove the camshaft sprocket if it has not been removed already.

28 Do not dispose of the old gasket until a new one has been obtained. The correct thickness of gasket is determined after measuring the protrusion of the pistons at TDC.

29 Clean the gasket faces of the cylinder head and cylinder block, preferably using a soft blunt instrument to prevent damage to the mating surfaces. Clean the bolt holes in the cylinder block.

Gasket selection

30 Check that the timing belt is clear of the injection pump sprocket, then turn the engine until pistons 1 and 4 are at TDC. Position a dial test indicator on the cylinder block and zero it on the block face. Transfer the probe to the centre of piston 1 then slowly turn the crankshaft back and forth past TDC noting the highest reading on the indicator **(see illustration)**. Record this reading.

31 Repeat this measurement procedure on piston 4 then turn the crankshaft half a turn (180°) and repeat the procedure on pistons 2 and 3.

32 If a dial test indicator is not available, piston protrusion may be measured using a straight-edge and feeler blades or vernier

calipers, however, these methods are inevitably less accurate and cannot therefore be recommended.

33 Ascertain the greatest piston protrusion measurement and use this to determine the correct cylinder head gasket, identified by the number of notches **(see illustration)**. Gaskets are supplied according to a three-class or five-class system as follows:

Three-class system

Piston protrusion	Gasket thickness	Notches
0.54 to 0.65 mm	1.49 mm	1*
0.65 to 0.77 mm	1.61 mm	2
0.77 to 0.82 mm	1.73 mm	3

*Original gasket fitted on new vehicle

Five-class system

Piston protrusion	Gasket thickness	Notches
0.56 to 0.67 mm	1.36 mm	1
0.67 to 0.71 mm	1.40 mm	2
0.71 to 0.75 mm	1.44 mm	3
0.75 to 0.79 mm	1.48 mm	4
0.79 to 0.83 mm	1.52 mm	5

34 The single notch on the gasket centre line indicates that the gasket is designed for use with the XUD 7 (161A) engine and has no significance for the gasket thickness. Make sure there is only one centre line notch.

Cylinder head bolt examination

35 Models from 1993 onwards are fitted with cylinder-head bolts that have to be measured

to determine whether or not they can be re-used. Measure the length of each bolt from the base of the head to the end of the shank **(see illustration)**. If any bolts exceed the maximum length, they must be renewed. It is considered good practice to renew the cylinder-head bolts anyway, regardless of their model year and apparent condition, because they are cheap compared with the cost of repairs following the failure of a gasket in service.

Refitting

36 Turn the crankshaft clockwise (viewed from the timing belt end) until pistons 1 and 4 pass bottom dead centre (BDC) and start to rise, then position them halfway up their bores. Pistons 2 and 3 will also be at their mid-way positions, but descending their bores.

37 Fit the correct gasket the right way round on the cylinder block with the identification notches or holes at the flywheel end **(see illustration)**. Make sure that the location dowel is in place at the timing end of the block.

38 Check that the Woodruff key is in place on the camshaft. Lower the cylinder head onto the block, and at the same time loosely fit the camshaft sprocket and bolt.

39 Apply Molykote G Rapide Plus or other suitable high-temperature grease to the threads and contact faces of the cylinder head bolts (cleaning them first if the old ones are being re-used). This will ensure that the correct torque is achieved by longitudinal stress, rather than friction on the bolt head and threads. Insert the bolts, together with their washers, and tighten them in the sequence shown **(see illustration)**, in stages as given in Specifications. Use an angle gauge for the final stage angle tightening **(see illustration)**.

40 Recheck the valve clearances, referring to Section 13 and adjust them as necessary. Do this even if the clearances have been adjusted with the cylinder head removed, as there may be minor differences.

41 Lubricate the exhaust manifold-to-downpipe contact surfaces with heat resistant grease, then reconnect them and fit the bolts, together with the springs, cups and self-locking nuts. Tighten the nuts progressively until approximately four threads are visible and the springs are compressed to 22.0 mm in length.

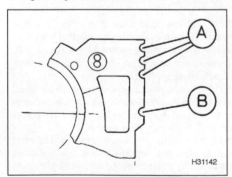

14.33 Head gasket thickness identification notches

a = gasket thickness identification
b = 161A engine identification

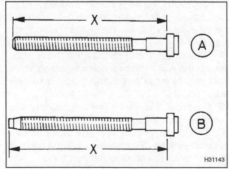

14.35 Maximum length (X) of cylinder head bolts - 1993 onwards

A Bolt without guide boss   X = 121.5 mm
B Bolt with guide boss   X = 125.5 mm

14.37 Cylinder head gasket identification notches (arrowed)

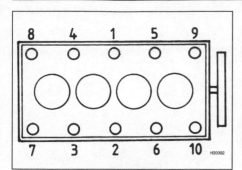

14.39a Cylinder head bolt tightening sequence

14.39b Angle-tightening a cylinder head bolt, using a commercially-available angle gauge

42 Tighten the camshaft sprocket bolt to the specified torque while holding the camshaft stationary with a spanner on the lug between the 3rd and 4th cams.

43 Turn the camshaft until the tips of cams 4 and 6 (counting from the flywheel end) are facing downwards.

44 Turn the crankshaft a quarter turn clockwise until pistons 1 and 4 are at TDC, and fit the TDC dowel rod to the flywheel. Do not turn the crankshaft anti-clockwise otherwise pistons 2 and 3 will pass their TDC positions and will strike valves 4 and 6.

45 Align the hole and refit the M8 timing bolt to the camshaft sprocket.

46 Refit the camshaft cover, using a new gasket, and reconnect the oil filler breather hose.

47 If the right-hand engine mounting-to-body bracket has been temporarily refitted, remove

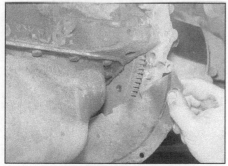

15.3 Remove the flywheel cover plate

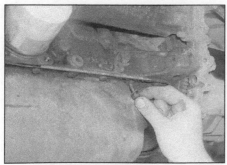

15.4 Remove the sump bolts

it again, supporting the engine with a hoist or trolley jack.

48 Apply locking fluid to the threads, then refit and tighten the two upper bolts to the right-hand engine mounting bracket. Also refit the tensioner adjustment bolt and tighten it. Loosen the tensioner pivot nut.

49 Refit and adjust the timing belt (see Section 7).

50 Reconnect the glow plug wiring.

51 Refit the left-hand engine lifting bracket.

52 On models with a brake vacuum pump driven directly from the camshaft, reconnect the brake vacuum hose.

53 On models with a belt-driven brake vacuum pump (and optionally a power steering pump), refit and tension the drivebelt(s) as described in Chapter 1B.

54 Clean the thermostat housing mating faces then refit it, using with a new gasket, and tighten the bolts. Refit any hoses that were removed.

55 Refit the fuel injection pipes and tighten the union nuts.

56 Refit the fuel injector leak-off hoses and the leak-off return hose. On models from February 1993 onwards, refit the fuel system automatic bleed valve to the cylinder head.

57 Reconnect the heater hose to the cylinder head.

58 Refit the left- and right-hand timing belt covers (see Section 6).

59 Refit the right-hand engine mounting-to-body bracket and tighten the nuts. Remove the hoist or trolley jack.

60 Refit the air cleaner, and reconnect the air inlet hose to the inlet manifold (see Chapter 4C).

15.6 Oil sump with gasket material removed

61 Reconnect the battery negative lead.

62 Refill the cooling system (see Chapter 1B).

63 Lower the vehicle to the ground.

64 Prime and bleed the fuel system (see Chapter 4C).

15 Sump - removal and refitting

Note: *A new sump gasket must be used on refitting.*

Removal

1 Apply the handbrake, chock the rear wheels, then jack up the front of the vehicle and support on axle stands (see "*Jacking and vehicle support*").

2 Position a container beneath the engine. Unscrew the drain plug and allow the oil to drain from the sump. Wipe clean the drain plug and refit it.

3 Unbolt the flywheel cover plate from the base of the transmission, and remove it **(see illustration)**.

4 Unscrew the sump bolts, noting the position of the various lengths and types of bolt **(see illustration)**. There are 23 bolts as follows:

a) 6 socket-head bolts
b) 15 bolts, 16 mm length
c) 2 bolts, 14 mm length

5 Remove the sump and gasket. The sump will probably be stuck in position in which case it will be necessary to cut it free using a thin knife.

Refitting

6 Clean all remains of gasket from the sump and block and wipe dry **(see illustration)**.

7 Apply a little sealing compound to the joints where the right-hand oil seal housing abuts the block on both sides **(see illustration)**.

8 Position a new gasket on the sump then lift the sump into position and insert the bolts in their correct locations.

9 Tighten the bolts evenly to the specified torque.

10 Lower the vehicle to the ground and refill the engine with the correct quantity and grade of oil (see Chapter 1B).

15.7 Apply sealing compound to the joints (arrowed) between the right-hand oil seal housing and block

16.2 Undo the oil pump bolts

16.3 Remove the oil pump and disengage it from the drive chain

16.4a Unscrew the oil pump bolts . . .

16.4b . . . separate the halves . . .

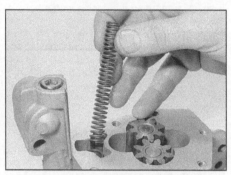

16.4c . . . and remove the relief valve spring . . .

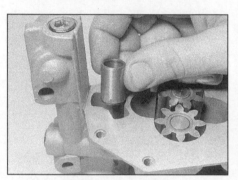

16.4d . . . and plunger

16 Oil pump, drive chain and sprockets - removal, inspection and refitting

Oil pump and sprocket

Removal

1 Remove the sump (see Section 15).
2 Undo the three bolts securing the oil pump to the crankcase. Identify them for position as all three are of different lengths **(see illustration)**.
3 Remove the pump and disengage it from the drive chain **(see illustration)**.

Separation and inspection

4 Remove the six bolts which hold the two halves of the oil pump together and separate the halves. Be prepared for the release of the relief valve spring and plungers, and remove them **(see illustrations)**.

5 If necessary remove the strainer by prising off the cap, then clean all components **(see illustrations)**.
6 Inspect the gears and the housings for wear and damage. Check the endfloat of the gears using a straight-edge and feeler blades; also check the clearance between the tip of the gear lobes and the housing **(see illustrations)**. If any of these clearances exceeds the specified limit, renew the pump. Note that except for the relief valve spring and plunger, individual components are not available.

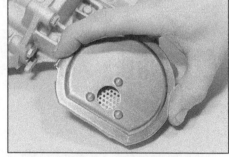

16.5a Removing the oil pump cap . . .

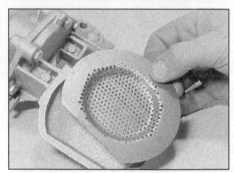

16.5b . . . and strainer

16.6a Oil pump rotors and housing

16.6b Checking the rotor endfloat

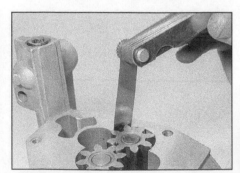

16.6c Checking the rotor side clearance

16.14 Remove the crankshaft sprocket

16.15 Undo the bolts and remove the crankshaft right-hand oil seal housing

7 Examine the teeth on the oil pump sprocket. If the teeth are "hooked", the complete oil pump must be renewed as the sprocket cannot be removed from the pump shaft.
8 If the pump is to be renewed, the chain and the crankshaft sprocket should also be renewed.

Refitting

9 Lubricate the gears with engine oil, then reassemble the oil pump in reverse order and tighten the six bolts evenly to the specified torque.
10 Refit the pump and engage the sprocket with the chain. Insert the bolts in their correct locations, according to their length, and tighten them evenly to the specified torque.
11 Refit the sump (see Section 15).

Oil pump drive chain and sprocket

Note: *A new crankshaft right-hand oil seal, and an oil seal housing gasket, must be used on refitting.*
Note: *Improved access to the timing belt system can be obtained by moving the engine to the left, as described in Section 19.*

Removal

12 Remove the oil pump as described above.
13 Remove the timing belt (see Section 7).

14 Slide the crankshaft timing belt sprocket off the shaft and recover the Woodruff key if it is loose **(see illustration)**.
15 Undo the bolts and remove the crankshaft right-hand oil seal housing **(see illustration)**. Recover the gasket. Remove the oil seal if it is still attached to the crankshaft.
16 Remove the chain and sprocket from the end of the crankshaft and recover the Woodruff key if it is loose.

Inspection

17 Examine the teeth on the sprockets, and renew them if the teeth are "hooked". If the oil pump sprocket is worn, the complete pump must be renewed as the sprocket cannot be removed from the pump shaft.
18 Examine the chain for wear. If the engine has completed a considerable mileage, or if the chain has a deeply bowed appearance when held horizontally (rollers vertical), renew the chain.

Refitting

19 Locate the Woodruff key on the nose of the crankshaft and refit the sprocket, teeth end first. Engage the chain with the sprocket.
20 Prise the oil seal from the crankshaft right-hand housing. Refit the housing to the cylinder block, together with a new gasket,

but without the oil seal, and tighten the bolts evenly to the specified torque.
21 Fit a new oil seal to the housing (see Section 17).
22 Refit the oil pump as described earlier, and refit the sump (see Section 15).
23 The remainder of refitting is the reverse of removal. Adjust the timing belt as described in Section 7.

17 Crankshaft oil seals - renewal

Timing belt end

Note: *Improved access to the timing belt system can be obtained by moving the engine to the left, as described in Section 19.*
1 Remove the timing belt (see Section 7).
2 Slide the timing belt sprocket from the crankshaft and recover the Woodruff key if it is loose.
3 Note the fitted depth, then pull the oil seal from the housing using a hooked instrument. Alternatively drill a small hole in the oil seal and use a self-tapping screw to remove it **(see illustration)**.
4 Clean the housing and crankshaft, then dip the new oil seal in engine oil and press it in (open end first) to the previously noted depth. A piece of thin plastic is useful to prevent damage to the oil seal **(see illustration)**.
5 Refitting is the reverse of removal. Adjust the timing belt as described in Section 7.

Flywheel end

6 Remove the flywheel as described in Section 18.
7 Using vernier calipers, measure the fitted depth of the oil seal and record it.
8 Pull out the oil seal using a hooked instrument. Alternatively, drill a small hole in the oil seal and use a self-tapping screw to remove it.

17.3 Using a self-tapping screw and a pair of pliers to remove the crankshaft right-hand oil seal

17.4 Fitting the timing belt end oil seal to the crankshaft with a plastic protector

17.10a Fitting the flywheel end oil seal to the crankshaft with a plastic protector

9 Clean the oil seal seating and crankshaft flange.

10 Dip the new oil seal in engine oil, locate it on the crankshaft open end first, and press it in squarely to the previously noted depth using a metal tube. A piece of thin plastic is useful to prevent damage to the oil seal. When fitted, note that the outer lip of the oil seal must point outwards; if it is pointing inwards use a piece of bent wire to pull it out **(see illustrations)**.

11 Refit the flywheel, referring to Section 18.

18 Flywheel -
removal, inspection and refitting

Note: *New flywheel bolts and a suitable locking compound are required on refitting.*

Removal

1 Remove the transmission as described in Chapter 7.

2 Remove the clutch as described in Chapter 6.

3 Hold the flywheel stationary with a suitable locking tool engaged in the ring gear teeth **(see illustration)**. Alternatively, insert a screwdriver between the ring gear teeth and the transmission location dowel, then unscrew and remove the bolts and lift the flywheel from the crankshaft. Take care, as the flywheel is heavy. Alignment marks are not required as there is a location dowel on the crankshaft flange.

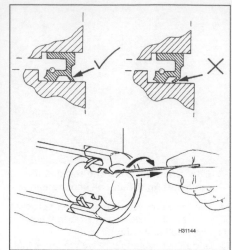

17.10b Correct fitting of the crankshaft flywheel end oil seal

Inspection

4 Examine the clutch friction disc contact surface of the flywheel for scoring, or signs of small hair cracks, caused by overheating. Light grooving or scoring may be ignored. Surface cracks or deep grooving can sometimes be removed by specialist machining, provided not too much metal is taken off, otherwise the flywheel must be renewed.

5 Inspect the flywheel ring gear for damaged or missing teeth. A ring gear, separate from the flywheel, is not available as a genuine Citroën part, although some specialist dealers may be able to fit one using a process of heat treatment which is beyond the scope of the DIY mechanic. Normally, if the ring gear is damaged, it will be necessary to renew the complete flywheel.

Refitting

6 Begin refitting by cleaning the mating surfaces of the crankshaft and flywheel.

7 Locate the flywheel on the crankshaft dowel.

8 Use new flywheel retaining bolts and apply locking fluid to the threads. Insert them and tighten them to the specified torque while holding the flywheel stationary **(see illustrations)**.

9 Refit the clutch driven and pressure plates (see Chapter 6).

10 Refit the transmission to the engine (see Chapter 7).

19 Engine/transmission mountings - inspection, removal and refitting

Note: *The procedure for moving the engine to the left, to prepare for work on the timing system, is given in paragraphs 14-21.*

Inspection

1 If improved access is required, apply the handbrake, jack up the front of the vehicle and support it securely on axle stands (see *"Jacking and vehicle support"*).

2 Check the mounting rubber to see if it is cracked, hardened or separated from the metal at any point. Renew the mounting if any such damage or deterioration is evident.

3 Check that all the mounting's fasteners are securely tightened. Use a torque wrench to check if possible.

4 Using a large screwdriver or a crowbar, check for wear in the mounting by carefully levering against it to check for free play. Where this is not possible, enlist the aid of an assistant to move the engine/transmission unit back and forth, or from side to side, while you watch the mounting. While some free play is to be expected even from new components, excessive wear should be obvious. If excessive free play is found, check first that the fasteners are correctly secured, then renew any worn components as described below.

Removal and refitting
Right-hand mounting-to-body bracket

5 Support the engine with a suitable overhead hoist and lifting tackle attached to the lifting bracket at the right-hand end of the engine. Alternatively, the engine can be supported using a trolley jack and interposed block of wood beneath the sump, in which case, be prepared for the engine to tilt backwards when the bracket is removed.

Note: *If the vertical right-hand mounting bracket and timing belt tensioner is also to be removed (see below), improved access can be obtained*

18.3 Locking tool engaged in flywheel ring gear teeth

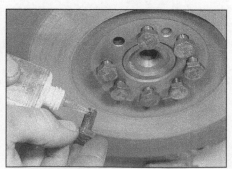

18.8a Apply locking fluid to the flywheel bolts . . .

18.8b . . . and tighten them to the specified torque

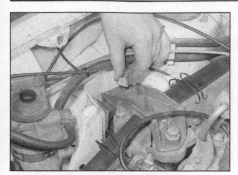

19.7a Undo the nuts . . .

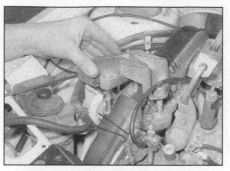

19.7b . . . and remove the right-hand engine mounting-to-body bracket

19.9a Right-hand engine mounting rubber and side stops

by disconnecting both the left and right mountings and moving the engine to the left. In this case, use an overhead hoist attached to both the left and right lifting brackets. It will not be sufficient just to use a trolley jack.

6 Release the fuel return hose from its clip on the bodywork and move it to one side. On models from February 1993 onwards with a fuel priming bulb, release the fuel feed hose from its clip and move it to one side. Move any other hoses and cables clear of the engine mounting.

7 Undo the nut securing the bracket to the mounting rubber, then undo the three remaining nuts and remove the bracket, noting the location of any shims **(see illustrations)**.

8 Refitting is the reverse of removal. Tighten the bracket securing nuts to the specified torque.

Right-hand mounting rubber

9 The right-hand engine mounting rubber is screwed into a housing on the bodywork. Make up a tool similar to that shown, to engage with the slots in the rim of the rubber **(see illustrations)**. Assuming that the rubber is being renewed, the new component can be used as a guide when making the tool.

10 Remove the right-hand mounting-to-body bracket as described above.

11 Insert the improvised tool and unscrew the rubber.

12 If required, undo the nuts and remove the rubber stops from each side of the housing. Recover any shims.

13 Refitting is the reverse of removal. Tighten the rubber firmly to the body using the tool. Refit the mounting-to-body bracket as described above. With the weight of the engine on the mounting, the clearance between the mounting-to-body bracket and each rubber stop should be 1.0 ± 0.7 mm. If necessary, adjust the clearance by means of shims positioned under the stops.

Right-hand mounting bracket and timing belt tensioner

Note: *This procedure may be required for removal and refitting a number of timing system components, to obtain extra space by moving the engine to the left.*

19.9b Home-made tool for unscrewing the right-hand engine mounting rubber

14 The vertical right-hand mounting bracket is bolted to the end face of the engine, and houses a spring and plunger which operates the timing belt tensioner **(see illustration)**.

15 Apply the handbrake, then jack up the front right-hand corner of the vehicle and support it on an axle stand (see *"Jacking and vehicle support"*). Engage 4th or 5th gear so that the engine may be rotated by turning the right-hand wheel (for timing purposes). Alternatively, engage neutral and remove the wheel so that the engine can be turned from the crankshaft pulley bolt.

16 Support the weight of the engine using an overhead hoist fitted to both the left and right-hand lifting lugs.

19.9c Right-hand engine mounting rubber

> ⚠ **Warning: Do not attempt to support the engine using only a trolley jack. The engine requires support from above, otherwise it is likely to fall over within the engine compartment when both the left and right mountings are disconnected.**

17 Remove the left-hand engine mounting as described below.

18 Remove the right-hand engine mounting-to-body bracket as described above.

19 Move the engine and transmission to the left as far as possible and support it in this position.

20 The remainder of the removal process, for the right-hand mounting bracket and timing

19.14 Right-hand engine mounting bracket and tensioner assembly

1 *Mounting bracket*
2 *Tensioner plunger*
3 *Tensioner roller*

19.25 Undo the bolts and remove the battery tray

19.26a Undo the nut (arrowed) from the left-hand engine/transmission mounting . . .

19.26b . . . then undo the two bolts and remove the mounting bracket

belt tensioner, is given in Section 9. If you are working on other components of the timing system, follow the appropriate Sections of this Chapter.

21 When finished, refit all relevant components, including the engine mountings, and lower the vehicle to the ground. Tighten the engine mounting nuts and bolts to the specified torque.

Left-hand mounting

22 Support the transmission with a hoist or with a trolley jack and block of wood.
23 Remove the air cleaner and trunking (see Chapter 4C).
24 Remove the battery (see Chapter 5A).
25 Undo the four bolts and remove the battery tray **(see illustration)**.
26 Slacken and remove the central nut and washer securing the mounting rubber to the transmission stud. Undo the two bolts securing the mounting bracket assembly to the vehicle body, then lower the engine slightly and remove the mounting bracket assembly. Recover the spacer from the stud **(see illustrations)**.
27 Undo the two nuts and bolts and remove the mounting rubber and support plate from

the bracket **(see illustration)**.
28 If required, unscrew the mounting stud from the transmission casing.
29 Refitting is the reverse of removal, noting the following points:
a) *Before fitting the mounting stud to the transmission, clean the threads and apply a little locking fluid.*
b) *Use a new self-locking nut to secure the mounting rubber to the transmission stud.*
c) *Tighten the nuts and bolts to torque where specified.*

Lower mounting
Removal

30 The lower mounting supports both the engine and the left-hand driveshaft intermediate bearing. The rubber bush is not available separately, and if it is damaged the complete mounting bracket must be renewed.
31 Apply the handbrake, then jack up the front of the vehicle and support securely on axle stands (see "*Jacking and vehicle support*").
32 Drain the transmission oil, with reference to Chapter 7, then perform the following operations, with reference to Chapter 8.
a) *Undo the clamp nut and bolt and disconnect the track control arm from the hub*

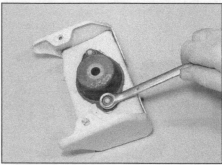

19.27 Removing the mounting rubber and support plate from the left-hand mounting bracket

carrier on the right-hand front suspension.
b) *Slacken the two nuts that secure the intermediate bearing to the mounting bracket, then push the bolts in and turn them by 90° to unlock the bearing from the bracket.*
33 Remove the nut and bolt securing the rubber bush to the torque link on the subframe **(see illustration)**.
34 Undo the four bolts connecting the engine mounting bracket to the cylinder block **(see illustration)**.

19.33 Remove the central nut and bolt from the lower engine mounting rubber

19.34 Undo the four bolts connecting the engine mounting bracket to the cylinder block

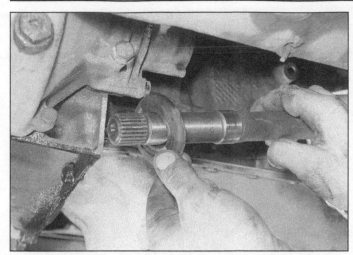

19.36a Pull the driveshaft out of the differential housing and remove the dust cover . . .

19.36b . . . then manoeuvre the bracket away from the torque link and slide it off the end of the shaft

35 Pull the bracket from the cylinder block, so that it comes off the two dowels. Take care not to strain the driveshaft joint.

36 Have an assistant pull the wheel hub outwards so that the intermediate bearing comes out of the bracket, and the end of the driveshaft comes out of the differential housing. Remove the dust cover from the end of the driveshaft, then manoeuvre the bracket away from the torque link and slide it off the end of the shaft. Support the driveshaft by suspending it with wire or string to avoid straining the joint (see illustrations).

37 Plug or tape over the differential aperture to prevent the entry of dirt.

38 If required, undo the nut and bolt and disconnect the torque link from the subframe (see illustration).

Refitting

39 Refitting is the reverse of removal. Refit the driveshaft to the transmission, using a new seal, with reference to Chapter 8. Refill the transmission with oil, with reference to Chapter 7. Tighten the mounting bolts to the specified torque.

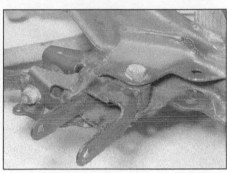

19.38 Torque link connected to subframe

Chapter 2 Part C:
Engine removal and overhaul procedures

Contents

Degrees of difficulty

Easy, suitable for novice with little experience	Fairly easy, suitable for beginner with some experience	Fairly difficult, suitable for competent DIY mechanic	Difficult, suitable for experienced DIY mechanic	Very difficult, suitable for expert DIY or professional 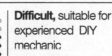

Specifications

TU series petrol engine - 954 cc and 1124 cc

Cylinder head

Warp limit	0.05 mm
Refinishing limit (1124 cc engines)*	0.20 mm

*Fill with a 0.20 mm repair seal

Valves

Valve head diameter:	
Inlet:	
954 cc engine	34.7 mm
1124 cc engine	36.7 mm
Exhaust:	
954 cc engine	27.7 mm
1124 cc engine	29.2 mm
Valve stem diameter:	
Inlet	6.965 to 6.980 mm
Exhaust:	
954 cc engine	6.960 to 6.975 mm
1124 cc engine	6.945 to 6.960 mm
Overall length:	
Inlet	112.76 mm
Exhaust	112.56 mm

Cylinder block

Cylinder bore diameter:	
954 cc engine:	
Size group A	70.000 to 70.010 mm
Size group B	70.010 to 70.020 mm
Size group C	70.020 to 70.030 mm
1124 cc engine:	
Size group A	72.000 to 72.010 mm
Size group B	72.010 to 72.020 mm
Size group C	72.020 to 72.030 mm
Liner protrusion above block mating surface:	
Standard	0.03 to 0.10 mm
Maximum difference between any two liners	0.05 mm

TU series petrol engine - 954 cc and 1124 cc (continued)

Pistons and piston rings

Piston diameter:
  954 cc engine:
    Size group A .... 69.960 to 69.970 mm
    Size group B .... 69.970 to 69.980 mm
    Size group C .... 69.980 to 69.990 mm
  1124 cc engine:
    Size group A .... 71.960 to 71.970 mm
    Size group B .... 71.970 to 71.980 mm
    Size group C .... 71.980 to 71.990 mm
Piston ring end gaps:
  Top compression ring .... 0.25 to 0.45 mm
  Second compression ring .... 0.25 to 0.45 mm
  Oil control ring* .... 0.3 to 0.5 mm
Maximum weight difference between any two pistons: .... 2.0 g
These are suggested figures, typical for this type of engine - no exact values are stated by Citroën.

Connecting rods

Maximum weight difference between any two connecting rods ...... 3.0 g

Crankshaft

Endfloat .... 0.1 to 0.3 mm
Main bearing journal diameter:
  Standard .... 49.965 to 49.981 mm
  Undersize .... 49.665 to 49.681 mm
Big-end bearing journal diameter:
  954 cc engine:
    Standard .... 38.000 ± 0.008 mm
    Undersize .... 37.700 ± 0.008 mm
  1124 cc engine:
    Standard .... 44.975 to 44.991 mm
    Undersize .... 44.675 to 44.691 mm
Maximum bearing journal out-of-round .... 0.007 mm
Main bearing running clearance:
  954 cc engines .... Not available at time of writing. Consult your Citroën dealer
  1124 cc engines*:
    Pre-February 1992 models .... 0.023 to 0.083 mm
    February 1992-on models .... 0.023 to 0.048 mm
Big-end bearing running clearance - all models** .... 0.025 to 0.050 mm
On 1124 cc models, the main bearing shells were modified in February 1992, resulting in a reduction in the specified running clearance - see text for further information.
**These are suggested figures, typical for this type of engine - no exact values are stated by Citroën.*

Torque wrench settings

Refer to Chapter 2A Specifications.

XUD7 diesel engine - 1769 cc

Cylinder head

Warp limit .... 0.07 mm subject to camshaft turning freely
Swirl chamber protrusion .... 0 to 0.03 mm
Cylinder head height (from gasket face to camshaft centre line):
  New .... 140.1 ± 0.15 mm
  Minimum after refinishing* .... 139.55 mm
Valve seats and swirl chambers must also be machined, and new washers fitted. A camshaft sprocket with the identification number 2 is required.

Valves

Seat angle (inclusive):
  Inlet .... 120°
  Exhaust .... 90°
Valve recess below cylinder head:
  Inlet .... 0.50 to 1.05 mm
  Exhaust .... 0.90 to 1.45 mm
Valve timing (at 1.0 mm clearance):
  Inlet opens .... 8° BTDC
  Inlet closes .... 40° ABDC
  Exhaust opens .... 56° BBDC
  Exhaust closes .... 12° ATDC

XUD7 diesel engine - 1769 cc (continued)

Cylinder block

Cylinder bore diameter:
Standard	80.000 to 80.018 mm
Oversize	80.030 to 80.048 mm

Pistons and piston rings

Piston diameter:
Standard	79.930 ± 0.008 mm
Oversize	76.960 ± 0.008 mm

Piston ring end gaps (fitted):
Top compression	0.20 to 0.40 mm
2nd compression	0.15 to 0.35 mm
Oil control ring	0.10 to 0.30 mm
Maximum weight difference between any two pistons	2.5 g
Maximum piston protrusion difference between any two pistons	0.12 mm

Connecting rods and gudgeon pins

Connecting rod length	145.0 ± 0.025 mm
Maximum difference in weight between connecting rods	4.0 g
Connecting rod small-end bush inner diameter	25.007 to 25.020 mm
Gudgeon pin diameter	25.0 + 0 - 0.006 mm
Gudgeon pin length	72.0 + 0 - 0.3 mm

Crankshaft

Endfloat	0.07 to 0.32 mm
Endfloat control thrustwasher thicknesses	2.30, 2.40, 2.45, and 2.50 mm

Main bearing journal diameter:
Standard	60.0 + 0 - 0.019 mm
Undersize	59.7 + 0 - 0.019 mm

Big-end bearing journal diameter:
Standard	50.0 + 0 - 0.016 mm
Undersize	49.7 + 0 - 0.016 mm
Maximum bearing journal out-of-round	0.007 mm
Main and big-end bearing running clearance **	0.025 to 0.050 mm

**These are suggested figures, typical for this type of engine - no exact values are stated by Citroën.*

Torque wrench settings

Refer to Chapter 2B Specifications.

1 General information

Included in this Part of Chapter 2 are details of removing the engine from the vehicle, and the general overhaul procedures for the cylinder head, cylinder block and all other engine internal components.

The information given ranges from advice concerning preparation for an overhaul and the purchase of replacement parts, to detailed step-by-step procedures covering removal, inspection, renovation and refitting of engine internal components.

After Section 5, all instructions are based on the assumption that the engine has been removed from the vehicle. For information concerning repairs with the engine in the vehicle, as well as the removal and refitting of those external components necessary for full overhaul, refer to Part A or B of this Chapter (as applicable) and to Section 5. Ignore any preliminary dismantling operations described in Part A (petrol engine) or Part B (diesel engine) that are no longer relevant once the engine has been removed from the vehicle.

Apart from torque wrench settings, which are given at the beginning of Part A or Part B, all specifications relating to engine overhaul are at the beginning of this Part of Chapter 2.

2 Engine overhaul - general information

It is not always easy to determine when, or if, an engine should be completely overhauled, as a number of factors must be considered.

High mileage is not necessarily an indication that an overhaul is needed, while low mileage does not preclude the need for an overhaul. Frequency of servicing is probably the most important consideration. An engine which has had regular and frequent oil and filter changes, as well as other required maintenance, should give many thousands of miles of reliable service. Conversely, a neglected engine may require an overhaul very early in its life.

Excessive oil consumption is an indication that piston rings, valve seals and/or valve guides are in need of attention. Make sure that oil leaks are not responsible before deciding that the rings and/or guides are worn. Perform a compression test, as described in Part A or B of this Chapter, to determine the likely cause of the problem.

Check the oil pressure with a gauge fitted in place of the oil pressure switch, and compare it with that specified in Parts A or B. If it is extremely low, the main and big-end bearings, and/or the oil pump, are probably worn out.

Loss of power, rough running, knocking or metallic engine noises, excessive valve gear noise, and high fuel consumption may also point to the need for an overhaul, especially if they are all present at the same time. If a complete service does not remedy the situation, major mechanical work is the only solution.

An engine overhaul involves restoring all internal parts to the specification of a new engine. During an overhaul, the pistons and the piston rings are renewed. On petrol engines which have aluminium cylinder blocks, the cast iron cylinder liners are also renewed. New main and big-end bearings are generally fitted. If necessary, the crankshaft may be renewed, to restore the journals. The valves are also serviced as well, since they are usually in less-than-perfect condition at this point. While the engine is being overhauled, other components, such as the distributor, starter and alternator, can be overhauled as well. The end result should be an as-new engine that will give many trouble-free miles.

Note: *Critical cooling system components such as the hoses, thermostat and water pump should be renewed when an engine is overhauled. The radiator should be checked carefully, to ensure that it is not clogged or leaking. Also, it is a good idea to renew the oil pump whenever the engine is overhauled.*

Before beginning the engine overhaul, read through the entire procedure, to familiarise yourself with the scope and requirements of the job. Overhauling an engine is not difficult if you carefully follow all of the instructions, have the necessary tools and equipment, and pay close attention to all specifications. It can, however, be time-consuming. Plan on the vehicle being off the road for a minimum of two weeks, especially if parts must be taken to an engineering works for repair or reconditioning. Check on the availability of parts, and make sure that any necessary special tools and equipment are obtained in advance. Most work can be done with typical hand tools, although a number of precision measuring tools are required for inspecting parts to determine if they must be renewed. Often, the engineering works will handle the inspection of parts, and can offer advice concerning reconditioning and renewal.

Note: *Always wait until the engine has been completely dismantled, and until all components (especially the cylinder block and the crankshaft) have been inspected, before deciding what service and repair operations must be performed by an engineering works. The condition of these components will be the major factor to consider when determining whether to overhaul the original engine, or to buy a reconditioned unit. Do not, therefore, purchase parts or have overhaul work done on other components until they have been thoroughly inspected. As a general rule, time is the primary cost of an overhaul, so it does not pay to fit worn or sub-standard parts.*

As a final note, to ensure maximum life and minimum trouble from a reconditioned engine, everything must be assembled with care, in a spotlessly-clean environment.

3 Engine removal - methods and precautions

If you have decided that the engine must be removed for overhaul or major repair work, several preliminary steps should be taken.

Locating a suitable place to work is extremely important. Adequate work space, along with storage space for the vehicle, will be needed. If a workshop or garage is not available, at the very least, a flat, level, clean work surface is required.

Cleaning the engine compartment and engine/transmission before beginning the removal procedure will help keep tools clean and organised.

An engine hoist or A-frame will also be necessary. Make sure that the equipment is rated in excess of the combined weight of the engine and transmission. Safety is of primary importance, considering the potential hazards involved in lifting the engine/transmission out of the vehicle.

If this is the first time you have removed an engine, an assistant should ideally be available. Advice and aid from someone more experienced would also be helpful. There are many instances when one person cannot simultaneously perform all of the operations required when lifting the engine out of the vehicle.

Plan the operation ahead of time. Before starting work, arrange to hire or obtain all the tools and equipment you will need. Some of the equipment necessary to perform engine/transmission removal and installation safely and with relative ease (in addition to an engine hoist) is as follows: a heavy-duty trolley jack, complete sets of spanners and sockets as described in the front of this manual, wooden blocks, and plenty of rags and cleaning solvent for mopping-up spilled oil, coolant and fuel. If the hoist must be hired, make sure that you arrange for it in advance, and perform all of the operations possible without it beforehand. This will save you time and money.

Plan for the vehicle to be out of use for quite a while. An engineering works will be required to perform some of the work which the do-it-yourselfer cannot accomplish without special equipment. These places often have a busy schedule, so it would be a good idea to consult them before removing the engine, in order to accurately estimate the amount of time required to rebuild or repair components that may need work.

Always be extremely careful when removing and refitting the engine/transmission. Serious injury can result from careless actions. Plan ahead and take your time, and a job of this nature, although major, can be accomplished successfully.

4 Engine and transmission - removal, separation and refitting

Note: *This Section describes the removal, separation and refitting of both petrol and diesel models. The illustrations are based on a diesel model, before the fuel filter/heater was relocated to the thermostat housing in February 1993.*

Removal

Note: *The engine can be removed from the vehicle only as a complete unit with the transmission; the two are then separated for overhaul. The engine is lifted out at an angle with the timing end upwards and the transmission end downwards. Make sure you have a hoist with sufficient height.*

1 Park the vehicle on firm, level ground, then open the bonnet and support it in the upright position.

2 Remove the complete air cleaner housing and duct assembly, as described in the relevant Part of Chapter 4. On diesel models, remove the air inlet duct from the inlet manifold and plug the manifold orifice with a clean rag **(see illustrations)**.

3 Remove the battery (see Chapter 5A), then undo the four bolts and remove the supporting baseplate.

4 Locate and disconnect the engine wiring harness connectors on the left-hand side of the engine compartment **(see illustration)**.

5 Disconnect the earth cable from the transmission **(see illustration)**.

4.2a Remove the air cleaner housing . . .

4.2b . . . and ducting

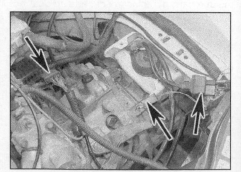

4.4 Disconnect the engine wiring harnesses (arrowed)

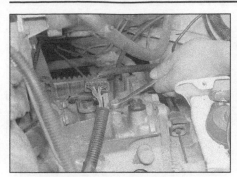

4.5 Disconnect the earth cable from the transmission

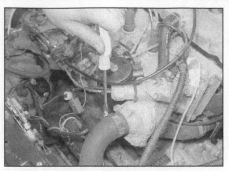

4.8a Disconnect the coolant hoses from the engine . . .

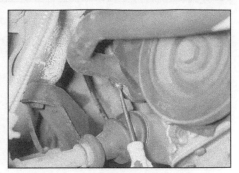

4.8b . . . including the return hose underneath the right wheel arch

6 Chock the rear wheels, then firmly apply the handbrake. Jack up the front of the vehicle, and securely support it on axle stands (see "*Jacking and vehicle support*"). Remove both front roadwheels.

7 Drain the cooling system (see Chapter 1A or 1B), saving the coolant if it is fit for re-use.

8 Disconnect the coolant hoses from the engine. Remove the cover panel from the right wheel arch to access the radiator bottom hose and disconnect it from the engine block (or from the fuel heater, as appropriate). Loosen the clips and disconnect the heater hoses from the bulkhead at the rear of the engine compartment **(see illustrations)**.

9 Although not essential, it may be worthwhile removing the radiator, which is easily damaged during engine removal (see Chapter 3). If you leave the radiator in position, place a piece of hardboard over the rear face to avoid damage.

4.15 Disconnect the clutch release cable from the transmission

10 Drain the transmission oil as described in Chapter 7. Refit the drain and filler plugs, and tighten them to their specified torque settings.

11 If the engine is to be dismantled, drain the oil, and if required remove the oil filter, working as described in Chapter 1A or 1B. Clean and refit the drain plug, tightening it securely.

12 Remove the bonnet (see Chapter 11).

13 Remove the wheel-changing jack from its support on the engine compartment bulkhead.

14 Disconnect the speedometer cable from the transmission (see Chapter 12).

15 Disconnect the clutch release cable from the transmission **(see illustration)**.

16 Disconnect the two gearchange link rods from their ball studs on the transmission, and move them to one side **(see illustration)**. Also disconnect the torque link rod, if it exists.

17 Disconnect the accelerator cable with reference to the relevant Part of Chapter 4.

Petrol engine models

18 On carburettor models, disconnect the choke cable with reference to the relevant Part of Chapter 4.

19 Disconnect the fuel supply hose from the fuel pump on the left-hand end of the cylinder head on carburettor models. On petrol injection models, disconnect the fuel inlet and return hoses with reference to the relevant Part of Chapter 4.

20 Disconnect the brake vacuum servo hose from the inlet manifold.

21 On petrol injection models, it may be

necessary to remove the ECU, although no specific information is available.

22 Undo the nuts and disconnect the exhaust front pipe from the manifold. Recover the gasket. If necessary, remove the heat shield to improve access to the nuts.

Diesel engine models

23 Disconnect the hose from the brake vacuum pump **(see illustration)**. The pump is either mounted directly on the left-hand end of the camshaft, or else mounted on the transmission and driven by a belt from the camshaft pulley.

24 On models with power steering, clamp the hoses underneath the power steering fluid reservoir to minimise fluid loss, then disconnect the fluid inlet and outlet hoses from the power steering pump. The pump is at the transmission end of the engine, driven by a belt from the camshaft pulley.

25 On models up to February 1993, disconnect the fuel supply and return hoses from the injection pump. Also disconnect the fuel hoses from the fuel heater at the back of the engine block on the right-hand side and remember which way round the hoses go for refitting **(see illustrations)**. The right hose is from the tank to the fuel heater. The left hose is from the fuel heater to the fuel filter. Plug the open ends of the pipes and fittings.

26 On models from February 1993 onwards there is no fuel heater on the back of the engine block. Instead there is a combined fuel filter and heater at the front of the engine, near the transmission. Disconnect the fuel inlet

4.16 Disconnect the gearchange link rods from their ball studs on the transmission

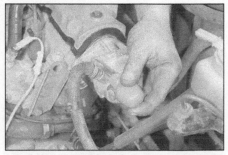

4.23 Disconnect the hose from the brake vacuum pump - model with pump driven directly from camshaft

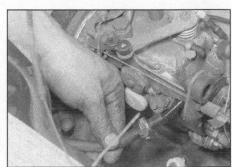

4.25a Disconnect the fuel supply hose from the injection pump . . .

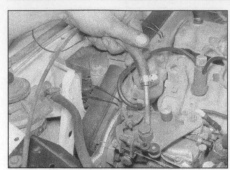

4.25b . . . and the fuel return hose . . .

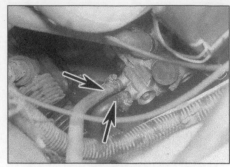

4.25c . . . and disconnect the fuel hoses (arrowed) from the fuel heater at the back of the engine block - diesel models up to February 1993

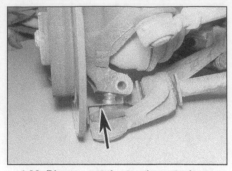
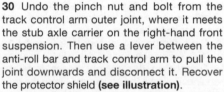

4.30 Disconnect the track control arm outer joint where it meets the stub axle carrier and recover the protector shield (arrowed)

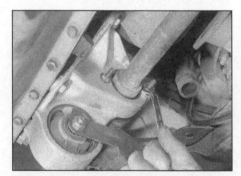

4.31 Slacken the nuts on the right-hand driveshaft intermediate bearing, then push the bolts in and turn them by 90° to unlock the bearing

4.32 Pull the driveshaft out of the differential housing and remove the dust cover (arrowed)

hose to the filter/heater unit at the joint where the flexible hose from the priming bulb meets the fixed hose, and disconnect the fuel return hose from the fuel pump, then plug the ends.

27 Disconnect the electrical cable from the No 1 glow plug at the transmission end of the engine.

28 Unscrew and remove the exhaust manifold-to-downpipe bolts, together with the springs and collars.

All models

29 All C15 models have an intermediate bearing on the right-hand driveshaft, supported by the rear engine mounting bracket. There are two different methods of disconnecting this type of right-hand driveshaft from the differential housing:

a) *Remove the complete right-hand driveshaft assembly (see Chapter 8), then go to paragraph 34. If you choose this method, you will need to slacken the hub nut first, before disconnecting any of the suspension components.*

⚠ ***Warning: The hub nut is very tight. Unless you have a suitable counterhold, you can pull the vehicle off the axle stands.*** *Alternatively, refit the roadwheel, lower the vehicle to the ground, apply the handbrake, then slacken the hub nut.*

b) *Leave the driveshaft connected to the wheel hub and pull it out of the differential housing as follows (with reference to Chapter 8):*

30 Undo the pinch nut and bolt from the track control arm outer joint, where it meets the stub axle carrier on the right-hand front suspension. Then use a lever between the anti-roll bar and track control arm to pull the joint downwards and disconnect it. Recover the protector shield **(see illustration)**.

31 Slacken the nuts that secure the right-hand driveshaft intermediate bearing to the rear engine mounting bracket, until the nuts are almost at the end of their threads. Then push the bolts in and turn them by 90° so that they can't be pulled out again. At this position, the intermediate bearing is unlocked from the mounting bracket **(see illustration)**.

32 Have an assistant pull the wheel hub outwards so that the bearing comes out of the bracket, and the end of the driveshaft comes out of the differential housing. It may be necessary to turn the steering wheel to get the required movement. Remove the dust cover that fits over the driveshaft housing oil seal, if it has come off with the driveshaft **(see illustration)**. Support the driveshaft so that it does not hang down.

33 Remove the central nut and bolt that goes through the mounting rubber, and the four bolts securing the mounting bracket to the engine block. Pull the bracket away from the engine block so it comes off the two dowels, then slide the bracket and mounting rubber off the end of the driveshaft **(see illustrations)**.

34 If you have chosen to remove the complete right-hand driveshaft, the rear

4.33a Remove the central nut and bolt that goes through the mounting rubber . . .

4.33b . . . and the four bolts securing the mounting bracket to the engine block . . .

4.33c . . . then slide the bracket and mounting rubber off the end of the driveshaft

4.37 With the hoist in place, remove the left-hand engine mounting . . .

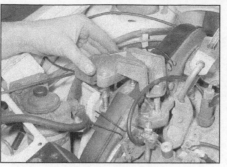

4.38 . . . and the right-hand mounting

4.40a Place a trolley jack under the sump and tilt the engine so that the transmission points downwards . . .

engine mounting will still be in place. Undo the single nut and bolt that goes through the mounting rubber.

35 Undo the pinch nut and bolt from the track control arm outer joint, where it meets the stub axle carrier on the left-hand suspension. Then use a lever between the anti-roll bar and track control arm to pull the joint downwards and disconnect it. Have an assistant pull the wheel hub outwards so that the end of the driveshaft comes out of the differential housing. It may be necessary to turn the steering wheel to get the required movement. Support the driveshaft so that it does not hang down.

36 Manoeuvre the engine hoist into position, and attach it to the lifting brackets bolted onto the cylinder head. Raise the hoist until it is supporting the weight of the engine.

37 Unscrew the nuts securing the left-hand engine mounting to the body, then unscrew and remove the centre bolt and remove the mounting **(see illustration)**.

38 Unscrew and remove the nuts securing the right-hand engine mounting upper bracket to the body mounting and bracket on the cylinder block. Withdraw the bracket **(see Illustration)**.

39 Make a final check that any components which would prevent the removal of the engine/transmission from the vehicle have been removed or disconnected. Ensure that components such as the gearchange selector rods are secured so that they cannot be damaged on removal.

40 Place a trolley jack under the sump, near the timing end, and raise it so that it takes the

weight of the engine. Then lower the hoist slightly so that the transmission end points downwards. With the engine tilted in this way, raise the hoist and carefully lift the engine up and over the front of the vehicle **(see illustrations)**. Enlist the help of an assistant during this procedure, to ensure that the engine remains clear of the body panels and other components. In particular, care must be taken to ensure that the following components are not damaged, and if necessary place a piece of board in front of them:

a) *The radiator, if it has not been removed.*
b) *The brake master cylinder and vacuum servo unit, which are at the left of the engine compartment on carburettor petrol models.*

Separation

41 With the engine/transmission assembly removed, support the assembly on suitable blocks of wood, on a workbench (or failing that, on a clean area of the workshop floor).

42 Disconnect the wire from the reversing light switch on the transmission. Undo the bolts and remove the engine wiring multi-plug bracket **(see illustration)**.

43 Both petrol and diesel models have a TDC sensor at the rear of the transmission, for diagnostic purposes. On petrol models, disconnect the wiring connector from the sensor. On diesel models the sensor is mounted on a sensor holder, bolted to the engine block. Slacken the bolts and remove the sensor, then undo the bolts and remove the sensor holder **(see illustration)**.

4.40b . . . then raise the hoist and lift out the engine

44 Unscrew the retaining bolts, and remove the starter motor from the transmission (see Chapter 5A).

45 On diesel engines, unbolt and remove the lower cover from the transmission **(see illustration)**.

46 Ensure that both the engine and transmission are adequately supported, then slacken and remove the bolts securing the transmission housing to the engine. Note the correct fitted positions of each bolt (and, where fitted, the relevant brackets) as they are removed, to use as a reference on refitting.

47 Carefully withdraw the transmission from the engine, ensuring that the weight of the transmission is not allowed to hang on the input shaft while it is engaged with the clutch friction disc.

48 If they are loose, remove the locating dowels from the engine or transmission, and keep them in a safe place.

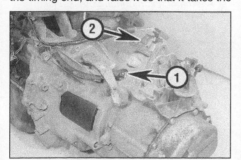

4.42 Disconnect the reversing light wire (1) and the engine wiring multi-plug bracket (2) from the transmission

4.43 Remove the TDC sensor

4.45 Remove the flywheel cover plate

Reconnection

49 Apply a smear of high-melting-point grease to the splines of the transmission input shaft. Do not apply too much, otherwise there is a possibility of the grease contaminating the clutch friction plate.

50 Ensure that the locating dowels are correctly positioned in the engine or transmission.

51 Carefully offer the transmission to the engine, until the locating dowels are engaged. Ensure that the weight of the transmission is not allowed to hang on the input shaft as it is engaged with the clutch friction disc.

52 Refit the transmission housing-to-engine bolts, ensuring that all the necessary brackets are correctly positioned, and tighten them to the specified torque (see Chapter 7).

53 Check that the release fork and bearing are correctly engaged.

 If necessary, the clutch cable can be refitted at this stage and the clutch pedal depressed in order to check the operation of the release fork. Should there be any doubt that the clutch is working properly, the gearbox should be removed and the clutch checked for correct assembly.

54 Reconnect the wire to the reversing light switch, and the engine wiring multi-plug bracket to the transmission. Refit the TDC sensor.

55 Refit the starter motor, and tighten the retaining bolts. On diesel engines, refit the lower cover to the transmission, and tighten the bolts.

Refitting

56 Examine the driveshaft oil seals in the left and right-hand differential housing for signs of wear, and if necessary, renew them as described in Chapter 7. They should normally be renewed as a matter of course.

57 If the hoist has been removed, reconnect it to the engine lifting lugs.

58 Raise the engine slightly and use a jack underneath the engine to tilt it so that the transmission points downwards, then raise the hoist sufficiently to clear the bodywork components. With the aid of an assistant, manoeuvre the engine over the front of the vehicle and carefully lower it into the engine compartment, making sure that it clears the surrounding components. In particular, take care not to damage the radiator, if it is fitted, or the brake master cylinder and vacuum servo unit on carburettor petrol models. Use a jack underneath the transmission to level the assembly and bring it in line with the left and right mounting points.

59 Refit the left-hand engine mounting to the bodywork using two bolts in the lower section of the bracket and tighten them securely. Insert the spindle on the transmission into the mounting rubber, then fit the nut and tighten it to the specified torque.

60 Refit the right-hand engine mounting bracket using three bolts on the engine and one nut on the mounting rubber on the bodywork, and tighten them to the specified torque.

61 Reconnect both driveshafts to the transmission and refit the rear engine mounting, using the reverse of the removal procedure, referring to Chapter 8 as required. Remember to refit the dust cover over the right-hand driveshaft, if it has been removed, and slide it along the shaft so that it covers the oil seal in the differential. Reconnect the track control arm ball joints to the stub axle carriers and remember to fit the protector plates. Fit the pinch nuts and bolts, using new nuts, and tighten them.

Diesel engine models

62 Lubricate the exhaust manifold-to-downpipe contact surfaces with heat resistant grease, then reconnect them and fit the bolts, together with the springs, cups and self-locking nuts. Tighten the nuts progressively until approximately four threads are visible and the springs are compressed to 22.0 mm in length.

63 Reconnect the electrical cable to the No 1 glow plug.

64 Reconnect the fuel hoses to the injection pump and fuel heater as appropriate, depending on model.

65 Reconnect the vacuum hose to the brake vacuum pump.

Petrol engine models

66 On petrol injection models, refit the ECU if it has been removed, and reconnect the wiring.

67 Reconnect the exhaust front pipe to the manifold, using a new gasket if required, and tighten the nuts to the specified torque, given in Chapter 4A or 4B.

68 Reconnect the brake vacuum servo hose to the inlet manifold.

69 Reconnect the fuel supply hose to the fuel pump on the left-hand end of the cylinder head on carburettor models. On petrol injection models, reconnect he fuel inlet and return hoses.

70 On carburettor models, reconnect and adjust the choke cable with reference to Chapter 4A.

All models

71 Reconnect the engine wiring harness connectors on the left-hand side of the engine compartment.

72 Refit the earth cable to the body on the left-hand side of the engine compartment, and tighten the bolt.

73 Reconnect the accelerator cable with reference to the relevant Part of Chapter 4.

74 Reconnect the two gearchange link rods to their ball studs on the transmission. Reconnect the torque link rod if it exists.

75 Reconnect and adjust the clutch release cable with reference to Chapter 6.

76 Reconnect the speedometer cable to the transmission.

77 Refit the wheel-changing jack to its support on the engine compartment bulkhead.

78 Refit the bonnet (see Chapter 11).

79 If applicable, refill the engine with the correct quantity and grade of oil.

80 Refill the transmission with the correct quantity and grade of oil.

81 Refit the radiator, if it has been removed.

82 Reconnect the coolant hoses to the engine, and the heater hoses to the bulkhead at the rear of the engine compartment. Refit the cover panel under the right wheel arch that was removed to access the radiator bottom hose.

83 Refill the cooling system with coolant (see Chapter 1A or 1B).

84 Refit the battery (Chapter 5A) and the supporting base plate.

85 Refit the air cleaner housing and duct assembly with reference to the relevant Part of Chapter 4.

86 Refit both front roadwheels, lower the vehicle to the ground, and tighten the wheel bolts to the specified torque.

87 Ensure that all wiring has been reconnected and all nuts and bolts have been tightened.

5 Engine overhaul - dismantling sequence

1 It is much easier to dismantle and work on the engine if it is mounted on a portable engine stand. These stands can often be hired from a tool hire shop. Before the engine is mounted on a stand, the flywheel should be removed, so that the stand bolts can be tightened into the end of the cylinder block.

2 If a stand is not available, it is possible to dismantle the engine with it blocked up on a sturdy workbench, or on the floor. Be extra-careful not to tip or drop the engine when working without a stand.

3 If you are going to obtain a reconditioned engine, all the external components must be removed first, to be transferred to the replacement engine (just as they will if you are doing a complete engine overhaul yourself). These components include the following:

a) *Alternator and mounting bracket (see Chapter 5A).*
b) *On petrol engines, the distributor, HT leads and spark plugs (see Chapters 1A and 5B).*
c) *On diesel engines, the fuel injection pump and mounting bracket, the fuel injectors and glow plugs (see Chapters 4C and 5C).*
d) *Thermostat and housing, including the fuel filter where applicable (see Chapter 3).*
e) *Coolant pump inlet manifold, where applicable (see Chapter 3).*
f) *Dipstick tube, where applicable.*
g) *On petrol engines, the carburettor/fuel injection system components (see Chapter 4A or 4B).*

6.4a Compressing the valve spring using a spring compressor - petrol engine

6.4b Remove the valve stem oil seal using a pair of pliers - petrol engine

6.6 Leak off pipe connected between two injectors - diesel engine

h) *All electrical switches and sensors, and the engine wiring harness.*
i) *Inlet and exhaust manifolds (see the relevant Part of Chapter 4).*
j) *Oil filter (see Chapter 1A or 1B).*
k) *On carburettor petrol engines, the fuel pump (see Chapter 4A).*
l) *Engine/transmission mountings (see Part A or B of this Chapter).*
m) *Engine lifting brackets*
n) *Flywheel (see Part A or B of this Chapter).*

Note: *When removing the external components from the engine, pay close attention to details that may be helpful or important during refitting. Note the fitted position of gaskets, seals, spacers, pins, washers, bolts, and other small items.*

4 If you are obtaining a "short" engine (which consists of the engine cylinder block, crankshaft, pistons and connecting rods all assembled), then the cylinder head, sump, oil pump, and timing belt will have to be removed also.

5 If you are planning a complete overhaul, the engine can be dismantled, and the internal components removed, in the order given below:

a) *Inlet and exhaust manifolds (see the relevant Part of Chapter 4).*
b) *Timing belt, sprockets and tensioner (see Part A or B of this Chapter).*
c) *Cylinder head (see Part A or B).*
d) *Flywheel (see Part A or B).*
e) *Sump (see Part A or B).*
f) *Oil pump (see Part A or B).*
g) *Piston/connecting rod assemblies (see Section 9).*

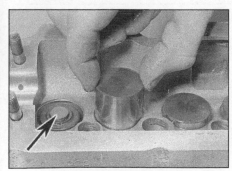

6.10 Remove the bucket tappets and shims (arrowed) - diesel engine

h) *Crankshaft (see Section 10).*

6 Before beginning the dismantling and overhaul procedures, make sure that you have all of the correct tools necessary. Refer to *"Tools and working facilities"* in the Reference Chapter of this manual for further information.

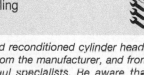

6 Cylinder head - dismantling

Note: *New and reconditioned cylinder heads are available from the manufacturer, and from engine overhaul specialists. Be aware that some specialist tools are required for the dismantling and inspection procedures, and new components may not be readily available. It may therefore be more practical and economical for the home mechanic to purchase a reconditioned head, rather than dismantle, inspect and recondition the original head.*

1 Remove the cylinder head as described in Part A or B of this Chapter (as applicable).
2 Remove the inlet and exhaust manifolds with reference to the relevant Part of Chapter 4 (if not already done). Remove the exhaust manifold gasket.

Petrol engines

3 Remove the camshaft as described in Part A of this Chapter.
4 Using a valve spring compressor, compress each valve spring in turn until the split collets can be removed. Release the compressor, and lift off the spring retainer, spring and spring seat. Using a pair of pliers, carefully extract the valve stem seal from the top of the guide **(see illustrations)**.

> **HAYNES HINT**
> *If, when the valve spring compressor is screwed down, the spring retainer refuses to free and expose the split collets, gently tap the top of the tool, directly over the retainer, with a light hammer. This will free the retainer.*

5 Withdraw the valve through the combustion chamber.

Diesel engines

6 Disconnect the leak off pipes (if not already removed) from between the injectors **(see illustration)**.
7 Remove the injectors, together with their copper washers, fire-seal washers and sleeves (see Chapter 4C).
8 Disconnect the wiring and unscrew the glow plugs.
9 Remove the camshaft as described in Part B of this Chapter.
10 Withdraw the bucket tappets, together with their respective shims, keeping them all identified for location **(see illustration)**.

> **HAYNES HINT**
> *Remove the tappets and shims as soon as the camshaft has been removed. Do not leave them in place while doing other jobs, otherwise they might fall out and get mixed up.*

11 Using a valve spring compressor, depress each valve spring retainer in turn to gain access to the collets. The valves are deeply recessed, so the end of the compressor may need to be extended with a tube or box section with a "window" for access. Remove the collets and release the compressor. Recover the retainer, large and small valve springs, and the spring seat, then withdraw the valve from the cylinder head **(see illustrations)**. From January 1992 onwards, diesel engines were fitted with valve stem oil seals for improved emission control. Using a pair of pliers, carefully extract the valve stem seal from the top of the guide.

6.11a Depress the retainer with a valve spring compressor and remove the collets, retainer . . .

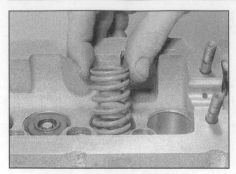

6.11b ... large valve spring ...

6.11c ... small valve spring ...

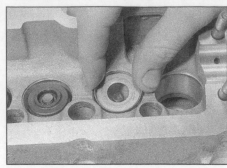

6.11d ... spring seat ...

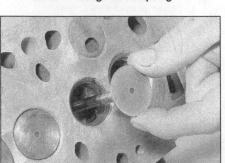

6.11e ... and valve - diesel engine

6.12 Remove the timing probe blanking plug from No 4 cylinder - diesel engine

12 If necessary, remove the timing probe blanking plug, which may be fitted next to the No. 4 injector at the timing belt end of the cylinder head **(see illustration)**. The timing probe was used until 1987 for static timing of the fuel injection pump, but is not used on any models covered in this manual.

All engines

13 It is essential that each valve is stored together with its collets, retainer, springs, and spring seat. Diesel engines have two springs for each valve **(see illustration)**. The valves should also be kept in their correct sequence, unless they are so badly worn that they are to be renewed. If they are going to be kept and used again, place each valve assembly in a labelled polythene bag or similar small container **(see illustration)**. Note that No 1 valve is nearest to the transmission end of the engine.

7 Cylinder head and valves - cleaning and inspection

1 Thorough cleaning of the cylinder head and valve components, followed by a detailed inspection, will enable you to decide how much valve service work must be carried out during the engine overhaul. **Note:** *If the engine has been severely overheated, it is best to assume that the cylinder head is warped - check carefully for signs of this.*

Cleaning

2 Scrape away all traces of old gasket material from the cylinder head. Take care not to damage the cylinder head surfaces.
3 Scrape away the carbon from the combustion chambers and ports, then wash the cylinder head thoroughly with paraffin or a suitable solvent.

4 Scrape off any heavy carbon deposits that may have formed on the valves, then use a power-operated wire brush to remove deposits from the valve heads and stems.
5 For complete cleaning, ideally the core plugs should be removed **(see illustration)**. Drill a small hole in each plug, then insert a self-tapping screw and pull out the plugs using a pair of grips or a slide hammer.
6 Remove all oil gallery plugs. If the plugs are tight, drill them out, then use new plugs and re-cut the threads on the cylinder head on refitting.
7 If the head is extremely dirty, it should be steam-cleaned.
8 If the head has been steam-cleaned, clean all oil holes and oil galleries one more time on completion. Flush all internal passages with warm water until the water runs clear. Dry the head thoroughly, and apply a light film of oil to all machined surfaces. If you have access to compressed air, use it to speed up the drying process, and to blow out all the oil holes and galleries.

⚠️ *Warning: Wear eye protection when using compressed air!*

9 If the head is not very dirty, you can do an adequate cleaning job with hot soapy water and a stiff brush. Take plenty of time, and do a thorough job.

HAYNES HINT *Regardless of the cleaning method used, be sure to clean all oil holes and galleries very thoroughly, dry the head completely, and coat all machined surfaces with light oil to prevent rusting.*

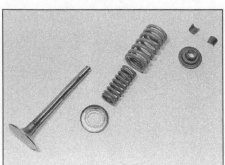

6.13a Valve components - diesel engine

6.13b Place each valve and its associated components in a labelled polythene bag

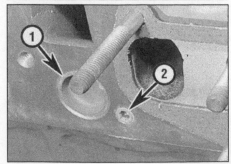

7.5 Core plug (1) and oil gallery plug (2) on manifold side of diesel engine

10 The threaded holes in the cylinder head must be cleaned out, to ensure accurate torque readings when tightening fixings during reassembly. Run the correct-size tap (which can be determined from the size of the relevant bolt which fits in the hole) into each of the holes to remove rust, corrosion, thread sealant or other contamination, and to restore damaged threads. If possible, use compressed air to clear the holes of debris produced by this operation. Do not forget to clean the threads of all nuts and bolts as well.

A good alternative is to inject aerosol-applied water-dispersant lubricant into each hole, using the long tube usually supplied.
Warning: Wear eye protection when cleaning out these holes in this way!

11 Apply suitable sealant to the new core plugs and fit them to the cylinder head. Make sure that they are driven in straight and seated correctly, or leakage could result. Special tools are available for this purpose, but a large socket, with an outside diameter that will just fit into the core plug, will work just as well.
12 Refit the oil gallery plugs, using new sealing washers where applicable, and tighten them securely.
13 If the engine is not going to be reassembled right away, cover the cylinder head with a large plastic bag to keep it clean.

Inspection

Note: *Be sure to perform all the following inspection procedures before concluding that the services of a machine shop or engine overhaul specialist are required. Make a list of all items that require attention.*

Cylinder head - all engines

14 Inspect the head very carefully for cracks, evidence of coolant leakage, and other damage. If cracks are found, a new cylinder head should be obtained.
15 Use a straight-edge and feeler blade, diagonally and along the edge, to check that the cylinder head surface is not distorted beyond the specified limit **(see illustration)**. On diesel engines, do not position the straight-edge over the swirl chambers, as they may be proud of the cylinder head face. If the cylinder head is distorted, it may be possible to have it machined, subject to the following conditions:
a) *On 1124 cc petrol engines, the cylinder head can be machined up to the specified limit and must be fitted with a repair seal.*
b) *On 954 cc petrol engines, no specific information is available. Consult your Citroën dealer.*
c) *On diesel engines, machining is possible, provided the camshaft turns freely. The cylinder head height must not be reduced below the specified limit. The valve seats and swirl chambers must also be*

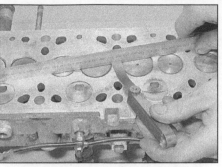

7.15 Checking the cylinder head gasket surface for distortion

machined, and new washers fitted. A camshaft sprocket with the identification number 2 is required (if not already fitted during production). The correct gasket must be fitted, depending on the measurement of piston protrusion above the cylinder head (see Part B of this Chapter).

16 Examine the valve seats in each of the combustion chambers. If they are severely pitted, cracked, or burned, they will need to be renewed or re-cut by an engine overhaul specialist. If they are only slightly pitted, this can be removed by grinding-in the valve heads and seats with fine valve-grinding compound, as described later in this Section.
17 Check the valve guides for wear by inserting the relevant valve, and checking for side-to-side motion of the valve. A very small amount of movement is acceptable. If the movement seems excessive, remove the valve. Measure the valve stem diameter (see later in this Section), and renew the valve if it is worn. If the valve stem is not worn, the wear must be in the valve guide, and the guide must be renewed. The renewal of valve guides is best carried out by a Citroën dealer or engine overhaul specialist, who will have the necessary tools available. Where no valve stem diameter is specified, seek the advice of a Citroën dealer on the best course of action.
18 If renewing the valve guides, the valve seats are to be re-cut or re-ground only *after* the guides have been fitted.

Cylinder head - diesel engines

19 Diesel engines have swirl chambers and bucket tappets, so that the following

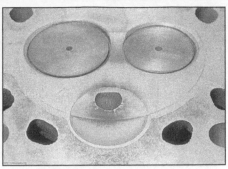

7.20 This swirl chamber shows the initial stages of cracking and burning

additional steps are required.
20 Inspect the swirl chambers for burning or damage such as cracking **(see illustration)**. Small cracks in the chambers are acceptable, but renewal of the chambers will be required if chamber tracts are badly burned and disfigured, or if they are no longer a tight fit in the cylinder head. If there is any doubt about the condition of the swirl chambers, consult a Citroën dealer or diesel engine repair specialist who will be able to renew them if necessary. Swirl chamber renewal is beyond the scope of the amateur mechanic.
21 Using a dial test indicator, check that the swirl chamber protrusion is within the limits given in the Specifications. Zero the dial test indicator on the gasket surface of the cylinder head, then measure the protrusion of the swirl chamber **(see illustrations)**. If the protrusion is not within the specified limits, consult a Citroën dealer or diesel engine repair specialist.
22 Check the tappet bores in the cylinder head for wear. If there is excessive wear, the cylinder head must be renewed.

Valves

Note: *From January 1992, modified exhaust valves were fitted to diesel engines. At the same time, modified exhaust valve seats were fitted, and valve stem oil seals were fitted to both the inlet and exhaust valves. The later exhaust valves can be fitted to the earlier cylinder head, provided the later exhaust valve seats are also fitted.*

23 Examine the head of each valve for pitting, burning, cracks, and general wear.

7.21a Zero the dial test indicator . . .

7.21b . . . then check the swirl chamber protrusion - diesel engine

7.24 Measuring a valve stem diameter

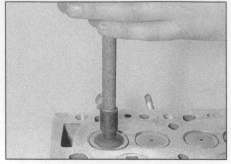

7.27 Grinding in the valves

7.28 Checking the valve head height

Check the valve stem for scoring and wear ridges. Rotate the valve, and check for any obvious indication that it is bent. Look for pits and excessive wear on the tip of each valve stem. Renew any valve that shows any such signs of wear or damage.

24 If the valve appears satisfactory at this stage, measure the valve stem diameter at several points using a micrometer **(see illustration)**. Any significant difference in the readings obtained indicates wear of the valve stem. Should any of these conditions be apparent, the valve(s) must be renewed.

25 If the valves are in satisfactory condition, they should be ground (lapped) into their respective seats, to ensure a smooth, gas-tight seal. If the seat is only lightly pitted, or if it has been re-cut, fine grinding compound *only* should be used to produce the required finish. Coarse valve-grinding compound should *not* be used, unless a seat is badly burned or deeply pitted. If this is the case, the cylinder head and valves should be inspected by an expert, to decide whether seat re-cutting, or even the renewal of the valve or seat (where possible) is required.

26 Valve grinding is carried out as follows. Place the cylinder head upside-down on a bench.

27 Smear a trace of the appropriate grade of valve-grinding compound on the seat face, and press a suction grinding tool onto the valve head. With a semi-rotary action, grind the valve head to its seat, lifting the valve occasionally to redistribute the grinding compound **(see illustration)**. A light spring placed under the valve head will greatly ease this operation.

28 If coarse grinding compound is being used, work only until a dull, matt even surface is produced on both the valve seat and the valve, then wipe off the used compound, and repeat the process with fine compound. When a smooth unbroken ring of light grey matt finish is produced on both the valve and seat, the grinding operation is complete. *Do not* grind-in the valves any further than absolutely necessary, or the seat will be prematurely sunk into the cylinder head. On diesel engines, use a dial-test indicator to check that the valve head height is within the specified limits **(see illustration)**.

29 When all the valves have been ground-in, carefully wash off *all* traces of grinding compound using paraffin or a suitable solvent, before reassembling the cylinder head.

Valve components

30 Examine the valve springs for signs of damage and discoloration. No minimum free length is specified by Citroën, so the only way of judging valve spring wear is by comparison with a new component.

31 Stand each spring on a flat surface, and check it for squareness. If any of the springs are damaged, distorted or have lost their tension, obtain a complete new set of springs. It is normal to renew the valve springs as a matter of course if a major overhaul is being carried out.

32 Renew the valve stem oil seals regardless of their apparent condition. On diesel engines, new seals are required for reassembly even if no seals were originally fitted.

Tappets - diesel engines

33 Examine the surfaces of the bucket tappets for wear or scoring. If there is excessive wear, the tappet should be renewed.

8 Cylinder head -
reassembly

1 With all the components cleaned, starting at one end of the cylinder head, fit the valve components as follows. If the original components are being refitted, all components must be refitted in their original positions. If new valves are being fitted, they must be fitted into the locations to which they have been ground.

Valve assembly - petrol engines

2 Lubricate the valve stem with clean engine oil, and insert the valve into its correct guide **(see illustration)**.

3 Refit the spring seat, then lubricate the new valve stem oil seal with clean engine oil and carefully locate it over the valve and onto the guide. Take care not to damage the oil seal as it is passed over the valve stem. Use a suitable socket or metal tube to press the seal firmly onto the guide **(see illustration)**.

4 Locate the valve spring on top of its seat, then refit the spring retainer.

5 Fit the spring compressor tool, and compress the spring until the retainer passes beyond the collet groove in the valve stem. Apply a little grease to the collet groove, and fit the split collets into the groove with the narrow ends nearest the spring. The grease should hold them in the groove. Slowly release the compressor tool, ensuring that the collets are not dislodged from the groove.

 HAYNES HINT *Use a little dab of grease to hold the collets in position on the valve stem while the spring compressor is released.*

6 Repeat the procedure on the remaining valves. With all the valves installed, place the cylinder head flat on the bench and, using a hammer and interposed block of wood, tap

8.2 Lubricate the valve stem before refitting the valve

8.3 Use a socket to fit the new valve stem oil seal

8.6 Tap the end of the valve stem to seat the collets in position

the end of each valve stem to settle the components **(see illustration)**.

Valve assembly - diesel engines

Note: *New valve stem oil seals should be fitted on reassembly, even if no seals were originally fitted.*

7 Lubricate the valve stem oil seal with clean engine oil, then fit the oil seal by pushing it into position in the cylinder head using a socket. Ensure that the seal is fully engaged with the cylinder head.

8 Lubricate the valve stem with clean engine oil, and insert the valve into its correct guide. To prevent damage to the oil seal as the valve is installed, wind a short length of tape round the top of the valve stem, to cover the collet groove. Remove the tape after the valve has been inserted.

9 Fit the spring seat, the small and large springs, and the spring retainer.

10 Proceed as described in paragraphs 5 and 6.

Diesel engines

11 Refit the timing probe blanking plug, if removed.

12 Insert and tighten the heater plugs to the specified torque (see Chapter 5C). Reconnect the wiring.

13 Insert and tighten the injectors, together with their copper washers, fire-seal washers and sleeves (see Chapter 4C). Reconnect the leak-off pipes.

14 Oil and insert the bucket tappets, together

with their respective shims, making sure that they are fitted in the correct locations, and with the size markings downwards **(see illustration)**. Make a note of the shim thickness fitted at each position, if not already done, for reference when checking the valve clearances.

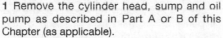

HAYNES HINT *Refit the tappets and shims immediately before refitting the camshaft. Do not refit them earlier, while doing other jobs, otherwise they might fall out and get mixed up.*

All engines

15 Refit the camshaft as described in Part A or B of this Chapter (as applicable).

16 Refit the inlet and exhaust manifolds with reference to the relevant Part of Chapter 4.

17 The cylinder head may now be refitted as described in Part A or B (as applicable).

9 Piston/connecting rod assembly - removal

1 Remove the cylinder head, sump and oil pump as described in Part A or B of this Chapter (as applicable).

2 If there is a pronounced wear ridge at the top of any bore, it may be necessary to remove it with a scraper or ridge reamer, to avoid piston damage during removal. Such a ridge may indicate that reboring is necessary, which will require new pistons (see also Section 11).

3 Using a hammer and centre-punch, paint or similar, mark each connecting rod big-end bearing cap with its respective cylinder number on the flat machined surface. If the engine has been dismantled before, note carefully any identifying marks made previously **(see illustration)**. Note that No 1 cylinder is at the transmission (flywheel) end of the engine.

4 Turn the crankshaft to bring pistons 1 and 4 to BDC (bottom dead centre).

5 Unscrew the nuts from No 1 piston big-end

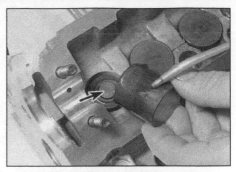

8.14 Oil and insert the bucket tappet, together with the shim (arrowed) - diesel engine

bearing cap. Take off the cap, and recover the bottom half bearing shell **(see illustration)**. If the bearing shells are to be re-used, tape the cap and the shell together.

6 To prevent the possibility of damage to the crankshaft bearing journals, tape over the connecting rod stud threads before removing the assemblies **(see illustration)**.

7 Using a hammer handle, push the piston up through the bore, and remove it from the top of the cylinder block. Recover the bearing shell, and tape it to the connecting rod for safe-keeping if it is to be re-used.

8 Loosely refit the big-end cap to the connecting rod, and secure it with the nuts, which will help to keep the components in their correct order.

9 Remove No 4 piston assembly in the same way.

10 Turn the crankshaft through 180° to bring pistons 2 and 3 to BDC (bottom dead centre), and remove them in the same way.

10 Crankshaft - removal

1 Remove the crankshaft timing sprocket, the flywheel and the oil pump as described in Part A or B of this Chapter.

2 Remove the pistons and connecting rods, as described in Section 9. **Note:** *If no work is to be done on the pistons and connecting*

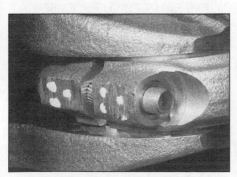

9.3 Connecting rod and big-end bearing cap marked for identification (No 3 cylinder shown)

9.5 Removing a big-end bearing cap and shell

9.6 To protect the crankshaft journals, tape over the connecting rod stud threads before removal

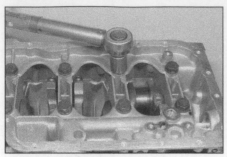

10.5 Progressively slacken and remove the ten large (11 mm) main bearing ladder retaining bolts . . .

10.6 . . . then lift the ladder away from the cylinder block - petrol engine

10.9 Remove the crankshaft right-hand oil seal housing - diesel engine

rods, there is no need to remove the cylinder head, or to push the pistons out of the cylinder bores. The pistons should just be pushed far enough up the bores that they are positioned clear of the crankshaft journals.

3 The crankshaft endfloat is controlled by thrustwashers on either side of No 2 main bearing. Check the endfloat as described in Section 13 and compare it with the specified limit, to determine whether new thrustwashers are required. Then proceed as follows.

Petrol engines

4 Work around the outside of the cylinder block, and unscrew all the small (6 mm) bolts securing the main bearing ladder to the base of the cylinder block. Note the correct fitted depth of both the left and right-hand crankshaft oil seals in the cylinder block/main bearing ladder.

5 Working in a diagonal sequence, evenly

and progressively slacken the ten large (11 mm) main bearing ladder retaining bolts by a turn at a time. Once all the bolts are loose, remove them from the ladder **(see illustration)**.

6 With all the retaining bolts removed, carefully lift the main bearing ladder casting away from the base of the cylinder block **(see illustration)**. Recover the lower main bearing shells, and tape them to their respective locations in the casting. If the two locating dowels are a loose fit, remove them and store them with the casting for safe-keeping.

7 Lift out the crankshaft, and discard both the oil seals. If not already removed, remove the oil pump drive chain from the end of the crankshaft, and slide off the drive sprocket. Recover the Woodruff key.

8 Recover the upper main bearing shells, and store them along with the relevant lower bearing shells. Also recover the two

thrustwashers (one fitted either side of No 2 main bearing) from the cylinder block.

Diesel engines

9 Undo the securing bolts and remove the crankshaft right-hand oil seal housing **(see illustration)**. Remove the gasket. Recover the oil seal if it is still attached to the crankshaft.

10 Remove the oil pump chain followed by the sprocket. Recover the Woodruff key if it is loose **(see illustrations)**.

11 Check that the main bearing caps are numbered 1 to 5 from the flywheel end **(see illustration)**. If not, mark them accordingly. Also note the fitted depth of the left-hand oil seal.

12 Invert the engine then unbolt and remove the main bearing caps. Recover the lower half bearing shells keeping them with their respective caps **(see illustration)**. Also recover the thrustwashers.

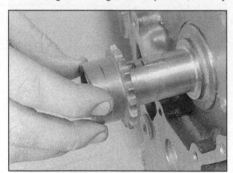

10.10a Slide off the oil pump sprocket . . .

10.10b . . . and remove the Woodruff key - diesel engine

10.11 Main bearing cap identification markings (arrowed) - diesel engine

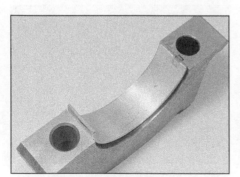

10.12 Main bearing cap and lower half bearing shell - diesel engine

10.13a Lift out the crankshaft . . .

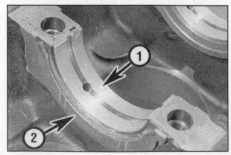

10.13b . . . and remove the upper half bearing shells (1) and thrustwashers (2) - diesel engine

11.1 Cylinder block core plugs (arrowed)

11.3 Remove the inspection plate (arrowed) from the flywheel end of the block - diesel engine

11.5a Removing an oil gallery plug

13 Lift out the crankshaft. Discard the left-hand oil seal. Recover the upper half bearing shells and keep them together with their respective caps, but identify them as the upper shells. Also recover and identify the upper thrustwashers **(see illustrations)**.

11 Cylinder block - cleaning and inspection

Cleaning

1 Remove all external components and electrical switches/sensors from the block. For complete cleaning, the core plugs should ideally be removed **(see illustration)**. Drill a small hole in the plugs, then insert a self-tapping screw into the hole. Pull out the plugs by pulling on the screw with a pair of grips, or by using a slide hammer.
2 On petrol engines, remove the liners as described in paragraph 12.
3 On diesel engines, remove the inspection plate from the flywheel end of the block, to enable cleaning of the coolant channels **(see illustration)**. Use a new rubber sealing ring when refitting the plate.
4 Scrape all traces of gasket from the cylinder block, and from the main bearing ladder (where fitted), taking care not to damage the gasket/sealing surfaces.
5 Remove all oil gallery plugs, where fitted. The plugs are usually very tight - they may have to be drilled out, and the holes re-

tapped. Use new plugs when the engine is reassembled. On diesel engines, one of the oil gallery plugs is on the flange beneath the oil filter mounting **(see illustrations)**.
6 If the block is extremely dirty, it should be steam-cleaned.
7 If the block has been steam-cleaned, clean all oil holes and oil galleries one more time. Flush all internal passages with warm water until the water runs clear. Dry thoroughly, and apply a light film of oil to all mating surfaces, to prevent rusting. On diesel engines, also oil the cylinder bores. If you have access to compressed air, use it to speed up the drying process, and to blow out all the oil holes and galleries.

⚠️ **Warning: Wear eye protection when using compressed air!**

8 If the castings are not very dirty, you can do an adequate cleaning job with hot soapy water and a stiff brush. Take plenty of time, and do a thorough job. Regardless of the cleaning method used, be sure to clean all oil holes and galleries very thoroughly, and to dry all components well. On diesel engines, protect the cylinder bores as described above, to prevent rusting.
9 All threaded holes must be clean, to ensure accurate torque readings during reassembly. To clean the threads, run the correct-size tap into each of the holes to remove rust, corrosion, thread sealant or sludge, and to restore damaged threads **(see illustration)**. If possible, use compressed air to clear the holes of debris produced by this operation.

HAYNES HINT *A good alternative is to inject aerosol-applied water-dispersant lubricant into each hole, using the long tube usually supplied.*
⚠️ *Warning: Wear eye protection when cleaning out these holes in this way!*

10 Apply suitable sealant to the new oil gallery plugs, and insert them into the holes in the block. Tighten them securely.
11 If the engine is not going to be reassembled right away, cover it with a large plastic bag to keep it clean; protect all mating surfaces and the cylinder bores as described above, to prevent rusting.

Inspection

Petrol engine

12 Petrol engines have aluminium alloy cylinder blocks with cast iron cylinder liners. Remove the liner clamps (where used), then use a hardwood drift to tap out each liner from inside the cylinder block. When all the liners are released, tip the cylinder block on its side and remove each liner from the top of the block. As each liner is removed, stick masking tape on its left-hand (transmission side) face, and write the cylinder number on the tape. No 1 cylinder is at the transmission end of the engine. Remove the O-ring from the base of each liner, and discard it **(see illustrations)**.
13 Check each cylinder liner for scuffing and

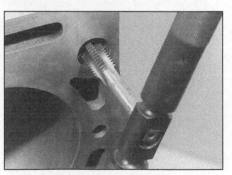

11.5b Oil gallery plug (arrowed) beneath oil filter mounting - diesel engine

11.9 Cleaning a cylinder block threaded hole using a suitable tap

11.12a Remove each cylinder liner . . .

11.12b . . . and recover the bottom O-ring seal (arrowed) - petrol engine

scoring. Check for signs of a wear ridge at the top of the liner, indicating that the bore is excessively worn.

14 If the necessary measuring equipment is available, measure the bore diameter of each cylinder liner at the top (just under the wear ridge), centre, and bottom of the cylinder bore, parallel to the crankshaft axis.

15 Next, measure the bore diameter at the same three locations, at right-angles to the crankshaft axis. Compare the results with the figures given in the Specifications.

16 Repeat the procedure for the remaining cylinder liners.

17 If the liner wear exceeds the permitted tolerances at any point, or if the cylinder liner walls are badly scored or scuffed, then renewal of the relevant liner assembly will be necessary. If there is any doubt about the condition of the cylinder bores, seek the advice of a Citroën dealer or engine reconditioning specialist.

18 If renewal is necessary, new liners, complete with pistons and piston rings, can be purchased from a Citroën dealer. Note that it is not possible to buy liners individually - they are supplied only as matched assemblies complete with pistons and rings. The gudgeon pins are an interference fit in the connecting rod small-end bearing. You will need to take the old pistons, complete with their connecting rods, to a Citroën dealer or engine repair specialist, who will have the necessary tools to remove the gudgeon pins and fit the connecting rods to the new pistons. See Section 12 for further details of pistons and connecting rods.

11.22 Check the cylinder liner protrusion - petrol engine

19 To allow for manufacturing tolerances, pistons and liners are separated into three size groups. The size group of each piston is indicated by a letter (A, B or C) stamped onto its crown, and the size group of each liner is indicated by a series of 1 to 3 notches on the upper lip of the liner; a single notch for group A, two notches for group B, and three notches for group C. Ensure that each piston and its respective liner are both of the same size group. It is permissible to have different size group piston and liner assemblies fitted to the same engine, but never fit a piston of one size group to a liner in a different group.

20 Before installing the liners, thoroughly clean the liner mating surfaces in the cylinder block, and use fine abrasive paper to polish away any burrs or sharp edges which might damage the liner O-rings. Clean the liners and wipe dry, then fit a new O-ring to the base of each liner. To aid installation, apply a smear of oil to each O-ring and to the base of the liner.

21 If the original liners are being refitted, use the marks made on removal to ensure that each is refitted the correct way round, and is inserted into its original position. Insert each liner into the cylinder block, taking care not to damage the O-ring, and press it home as far as possible by hand. Using a hammer and a block of wood, tap each liner lightly but fully onto its locating shoulder. Wipe clean, then lightly oil, all exposed liner surfaces, to prevent rusting.

22 With all four liners correctly installed, use a dial gauge (or a straight-edge and feeler blade) to check that the protrusion of each liner above the upper surface of the cylinder block is within the limits given in the Specifications **(see illustration)**. The maximum difference between any two liners must not be exceeded.

23 If new liners are being fitted, it is permissible to interchange them to bring the difference in protrusion within limits. Remember to keep each piston with its respective liner.

24 If liner protrusion cannot be brought within limits, seek the advice of a Citroën dealer or engine reconditioning specialist before proceeding with the engine rebuild.

Diesel engine

25 Diesel engines have cast iron cylinder blocks and there are no cylinder liners. Visually check the block for cracks, rust and corrosion. Look for stripped threads in the threaded holes (it may be possible to re-cut stripped threads using a suitable tap). If there has been any history of internal coolant leakage, it may be worthwhile asking an engine overhaul specialist to check the block for cracks using special equipment. If defects are found, have the block repaired if possible, or renew the assembly.

26 Examine the cylinder bores for taper, ovality, scoring and scratches. Start by carefully examining the top of the cylinder bores. If they are at all worn, a very slight

ridge will be found on the thrust side. This marks the top of the piston ring travel.

27 Measure the bore diameter of each cylinder at the top (just under the wear ridge), and at the centre and bottom of the cylinder bore, parallel to the crankshaft axis. For accurate assessment, a bore micrometer is required. However, a rough measurement can be made by inserting feeler blades between a piston (without rings) and the bore wall.

28 Next, measure the bore diameter at the same three locations, at right-angles to the crankshaft axis. Compare the results with the figures given in the Specifications.

29 Repeat the procedure for the remaining cylinders.

30 If the cylinder wear exceeds the permitted tolerances, or if the cylinder walls are badly scored or scuffed, then the cylinders will have to be rebored by a Citroën dealer or engineering workshop who will be able to supply suitable oversize pistons.

31 If the bore wear is marginal, and within the specified tolerances, special piston rings can be fitted to offset the wear. In this case, the bores should be honed to allow the new rings to bed-in correctly and provide the best possible seal. The conventional type of hone has spring-loaded stones, and is used with a power drill. You will also need some paraffin or honing oil and rags. The hone should be moved up and down the cylinder bore to produce a crosshatch pattern, and plenty of honing oil should be used. Ideally, the crosshatch lines should intersect at approximately a 60° angle. Do not remove more material than is necessary to produce the required finish. If new pistons are being fitted, the piston manufacturers may specify a finish with a different angle, so their instructions should be followed. Do not withdraw the hone from the cylinder while it is still being turned - stop it first. After honing a cylinder, wipe out all traces of the honing oil. An engine overhaul specialist will be able to carry out this work at moderate cost.

32 If the cylinder bores are within the specified tolerances, but have developed a 'glazed' appearance, they will need to be honed as described above, and new piston rings fitted. Glazing of cylinder bores can occur due to long periods of idling, incorrect running-in procedures or use of the incorrect grade of oil during running-in, and will cause the engine to consume excessive amounts of oil.

12 Piston/connecting rod assembly - inspection

1 Before the inspection process can begin, the piston/connecting rod assemblies must be cleaned, and the original piston rings removed from the pistons.

2 Carefully expand the old rings over the top of the pistons. The use of two or three old feeler blades will be helpful in preventing the rings dropping into empty grooves **(see illustration)**. Be careful not to scratch the piston with the ends of the ring. The rings are brittle, and will snap if they are spread too far. They're also very sharp - protect your hands and fingers. Note that the third ring incorporates an expander. Always remove the rings from the top of the piston. Keep each set of rings with its piston if the old rings are to be re-used (although it is not recommended).

3 Scrape away all traces of carbon from the top of the piston. A hand-held wire brush or a piece of fine emery cloth can be used, once the majority of the deposits have been scraped away.

4 Remove the carbon from the ring grooves in the piston, using an old ring. Break the ring in half to do this (be careful not to cut your fingers - piston rings are sharp). Be careful to remove only the carbon deposits - do not remove any metal, and do not nick or scratch the sides of the ring grooves.

5 Once the deposits have been removed, clean the piston/connecting rod assembly with paraffin or a suitable solvent, and dry thoroughly. Make sure that the oil return holes in the ring grooves are clear.

6 If the pistons and cylinder bores are not damaged or worn excessively, and if the cylinder block (on diesel engines) does not need to be rebored, the original pistons can be refitted. Normal piston wear shows up as even vertical wear on the piston thrust surfaces, and slight looseness of the top ring in its groove. New piston rings should always be used when the engine is reassembled.

7 Carefully inspect each piston for cracks around the skirt, around the gudgeon pin holes, and at the piston ring "lands" (between the ring grooves).

8 Look for scoring and scuffing on the piston skirt, holes in the piston crown, and burned areas at the edge of the crown. If the skirt is scored or scuffed, the engine may have been suffering from overheating, and/or abnormal combustion which caused excessively high operating temperatures. The cooling and lubrication systems should be checked thoroughly. Scorch marks on the sides of the pistons show that blow-by has occurred. A hole in the piston crown, or burned areas at the edge of the piston crown, indicates that abnormal combustion (pre-ignition, knocking, or detonation) has been occurring. If any of the above problems exist, the causes must be investigated and corrected, or the damage will occur again. The causes may include incorrect ignition/injection pump timing, or a faulty injector (as applicable).

9 Corrosion of the piston, in the form of pitting, indicates that coolant has been leaking into the combustion chamber and/or the crankcase. Again, the cause must be corrected, or the problem may persist in the rebuilt engine.

10 Examine each connecting rod carefully for signs of damage, such as cracks around the

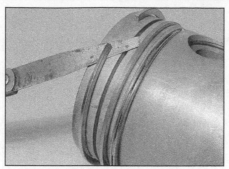

12.2 Removing a piston ring with a feeler blade

big-end and small-end bearings. Check that the rod is not bent or distorted. Damage is highly unlikely, unless the engine has been seized or badly overheated. Detailed checking of the connecting rod assembly can be carried out by a Citroën dealer or engine repair specialist with the necessary equipment.

11 Examine the mating faces of the big-end caps and connecting rods, to see if they have ever been filed, in a mistaken attempt to take up bearing wear. This is extremely unlikely, but if evident, the offending connecting rods and caps must be renewed.

Petrol engines

12 On petrol engines with cylinder liners, it is not possible to renew the pistons separately. Pistons are only supplied with piston rings and a liner, as a part of a matched assembly (see Section 11).

13 The gudgeon pins are an interference fit in the connecting rod small-end bearing. Therefore, you will need to take the old pistons, complete with their connecting rods, to a Citroën dealer or engine repair specialist, who will have the necessary tools to remove the gudgeon pins and fit the connecting rods to the new pistons.

Diesel engines

14 On diesel engines, the gudgeon pins are the floating type, secured in position by two circlips. Pistons are supplied complete with new gudgeon pins and circlips. Circlips can also be purchased individually.

15 Measure the piston diameters, and check that they are within limits for the corresponding

12.17a Prise out the gudgeon pin circlip - diesel engine

12.15 Measuring a piston diameter using a micrometer

bore diameters **(see illustration)**. If the piston-to-bore clearance is excessive, the block will have to be rebored, and new pistons and rings fitted.

16 If renewing pistons without reboring make sure that the correct size is obtained. Piston class is denoted by an "A" or "A1" mark, or no mark at all on the centre of the crown. The identical code appears also on the corner of the cylinder block at the timing belt end. The piston weight class is stamped on the crown and must be identical on all pistons in the same engine. **Note:** *The marks "R1", "R2" etc. generally denote rebore sizes on XUD series engines. There is only one rebore size on the XUD7 engine covered in this manual.*

17 To separate a piston from its connecting rod, use a small flat-bladed screwdriver to prise out the circlips, then push out the gudgeon pin. Hand pressure should be sufficient to remove the pin. Identify the piston and rod to ensure correct reassembly **(see illustrations)**.

18 Examine the gudgeon pin and connecting rod small-end bearing for signs of wear or damage. Wear can be cured by renewing both the pin and bush. Bush renewal, however, is a specialist job, because press facilities are required, and the new bush must be reamed accurately. Examine the connecting rods carefully as described in paragraph 10, and if necessary take them to an engine overhaul specialist for a detailed check.

19 Reassemble the pistons and rods, using new components as appropriate. Make sure the pistons are fitted the right way round. The cloverleaf recess on the piston crown must

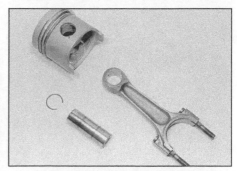

12.17b Piston and connecting rod components - diesel engine

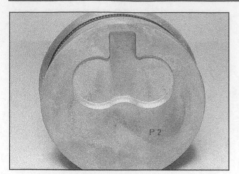

12.19a Cloverleaf recess on the piston crown - diesel engine

face the same way as the big-end bearing shell cut-out in the connecting rod. Apply a smear of clean engine oil to the gudgeon pin. Slide it into the piston and through the connecting rod small-end **(see illustrations)**. Check that the piston pivots freely on the rod, then secure the gudgeon pin in position with two new circlips. Ensure that each circlip is correctly located in its groove in the piston.

20 See Section 16 for details of refitting piston rings.

13 Crankshaft - inspection

Checking crankshaft endfloat

1 If the crankshaft endfloat is to be checked, this must be done when the crankshaft is still installed in the cylinder block, but is free to move.

2 Check the endfloat using a dial gauge in contact with the end of the crankshaft. Push the crankshaft fully one way, and then zero the gauge. Push the crankshaft fully the other way, and check the endfloat. The result can be compared with the specified amount, and will give an indication as to whether new thrustwashers are required **(see illustration)**.

3 If a dial gauge is not available, feeler blades can be used. First push the crankshaft fully towards the flywheel end of the engine, then use feeler blades to measure the gap between the No 1 crankpin web and No 2 main bearing thrustwasher **(see illustration)**.

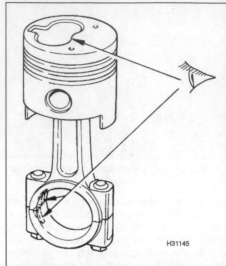

H31145

12.19b The cloverleaf recess must face the same way as the bearing shell cut-outs in the connecting rod - diesel engine

Inspection

4 Clean the crankshaft using paraffin or a suitable solvent, and dry it, preferably with compressed air if available.

⚠️ *Warning: Wear eye protection when using compressed air! Be sure to clean the oil holes with a pipe cleaner or similar probe, to ensure that they are not obstructed.*

5 Check the main and big-end bearing journals for uneven wear, scoring, pitting and cracking.

6 Big-end bearing wear is accompanied by distinct metallic knocking when the engine is running (particularly noticeable when the engine is pulling from low revs) and some loss of oil pressure.

7 Main bearing wear is accompanied by severe engine vibration and rumble - getting progressively worse as engine revs increase - and again by loss of oil pressure.

8 Check the bearing journal for roughness by running a finger lightly over the bearing surface. Any roughness (which will be accompanied by obvious bearing wear) indicates that the crankshaft requires regrinding (where possible) or renewal.

12.19c Push the gudgeon pin into the piston - diesel engine

9 If the crankshaft has been reground, check for burrs around the crankshaft oil holes (the holes are usually chamfered, so burrs should not be a problem unless regrinding has been carried out carelessly). Remove any burrs with a fine file or scraper, and thoroughly clean the oil holes as described previously.

10 Using a micrometer, measure the diameter of the main and big-end bearing journals, and compare the results with the Specifications **(see illustration)**. By measuring the diameter at a number of points around each journal's circumference, you will be able to determine whether or not the journal is out-of-round. Take the measurement at each end of the journal, near the webs, to determine if the journal is tapered. Compare the results obtained with those given in the Specifications. Where no specified journal diameters are quoted, seek the advice of a Citroën dealer.

11 Check the oil seal contact surfaces at each end of the crankshaft for wear and damage. If the seal has worn a deep groove in the surface of the crankshaft, consult an engine overhaul specialist. Repair may be possible, but otherwise a new crankshaft will be required.

12 Note that Citroën produce a set of undersize bearing shells for both the main bearings and big-end bearings. If the crankshaft has worn beyond the specified limits, and the crankshaft journals have not already been reground, it may be possible to have the crankshaft reconditioned, and to fit the undersize shells. Seek the advice of your Citroën dealer or engine specialist on the best course of action.

13.2 Checking crankshaft endfloat using a dial gauge

13.3 Checking crankshaft endfloat using feeler blades

13.10 Measuring a crankshaft big-end journal diameter using a micrometer

14 Main and big-end bearings - inspection

1 Even though the main and big-end bearings should be renewed during the engine overhaul, the old bearings should be retained for close examination, as they may reveal valuable information about the condition of the engine.

2 Bearing failure can occur due to lack of lubrication, the presence of dirt or other foreign particles, overloading the engine, or corrosion (see illustration). Regardless of the cause of bearing failure, the cause must be corrected (where applicable) before the engine is reassembled, to prevent it from happening again.

3 When examining the bearing shells, remove them from the cylinder block, the main bearing ladder/caps (as appropriate), the connecting rods and the connecting rod big-end bearing caps. Lay them out on a clean surface in the same general position as their location in the engine. This will enable you to match any bearing problems with the corresponding crankshaft journal. Do not touch any shell's bearing surface with your fingers while checking it, or the delicate surface may be scratched.

4 Dirt and other foreign matter gets into the engine in a variety of ways. It may be left in the engine during assembly, or it may pass through filters or the crankcase ventilation system. It may get into the oil, and from there into the bearings. Metal chips from machining operations and normal engine wear are often present. Abrasives are sometimes left in engine components after reconditioning, especially when parts are not thoroughly cleaned using the proper cleaning methods. Whatever the source, these foreign objects often end up embedded in the soft bearing material, and are easily recognised. Large particles will not embed in the bearing, and will score or gouge the bearing and journal. The best prevention for this cause of bearing failure is to clean all parts thoroughly, and keep everything spotlessly-clean during engine assembly. When the engine has been installed in the vehicle, make sure the engine oil and filter changes are carried out at the regular service intervals.

5 Lack of lubrication (or lubrication breakdown) has a number of interrelated causes. Excessive heat (which thins the oil), overloading (which squeezes the oil from the bearing face) and oil leakage (from excessive bearing clearances, worn oil pump or high engine speeds) all contribute to lubrication breakdown. Blocked oil passages, which usually are the result of misaligned oil holes in a bearing shell, will also oil-starve a bearing, and destroy it. When lack of lubrication is the cause of bearing failure, the bearing material is wiped or extruded from the steel backing of the bearing. Temperatures may increase to the point where the steel backing turns blue from overheating.

6 Driving habits can have a definite effect on bearing life. Full-throttle, low-speed operation (labouring the engine) puts very high loads on bearings, tending to squeeze out the oil film. These loads cause the bearings to flex, which produces fine cracks in the bearing face (fatigue failure). Eventually, the bearing material will loosen in pieces, and tear away from the steel backing.

7 Short-distance driving leads to corrosion of bearings, because insufficient engine heat is produced to drive off the condensed water and corrosive gases. These products collect in the engine oil, forming acid and sludge. As the oil is carried to the engine bearings, the acid attacks and corrodes the bearing material.

8 Incorrect bearing installation during engine assembly will also lead to bearing failure. Tight-fitting bearings leave insufficient bearing running clearance, and will result in oil starvation. Dirt or foreign particles trapped behind a bearing shell result in high spots on the bearing, which lead to failure.

9 Do not touch any shell's bearing surface with your fingers during reassembly; there is a risk of scratching the delicate surface, or of depositing particles of dirt on it.

10 As mentioned at the beginning of this Section, the bearing shells should be renewed as a matter of course during engine overhaul; to do otherwise is false economy. Refer to Sections 17 and 18 for details of main and big-end bearing shell selection.

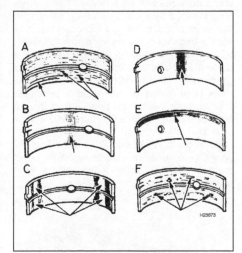

14.2 Typical bearing failures

A Scratched by dirt; dirt embedded in bearing material
B Lack of oil; overlay wiped out
C Improper seating; bright (polished) sections
D Tapered journal; overlay gone from entire surface
E Radius ride
F Fatigue failure; craters or pockets

15 Engine overhaul - reassembly sequence

1 Before reassembly begins, ensure that all new parts have been obtained, and that all necessary tools are available. Read through the entire procedure to familiarise yourself with the work involved, and to ensure that all items necessary for reassembly of the engine are at hand. In addition to all normal tools and materials, thread-locking compound will be needed. A suitable tube of liquid sealant will also be required for the joint faces that are fitted without gaskets. It is recommended that Citroën's own product(s) are used, which are specially formulated for this purpose; the relevant product names are quoted in the text of each Section where they are required.

2 In order to save time and avoid problems, engine reassembly can be carried out in the following order:

a) Piston rings (Section 16)
b) Crankshaft (Section 17).
c) Piston/connecting rod assemblies (Section 18).
d) Oil pump (See Part A or B - as applicable).
e) Sump (See Part A or B - as applicable).
f) Flywheel (See Part A or B - as applicable).
g) Cylinder head (See Part A or B - as applicable).
h) Timing belt tensioner and sprockets, and timing belt (See Part A or B - as applicable).
i) Engine external components.

3 At this stage, all engine components should be absolutely clean and dry, with all faults repaired. The components should be laid out (or in individual containers) on a completely clean work surface.

16 Piston rings - refitting

1 Before fitting new piston rings, the ring end gaps must be checked as follows.

2 Lay out the piston/connecting rod assemblies and the new piston ring sets, so that the ring sets will be matched with the same piston and cylinder/liner during the end gap measurement and subsequent engine reassembly.

3 Insert the top ring into the first cylinder/liner, and push it down the bore using the top of the piston. This will ensure that the ring remains square with the cylinder walls. Position the ring near the bottom of the cylinder bore, at the lower limit of ring travel. Note that the top and second compression rings are different.

4 Measure the end gap using feeler blades.

16.5 Measuring a piston ring end gap

5 Repeat the procedure with the ring at the top of the cylinder bore, at the upper limit of its travel, and compare the measurements with the figures given in the Specifications **(see illustration)**.

6 If the gap is too small (unlikely if genuine Citroën parts are used), it must be enlarged, or the ring ends may contact each other during engine operation, causing serious damage. Ideally, new piston rings providing the correct end gap should be fitted. As a last resort, the end gap can be increased by filing the ring ends very carefully with a fine file. Mount the file in a vice equipped with soft jaws, slip the ring over the file with the ends contacting the file face, and slowly move the ring to remove material from the ends. Take care, as piston rings are sharp, and are easily broken.

7 With new piston rings, it is unlikely that the end gap will be too large. If the gaps are too large, check that you have the correct rings for your engine and for the particular cylinder bore size.

8 Repeat the checking procedure for each ring in the first cylinder, and then for the rings in the remaining cylinders. Remember to keep rings, pistons and cylinders matched up.

9 Once the ring end gaps have been checked and if necessary corrected, the rings can be fitted to the pistons.

10 Fit the piston rings using the same technique as for removal. Fit the bottom (oil control) ring first, and work up. Note that the oil control ring may incorporate an expander. First insert the expander, then fit the oil control ring with its gap positioned 180° from

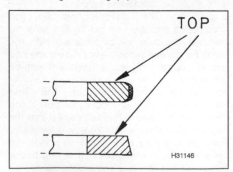

16.10a Piston ring cross sections

the expander gap. Ensure that the second compression ring is fitted the correct way up. If a stepped ring is being fitted, fit the ring with the smaller diameter of the step uppermost **(see illustrations)**. Arrange the gaps of the top and second compression rings 120° either side of the oil control ring gap. **Note:** *Always follow any instructions supplied with the new piston ring sets - different manufacturers may specify different procedures. Do not mix up the top and second compression rings, as they have different cross-sections.*

17 Crankshaft - refitting and main bearing running clearance check

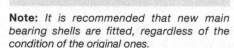

Note: *It is recommended that new main bearing shells are fitted, regardless of the condition of the original ones.*

Selection of new bearing shells

Petrol engines

1 On early petrol engines, both upper and lower main bearing shells were of the same thickness, with only two sizes of bearing shells being available: a standard size for use with the standard crankshaft, and a set of undersize bearing shells for use once the crankshaft bearing journals have been reground.

2 However, since February 1992, the specified main bearing running clearance tolerance has been significantly reduced. This has been achieved by the introduction of three different grades of bearing shell, in both standard sizes and undersizes. The grades are indicated by a colour-coding marked on the edge of each shell, which denotes the shell's thickness, as listed in the following table. The upper shell on all bearings is of the same size (class B, colour code black), and the running clearance is controlled by fitting a lower bearing shell of the required thickness. This arrangement has been fitted to all engines produced since February 1992 and, if possible, should also be fitted to earlier engines during overhaul. Seek the advice of your Citroën dealer on parts availability and the best course of action when ordering new bearing shells.

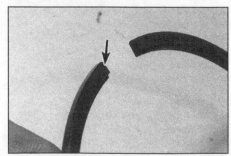

16.10b Stepped rings should be fitted with the smaller diameter of the step (arrowed) uppermost

Bearing colour code	Thickness (mm)	
	Standard	Undersize
Blue (class A)	1.823	1.973
Black (class B)	1.835	1.985
Green (class C)	1.848	1.998

3 On early engines, the correct size bearing shell must be selected by measuring the running clearance, as described in the section beginning at paragraph 10.

4 On engines produced since February 1992, when the new bearing shell sizes were introduced, the crankshaft and cylinder block have reference marks on them, to identify the size of the journals and bearing bores.

5 The cylinder block reference marks are on the right-hand (timing belt) end of the block, and the crankshaft reference marks are on the right-hand (timing belt) end of the crankshaft, on the right-hand web of No 4 crankpin. These marks can be used to select bearing shells of the required thickness as follows.

6 On both the crankshaft and block, there are two lines of identification: a bar code, which is used by Citroën during production, and a row of five letters. The first letter in the sequence refers to the size of No 1 bearing (at the flywheel end). The last letter in the sequence (which is followed by an arrow) refers to the size of No 5 main bearing.

7 Obtain the identification letter of both the relevant crankshaft journal and the cylinder block bearing bore. These letters can now be used with the accompanying chart, to select the bearing shell grade. Noting that the cylinder block letters are listed across the top of the chart, and the crankshaft letters down the side, trace a vertical line down from the relevant cylinder block letter, and a horizontal line across from the relevant crankshaft letter, and find the point at which both lines cross. This crossover point will indicate the grade of lower bearing shell required to give the correct main bearing running clearance. For example, the illustration shows cylinder block reference G, and crankshaft reference T, crossing at a point within the area of Class A, indicating that a blue-coded (Class A) lower bearing shell is required to give the correct main bearing running clearance **(see illustration)**.

8 Repeat this procedure so that the required bearing shell grade is obtained for each of the five main bearing journals.

Diesel engines

9 There are no colour-coded main bearing shells on the XUD 7 diesel engine. Instead, the bearing shells have codes marked on the back, to denote their size. If the shells are to be renewed, the old shells should be taken to your dealer to ensure that the correct new shells are obtained. If the crankshaft has been reground, the engineering works will advise you about the correct size shells. If there is any doubt about which bearing shells should be used, consult your Citroën dealer.

Main bearing running clearance check

Petrol engines

10 On early petrol engines, if the modified bearing shells are to be fitted, obtain a set of new black (Class B) upper bearing shells and new blue (Class A) lower bearing shells. On later (February 1992-on) engines where the modified bearing shells are already fitted, the running clearance check can be carried out using the original bearing shells. However, it is preferable to use a new set, since the results obtained will be more conclusive.

11 Clean the backs of the bearing shells, and the bearing locations in both the cylinder block and the main bearing ladder.

12 Press the bearing shells into their locations, ensuring that the tab on each shell engages in the notch in the cylinder block or main bearing ladder location. Take care not to touch any shell's bearing surface with your fingers. Note that the grooved bearing shells, both upper and lower, are fitted to Nos 2 and 4 main bearings **(see illustration)**. If the original bearing shells are being used for the check, ensure that they are refitted in their original locations. The clearance can be checked in either of two ways.

13 One method (which will be difficult to achieve without a range of internal micrometers or internal/external expanding calipers) is to refit the main bearing ladder casting to the cylinder block, with the bearing shells in place. With the casting retaining bolts correctly tightened, measure the internal diameter of each assembled pair of bearing shells. If the diameter of each corresponding crankshaft journal is measured and then subtracted from the bearing internal diameter, the result will be the main bearing running clearance.

14 The second (and more accurate) method is to use an American product known as "Plastigage". This consists of a fine thread of perfectly-round plastic, which is compressed between the bearing shell and the journal. When the shell is removed, the plastic is deformed, and can be measured with a special card gauge supplied with the kit. The running clearance is determined from this gauge. Plastigage should be available from

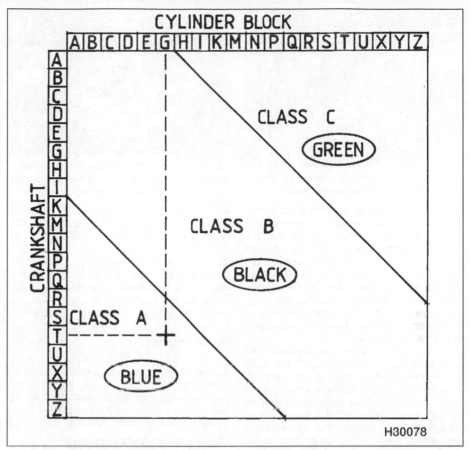

17.7 Main bearing shell selection chart for use with petrol engines from February 1992

your dealer (reference number OUT 30 4133 T). Otherwise, enquiries at one of the larger specialist motor factors should produce the name of a stockist in your area. The procedure for using Plastigage is as follows.

15 With the main bearing upper shells in place, carefully lay the crankshaft in position. Do not use any lubricant; the crankshaft journals and bearing shells must be perfectly clean and dry.

16 Cut several lengths of the appropriate-size Plastigage (they should be slightly shorter than the width of the main bearings), and place one length on each crankshaft journal axis **(see illustration)**.

17 With the main bearing lower shells in position, refit the main bearing ladder casting, tightening its retaining bolts as described in paragraph 31. Take care not to disturb the Plastigage, and *do not* rotate the crankshaft at any time during this operation.

18 Remove the main bearing ladder casting, again taking great care not to disturb the Plastigage or rotate the crankshaft.

19 Compare the width of the crushed Plastigage on each journal to the scale printed on the Plastigage envelope, to obtain the main bearing running clearance **(see illustration)**. Compare the clearance measured with that given in the Specifications at the start of this Chapter.

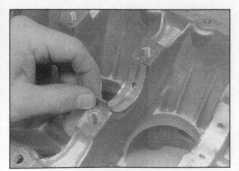

17.12 On petrol engines, the grooved bearing shells are fitted to Nos 2 and 4 main bearing journals

17.16 Plastigage in place on a crankshaft main bearing journal

17.19 Measuring the width of the deformed Plastigage using the scale on the card provided

17.26 Refit the upper thrustwashers to No 2 main bearing - petrol engine

17.27 Ensure that each bearing shell tab (arrowed) is correctly located and lubricate the shell with clean engine oil

17.28 Refit the oil pump drive chain and sprocket - petrol engine

20 If the clearance is significantly different from that expected, the bearing shells may be the wrong size (or excessively worn, if the original shells are being re-used). Before deciding that different-size shells are required, make sure that no dirt or oil was trapped between the bearing shells and the main bearing ladder or block when the clearance was measured. If the Plastigage was wider at one end than at the other, the crankshaft journal may be tapered.

21 If the clearance is not as specified, use the reading obtained, along with the shell thicknesses quoted above, to calculate the necessary grade of bearing shells required. When calculating the bearing clearance required, bear in mind that it is always better to have the running clearance towards the lower end of the specified range, to allow for wear in use.

22 Where necessary, obtain the required grades of bearing shell, and repeat the running clearance checking procedure as described above.

23 On completion, carefully scrape away all traces of the Plastigage material from the crankshaft and bearing shells. Use your fingernail, or a wooden or plastic scraper which is unlikely to score the bearing surfaces.

Diesel engines

24 The procedure for measuring the main bearing running clearance check is similar to that described in the previous paragraphs for petrol engines, except that each main bearing has a separate cap which must be fitted and the bolts tightened to the specified torques.

Final crankshaft refitting

Petrol engines

25 Carefully lift the crankshaft out of the cylinder block once more.

26 Using a little grease, stick the upper thrustwashers to each side of the No 2 main bearing upper location. Ensure that the oilway grooves on each thrustwasher face outwards (away from the cylinder block) **(see illustration)**.

27 Place the bearing shells in their locations as described earlier. If new shells are being fitted, ensure that all traces of protective grease are cleaned off using paraffin. Wipe dry the shells and connecting rods with a lint-free cloth. Liberally lubricate each bearing shell in the cylinder block with clean engine oil **(see illustration)**.

28 Refit the Woodruff key, then slide on the oil pump drive sprocket, and locate the drive chain on the sprocket **(see illustration)**. Lower the crankshaft into position so that Nos 2 and 3 cylinder crankpins are at TDC; Nos 1 and 4 cylinder crankpins will be at BDC ready for fitting No 1 piston. Check the crankshaft endfloat as described in Section 13.

29 Thoroughly degrease the mating surfaces of the cylinder block and the main bearing ladder. Apply a thin bead of suitable sealant to the cylinder block mating surface of the main bearing ladder casting, then spread to an even film **(see illustration)**. Citroën recommend the use of Auto Joint E10 sealant. In the absence of the specified sealant, any suitable good-quality sealant may be used.

30 Lubricate the lower bearing shells with clean engine oil, then refit the main bearing ladder, ensuring that the shells are not displaced, and that the locating dowels engage correctly **(see illustration)**.

31 Install the ten 11 mm main bearing ladder retaining bolts, and tighten them all by hand only. Working progressively outwards from the centre bolts, tighten the ten bolts, by a turn at a time, to the specified Stage 1 torque. Once all the bolts have been tightened to the Stage 1 torque, angle-tighten them to the specified Stage 2 angle using a socket and extension bar. It is recommended that an angle-measuring gauge is used during this stage of tightening, to ensure accuracy **(see illustrations)**. If a gauge is not available, use a dab of white paint to make alignment marks between the bolt head and casting, to mark out the required angle.

17.29 Apply a thin film of suitable sealant to the cylinder block/crankcase mating surface . . .

17.30 . . . then lower the main bearing ladder into position - petrol engine

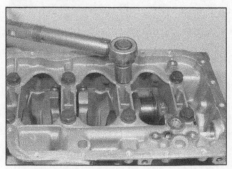

17.31a Tighten the ten 11 mm main bearing bolts to the Stage 1 torque . . .

17.31b . . . then angle-tighten them through the specified Stage 2 angle - petrol engine

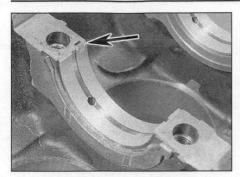

17.42 No 2 main bearing and thrustwashers. Make sure the tab (arrowed) engages with the recess in the seat

17.43 Lower the crankshaft into position - diesel engine

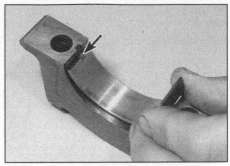

17.44 Fitting a bearing shell to a main bearing cap. Make sure the tab (arrowed) engages with the recess in the cap

32 Refit all the 6 mm bolts securing the main bearing ladder to the base of the cylinder block, and tighten them to the specified torque. Check that the crankshaft rotates freely.

33 Refit the piston/connecting rod assemblies to the crankshaft as described in Section 18.

34 Ensuring that the drive chain is correctly located on the sprocket, refit the oil pump and sump as described in Part A of this Chapter.

35 Fit two new crankshaft oil seals as described in Part A.

36 Refit the flywheel as described in Part A.

37 Where removed, refit the cylinder head as described in Part A.

38 Refit the crankshaft timing sprocket and timing belt as described in Part A.

Diesel engines

39 Carefully lift the crankshaft out of the cylinder block once more.

40 If not already done, wipe clean the main bearing shell seats in the block and caps. Wipe any protective coating from the new bearing shells.

41 Fit the upper main bearing shells (with the oil grooves) to their seats in the block. Make sure that the locating tabs on the shells engage with the recesses in the seats.

42 Fit the thrustwashers on each side of No 2 main bearing, grooved side outwards. Use a smear of grease to hold them in position (see illustration).

43 Lubricate the upper main bearing shells

17.47 Fitting No 5 main bearing cap - diesel engine

and lower the crankshaft into position (see illustration).

44 Fit the plain bottom half main bearing shells to their respective caps. Make sure that the locating tabs on the shells engage with the recesses in the caps (see illustration). Oil the shells.

45 Fit the thrustwashers on each side of No 2 main bearing cap using a smear of grease to hold them in position.

46 Before fitting the bearing caps, check that the crankshaft endfloat is within the specified limits, with reference to Section 13.

47 Fit the main bearing caps Nos 2 to 5 to their correct locations (see illustration) ensuring that they are the right way round (the bearing shell tab recesses in the block and caps must be on the same side). Insert the bolts loosely.

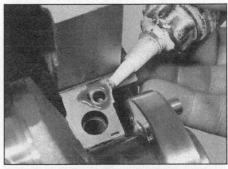

17.48 Apply thread locking fluid to No 1 main bearing cap joint face - diesel engine

48 Apply a small amount of thread locking fluid to the No 1 main bearing cap mating face on the cylinder block, around the sealing strip holes (see illustration).

49 Press the sealing strips into the grooves on each side of No 1 main bearing cap. A method is now required to hold the sealing strips in position while the bearing cap is being fitted. Citroën garages use a special clamp, but you can use two thin metal plates of 0.25 mm thickness or less (such as old feeler blades) provided all burrs that may damage the sealing strips are first removed. Oil both sides of the metal plates and hold them in position on the sealing strips. Fit the No 1 main bearing cap, insert the bolts loosely, then carefully pull out the metal plates with a pair of pliers in a horizontal direction (see illustrations).

17.49a Press the sealing strips onto both sides of No 1 main bearing cap . . .

17.49b . . . and use thin metal plates to hold them in position while fitting the cap . . .

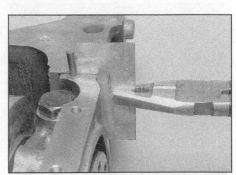

17.49c . . . then insert the bolts and carefully pull out the metal plates- diesel engine

17.50 Tighten the main bearing cap retaining bolts to the specified torque - diesel engine

17.53 Cut the sealing strips on No 1 bearing cap to the correct height using feeler blades and a knife

50 Tighten the main bearing cap bolts evenly to the specified torque **(see illustration)**.
51 Check that the crankshaft rotates freely - there must be no tight spots or binding.
52 Fit a new crankshaft left-hand oil seal, with reference to Part B.
53 Refit the piston/connecting rod assemblies as described in Section 18, but before refitting the oil pump and sump, use feeler blades and a knife to cut the sealing strips on No 1 bearing cap to 1.0 mm above the sump gasket mating surface **(see illustration)**.
54 Refit the oil pump, drive chain, right-hand oil seal housing and sump as described in Part B.
55 Refit the flywheel as described in Part B.
56 Where removed, refit the cylinder head as described in Part B.
57 Fit a new crankshaft right-hand oil seal, then refit the crankshaft timing sprocket and timing belt as described in Part B.

18 Piston/connecting rod assembly - refitting and big-end bearing running clearance check

Note: *This Section describes how to check the big-end bearing clearances and then fit the pistons to the cylinders. If you like, you can fit the pistons first and then check the bearing clearances. If the pistons are already fitted, there is no need to take them out.*

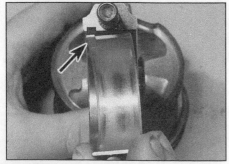

18.4 Fitting a bearing shell to a connecting rod - ensure that the tab (arrowed) engages with the recess in the connecting rod

Selection of bearing shells

1 There are two sizes of big-end bearing shell produced by Citroën; a standard size for use with the standard crankshaft, and an undersize for use once the crankshaft journals have been reground. When ordering shells, quote the diameter of the crankshaft big-end crankpins, to ensure that the correct set of shells are purchased.
2 Prior to refitting the piston/connecting rod assemblies, it is recommended that the big-end bearing running clearance is checked as follows:

Big-end bearing running clearance check

3 Clean the backs of the bearing shells, and the bearing locations in both the connecting rod and bearing cap.
4 Press the bearing shells into their locations, ensuring that the tab on each shell engages in the notch in the connecting rod and cap **(see illustration)**. Take care not to touch any shell's bearing surface with your fingers. If the original bearing shells are being used for the check, ensure they are refitted in their original locations. The clearance can be checked in either of two ways.
5 One method is to refit the big-end bearing cap to the connecting rod, ensuring they are fitted the correct way round (see paragraph 20), with the bearing shells in place. With the cap retaining nuts correctly tightened, use an internal micrometer or vernier caliper to measure the internal diameter of each assembled pair of bearing shells. If the diameter of each corresponding crankshaft journal is measured and then subtracted from the bearing internal diameter, the result will be the big-end bearing running clearance.
6 The second and more accurate method is to use Plastigage (see Section 17).
7 Ensure that the bearing shells are correctly fitted. Place a strand of Plastigage on each (cleaned) crankpin journal.
8 Refit the (clean) piston/connecting rod assemblies to the crankshaft, and refit the big-end bearing caps, using the marks made or noted on removal to ensure they are fitted the correct way around.
9 Tighten the bearing cap nuts as described

below in paragraph 21. Take care not to disturb the Plastigage or rotate the connecting rod during the tightening sequence.
10 Dismantle the assemblies without rotating the connecting rods. Use the scale printed on the Plastigage envelope to obtain the big-end bearing running clearance.
11 If the clearance is significantly different from that expected, the bearing shells may be the wrong size (or excessively worn, if the original shells are being re-used). Make sure that no dirt or oil was trapped between the bearing shells and the caps or connecting rods when the clearance was measured. If the Plastigage was wider at one end than at the other, the crankpins may be tapered.
12 Note that Citroën do not specify a recommended big-end bearing running clearance. The figure given in the Specifications is a guide figure, which is typical for this type of engine. Before condemning the components concerned, refer to your Citroën dealer or engine reconditioning specialist for further information on the specified running clearance. Their advice on the best course of action to be taken can then also be obtained.
13 On completion, carefully scrape away all traces of the Plastigage material from the crankshaft and bearing shells. Use your fingernail, or some other object which is unlikely to score the bearing surfaces.

Final piston/connecting rod refitting

14 Note that the following procedure assumes that on petrol engines the cylinder liners are in position in the cylinder block as described in Section 11, and that the crankshaft and main bearing ladder/caps are in place (see Section 17). It is possible to fit the pistons to the liners before fitting the liners to the cylinder block. The advantage of using this method is that the pistons enter the liner from the bottom end, which is tapered to allow easy entry of the piston rings (a piston ring compressor will still be required).
15 Ensure that the bearing shells are correctly fitted as described in paragraphs 3 and 4. If new shells are being fitted, ensure that all traces of the protective grease are cleaned off using paraffin. Wipe dry the shells and connecting rods with a lint-free cloth.
16 Lubricate the cylinder bores/liners, the pistons, and piston rings, then lay out each piston/connecting rod assembly in its respective position.
17 Start with assembly No 1. Make sure that the piston rings are still spaced as described in Section 16, then clamp them in position with a piston ring compressor.
18 Insert the piston/connecting rod assembly into the top of cylinder/liner No 1, correctly oriented as follows:
a) On petrol engines, the arrow on the piston crown must point towards the timing belt end of the engine.

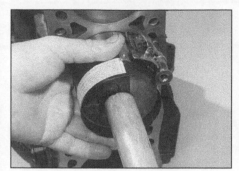

18.19 Tap the piston into the bore using a hammer handle

18.21 Tighten the big-end bearing cap nuts to the specified torque

18.24 Checking the crankshaft rotational torque

b) *On diesel engines, the cloverleaf recess on the piston crown must point towards the oil filter side of the engine. If there is also an arrow on the piston crown, it should point towards the timing belt end of the engine.*

19 Using a block of wood or hammer handle against the piston crown, tap the assembly into the cylinder/liner until the piston crown is flush with the top of the cylinder/liner **(see illustration)**.

20 Ensure that the bearing shell is still correctly installed. Liberally lubricate the crankpin and both bearing shells. Taking care not to mark the cylinder/liner bores, tap the piston/connecting rod assembly down the bore/liner and onto the crankpin. Refit the big-end bearing cap, tightening its retaining nuts finger-tight at first. Note that the faces with the identification marks must match (which means that the bearing shell locating tabs abut each other).

21 Tighten the bearing cap retaining nuts evenly and progressively to the specified torque **(see illustration)**.

22 Rotate the crankshaft. Check that it turns freely; some stiffness is to be expected if new components have been fitted, but there should be no signs of binding or tight spots.

23 Refit the remaining three piston /connecting rod assemblies in the same way.

24 On diesel engines, temporarily refit the pulley bolt to the nose of the crankshaft then, using a torque wrench, check that the torque required to turn the crankshaft does not exceed 40 Nm (30 lbf ft) **(see illustration)**. Any excessive tightness must be investigated

before proceeding. Dirt or grit trapped behind a bearing shell is one possible cause.

25 Refit the cylinder head, oil pump and sump as described in Part A or B of this Chapter (as applicable).

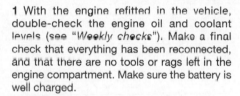

19 Engine -
initial start-up after overhaul

1 With the engine refitted in the vehicle, double-check the engine oil and coolant levels (see "*Weekly checks*"). Make a final check that everything has been reconnected, and that there are no tools or rags left in the engine compartment. Make sure the battery is well charged.

Petrol engine models

2 Remove the spark plugs. Disable the ignition system by disconnecting the ignition HT coil lead from the distributor cap, and earthing it on the cylinder block. Use a jumper lead or similar wire to make a good connection.

3 On fuel injection models, disconnect the fuel injection system relay unit (see Chapters 4B and 12) to disable the fuel pump and prevent contamination of the catalytic converter with unburnt fuel.

4 Turn the engine on the starter until the oil pressure warning light goes out.

5 Refit the spark plugs, and reconnect the spark plug (HT) leads, referring to Chapter 1A for further information. Reconnect any HT leads or wiring which was disconnected in

paragraph 2. Refit the fuel pump relay, if removed.

Diesel engine models

6 Prime the fuel system as described in Chapter 4C.

All models

7 Start the engine, noting that this may take a little longer than usual, due to the fuel system components having been disturbed.

8 While the engine is idling, check for fuel, water and oil leaks. Don't be alarmed if there are some odd smells and smoke from parts getting hot and burning off oil deposits.

9 Assuming all is well, keep the engine idling until hot water is felt circulating through the top hose, then switch off the engine.

10 After a few minutes, recheck the oil and coolant levels as described in *"Weekly checks"*, and top-up as necessary.

11 Check the ignition timing (petrol engines) or injection pump timing (diesel engines), and the idle speed settings (as appropriate), then switch the engine off.

12 If the cylinder head bolts were tightened as described, there is no need to re-tighten them once the engine has first run after reassembly.

13 If new pistons, rings or crankshaft bearings have been fitted, the engine must be treated as new, and run-in for the first 500 miles (800 km). *Do not* operate the engine at full-throttle, or allow it to labour at low engine speeds in any gear. It is recommended that the oil and filter be changed at the end of this period.

Chapter 3
Cooling, heating and ventilation systems

Contents

Degrees of difficulty

Easy, suitable for novice with little experience	**Fairly easy,** suitable for beginner with some experience	**Fairly difficult,** suitable for competent DIY mechanic 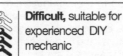	**Difficult,** suitable for experienced DIY mechanic	**Very difficult,** suitable for expert DIY or professional

Specifications

System

Type .... Pressurised, front-mounted radiator (with integral expansion tank up to January 1994), coolant pump and thermostat. Electric cooling fan(s)

Expansion tank

Cap pressure*: .... 1.0 bar (14.5 psi) or 1.4 bars (20.3 psi)
*Depends on production date. Value is normally stamped on cap.

Thermostat

Opening temperature (stamped on thermostat):
   Petrol engine models (from July 1988 onwards) .... 88°C
   Diesel engine models .... 83°C

Temperature senders

Temperature switch operating temperature:
   Petrol models .... 110°C
   Diesel models:
     Warning switch .... 105°C
     Emergency warning switch (if fitted) .... 112°C

Cooling fan(s)

Petrol engine models:
   Cut-in temperature .... 90°C
Diesel engine models:
   1st speed cuts in at .... 84°C
   2nd speed cuts in at .... 88°C

Torque wrench settings	Nm	lbf ft
Coolant pump-to-cylinder block (petrol engines)		
Smaller coolant pump housing securing bolts.	30	22
Larger coolant pump housing securing bolts	50	37
Coolant pump-to-housing (petrol engines)		
Upper bolt	16	12
Lower bolt	7	5
Coolant pump rear cover-to-housing (petrol engines)	7	5
Coolant pump-to-cylinder block (diesel engines)	12	9
Diesel post-heating system temperature sensor	18	13
Temperature warning light sender	18	13

1 General information and precautions

General information

1 The cooling system is the pressurised type, comprising a coolant pump driven by the timing belt, an aluminium crossflow radiator, electric cooling fan, a thermostat, heater matrix, expansion tank and all associated hoses and switches.

2 The system functions as follows. Cold coolant in the bottom of the radiator passes through the bottom hose to the coolant pump, where it is pumped around the cylinder block, head passages and (on petrol models only) the inlet manifold. After cooling the cylinder bores, combustion surfaces and valve seats, the coolant reaches the underside of the thermostat. When the engine is cold, the thermostat is closed and the coolant is returned to the coolant pump, either directly or via the heater matrix, according to model. When the coolant reaches a predetermined temperature, the thermostat opens, and the coolant passes through the top hose to the radiator. As the coolant circulates through the radiator, it is cooled by the inrush of air when the van is in forward motion, and also by the action of the electric cooling fan(s) when necessary. Upon reaching the bottom of the radiator, the coolant has now cooled, and the cycle is repeated.

3 When the engine warms up, the coolant expands, and some of it is displaced into the expansion tank. Coolant collects in the tank, and is returned to the radiator when the system cools.

4 The electric cooling fans mounted in front of the radiator are controlled by a thermostatic switch. At a predetermined coolant temperature, the switch/sensor actuates the fans. On models with two fans, the switch operates in two stages, actuating the fans in series at low speed or in parallel at high speed, depending on the temperature.

5 On petrol models, the expansion tank is integral with the radiator. The low temperature by-pass from the thermostat goes to the heater matrix and then to the coolant pump. On carburettor petrol models, the carburettor is heated from a hose which comes off the heater matrix return line **(see illustration)**.

6 On diesel models, the low temperature by-pass from the thermostat goes directly to the coolant pump, and the heater matrix receives water from a hose on the rear of the engine block. The fuel is heated by the cooling water on all diesel models. The fuel heater was mounted behind the cylinder block, integral with the coolant pump inlet manifold, until early 1993 when a new type of fuel filter with integral heater was mounted on top of the thermostat housing in front of the cylinder head **(see illustrations)**. The expansion tank was integral with the radiator, the same as on petrol models, until January 1994 when a separate expansion tank was introduced.

Precautions

⚠️ *Warning: Do not attempt to remove the expansion tank filler cap, or to disturb any part of the cooling system, while the engine is hot, as there is a high risk of scalding. If the filler cap must be removed before the engine and radiator have fully cooled (even though this is not recommended), the pressure in the cooling system must first be relieved. Cover the cap with a thick layer of cloth, to avoid scalding, and slowly unscrew the filler cap until a hissing sound is heard. When the hissing has stopped, indicating that the pressure has reduced, slowly unscrew the filler cap until it can be removed. If more hissing sounds are heard, wait until they have stopped before unscrewing the cap completely. At all times, keep well away from the filler cap opening, and protect your hands.*

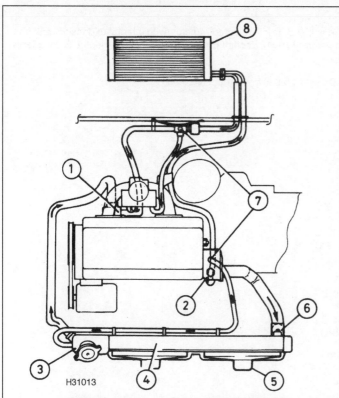

1.5 Cooling system circuit - petrol models

1 *Coolant pump*
2 *Thermostat housing*
3 *Expansion tank*
4 *Radiator*
5 *Electric cooling fan*

6 *Thermocontact (at bottom corner of radiator)*
7 *Bleed points*
8 *Heater matrix*

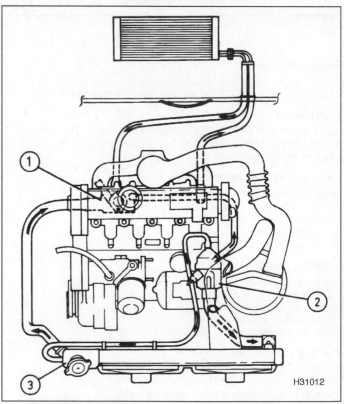

1.6a Cooling system circuit - diesel models to early 1993

1 *Integral fuel heater and coolant pump inlet manifold*

2 *Thermostat housing*
3 *Expansion tank*

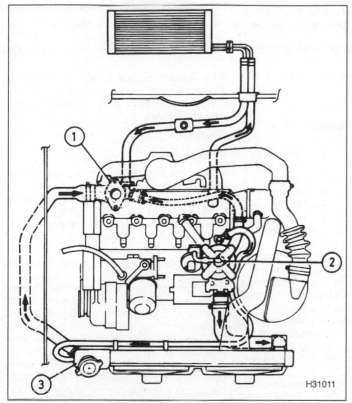

**1.6b Cooling system circuit -
diesel models from early 1993 to January 1994**

1 Coolant pump inlet manifold
2 Integral fuel heater and fuel filter mounted on thermostat housing
3 Expansion tank

⚠️ **Warning: Do not allow antifreeze to come into contact with your skin, or with the painted surfaces of the vehicle. Rinse off spills immediately, with plenty of water. Never leave antifreeze** lying around in an open container, or in a puddle in the driveway or on the garage floor. Children and pets are attracted by its sweet smell, but antifreeze can be fatal if ingested.

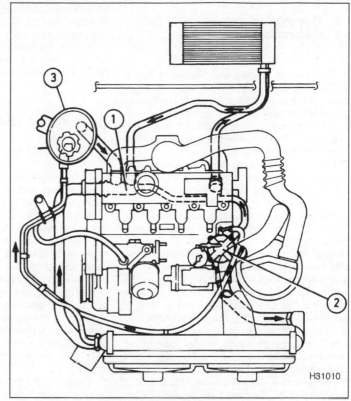

**1.6c Cooling system circuit -
diesel models from January 1994**

1 Coolant pump inlet manifold
2 Integral fuel heater and fuel filter mounted on thermostat housing
3 Expansion tank

⚠️ **Warning: If the engine is hot, the electric cooling fan may start rotating even if the engine is not running. Be careful to keep your hands, hair and any loose clothing well clear when working in the engine compartment.**

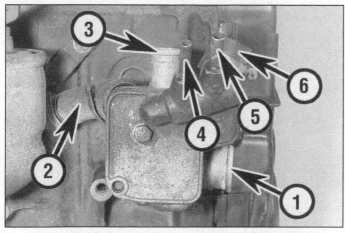

**1.6d Integral fuel heater and coolant pump inlet manifold -
diesel models to early 1993**

1 Return from radiator
2 Low temperature by-pass from thermostat
3 Return from heater matrix
4 Fuel outlet to filter
5 Fuel inlet from tank
6 Inlet from expansion tank (not used on this model)

1.6e Integral fuel heater and fuel filter mounted on thermostat housing - diesel models from early 1993

2 Cooling system hoses - disconnection and renewal

1 The number, routing and pattern of hoses will vary according to model, but the same basic procedure applies. Before commencing work, make sure that the new hoses are to hand, along with new hose clips if needed. It is good practice to renew the hose clips at the same time as the hoses.

2 Drain the cooling system, as described in Chapter 1A or 1B, saving the coolant if it is fit for re-use. Squirt a little penetrating oil onto the hose clips if they are corroded.

3 Release the hose clips from the hose concerned. Three types of clip are used; worm-drive, spring and "sardine-can". The worm-drive clip is released by turning its screw anti-clockwise. The spring clip is released by squeezing its tags together with pliers, at the same time working the clip away from the hose stub. The "sardine-can" clip is not re-usable, and is best cut off with snips or side cutters.

4 Unclip any wires, cables or other hoses which may be attached to the hose being removed. Make notes for reference when reassembling if necessary.

5 Release the hose from its stubs with a twisting motion. Be careful not to damage the stubs on delicate components such as the radiator. If the hose is stuck fast, try carefully prising the end of the hose with a screwdriver or similar, taking care not to use excessive force. The best course is often to cut off a stubborn hose using a sharp knife, but again be careful not to damage the stubs.

6 Before fitting the new hose, smear the stubs with washing-up liquid or a suitable rubber lubricant to aid fitting. Do not use oil or grease, which may attack the rubber.

7 Fit the hose clips over the ends of the hose, then fit the hose over its stubs. Work the hose into position. When satisfied, locate and tighten the hose clips.

8 Refill the cooling system as described in Chapter 1A or 1B. Run the engine, and check that there are no leaks.

9 Recheck the tightness of the hose clips on any new hoses after a few hundred miles.

10 Top-up the coolant level if necessary (see "Weekly checks").

3 Cooling system pressure - testing

1 In cases where leakage is difficult to trace a pressure test can prove helpful. The test involves pressurising the system by means of a hand pump and an adapter, which is fitted to the expansion tank or radiator in place of the filler cap. If you do not have the correct equipment, take the vehicle to a Citroën dealer for testing.

2 Fit the test equipment to the expansion tank or radiator, then run the engine to normal operating temperature and switch it off.

3 Apply 1.4 bar pressure and check that this pressure is held for at least 10 seconds. If the pressure drops prematurely there is a leak in the cooling system which must be traced and rectified.

4 Besides leaks from hoses, pressure can also be lost through leaks in the radiator and heater matrix. A blown head gasket or a cracked head or block can cause an "invisible" leak, but there are usually other clues to this condition such as poor engine performance, regular misfiring or combustion gases entering the coolant.

5 After completing the test, allow the engine to cool, then remove the test equipment.

6 The condition of the filler cap must not be overlooked. Normally it is tested with similar equipment to that used for the pressure test. The release pressure is given in the Specifications and is also usually stamped on the cap. Renew the cap with a replacement of the same rating, if it is faulty.

4 Radiator and expansion tank - removal, inspection and refitting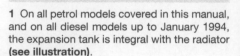

1 On all petrol models covered in this manual, and on all diesel models up to January 1994, the expansion tank is integral with the radiator **(see illustration)**.

Radiator

Removal

2 Drain the cooling system as described in Chapter 1A or 1B.

3 On diesel models, remove the air cleaner and inlet duct as described in Chapter 4C.

4 Disconnect the upper hose and bypass hose from the radiator. Alternatively, disconnect them from the thermostat housing.

5 Disconnect the lower hose from the radiator.

6 Disconnect the wiring from the thermal switch on the bottom left of the radiator, and from the level detector on the expansion tank if fitted.

7 Remove the radiator grille and cross panel (see Chapter 11).

8 Lift the radiator slightly so that it disengages from the lower studs, then disengage it from the upper plastic moulding on the fan assembly. Carefully lift the radiator from the engine compartment.

Inspection

9 If the radiator has been removed because of blockage (causing overheating), then try reverse-flushing or, in severe cases, use a radiator cleanser strictly in accordance with the manufacturer's instructions. Refer to Chapter 1A or 1B for further information.

10 Use a soft brush and an air line or garden hose to clear the radiator matrix of leaves, insects etc.

> **HAYNES HINT** *Minor leaks from the radiator can be cured using a suitable sealant with the radiator in situ.*

11 Major leaks or extensive damage should be repaired by a specialist, or the radiator should be renewed or exchanged for a reconditioned unit.

Caution: Do not attempt to weld or solder a leaking radiator, as damage to the plastic components may result.

12 Examine the radiator mounting rubbers for signs of damage or deterioration and renew if necessary.

Refitting

13 Refitting is the reverse of removal. If a new filler cap is fitted, make sure it has the correct pressure rating, according to the Specifications or the value stamped on the old cap. On completion, refill the system as described in Chapter 1A or 1B.

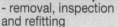

4.1 Radiator with integral expansion tank

1 Drain plug
2 Thermocontact for cooling fan
3 Inlet hose from thermostat
4 Bypass hose from thermostat
5 Expansion tank
6 Coolant level detector (if fitted) or blanking plug
7 Outlet hose to cooling pump

Expansion tank

14 On diesel models from January 1994 onwards, a separate expansion tank is mounted on the right-hand side of the engine compartment, near the bulkhead (**see illustration**).

Removal

15 With the engine cold, unscrew and remove the filler cap from the expansion tank.
16 Slacken the drain plug on the side of the tank and drain the fluid into a suitable container. Disconnect the wiring plug if there is a coolant level detector.
17 Disconnect the hoses from the bottom of the expansion tank and collect any remaining fluid.
18 Release the expansion tank from its wire bracket and withdraw it from the vehicle.

Refitting

19 Refitting is the reverse of removal. If a new filler cap is fitted, make sure it has the correct pressure rating, according to the Specifications or the value stamped on the old cap. On completion, refill the system as described in Chapter 1A or 1B.

---

5 Thermostat - removal, testing and refitting

Removal

1 The thermostat is located on the left-hand end of the cylinder head, within a thermostat

5.1 Thermostat location on petrol models (arrowed)

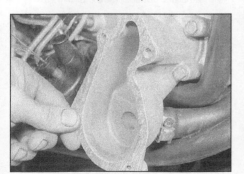

5.4b ... then remove the gasket - diesel model up to early 1993

4.14 Separate expansion tank - diesel models from January 1994

housing, and there are three different designs:
a) *Petrol models (see illustration).*
b) *Diesel models up to early 1993 (see illustrations 5.4a and 5.4b).*
c) *Diesel models from early 1993 onwards. A combined fuel filter and heater is mounted on top of the housing (see Section 1).*

In all cases, the thermostat is behind the housing cover that includes the outlet pipe to the radiator, and the method of removal is the same.
2 Drain the cooling system as described in Chapter 1A or 1B.
3 If desired for improved access, remove the air cleaner and/or the air intake ducting as described in the relevant part of Chapter 4. Disconnect the radiator upper hose from the thermostat housing if it gives better access. Position any other wiring and hoses clear of the thermostat housing. On diesel models there is no need to disconnect the fast idle thermostatic sensor.

5.4a Unscrew the retaining bolts and remove the thermostat housing cover . . .

5.5a Thermostat in housing cover held by circlip

4 Unscrew the retaining bolts, and carefully withdraw the thermostat housing cover. Remove the gasket, if fitted (**see illustrations**).
5 The thermostat is located in the housing or the cover, according to model. Use pliers to remove a circlip, if fitted, then remove the thermostat, noting which way round it is fitted, and recover the sealing ring(s) (**see illustrations**).

Testing

6 To test the thermostat, suspend it on a piece of string in a pan of cold water and check that it is initially closed. Heat the water and check that it begins to open at the temperature given in the Specifications, or the value stamped on the thermostat. Continue to heat the water and check the fully open temperature. Finally allow the water to cool and check that it fully closes.

Refitting

7 Refitting is the reverse of removal. Make sure the thermostat is fitted the correct way round. Use new sealing ring(s) if necessary, and a new gasket (if fitted). On completion, refill the cooling system as described in Chapter 1A or 1B.

---

6 Elcotric cooling fan - testing, removal and refitting

Testing

Petrol models

1 Current supply to the cooling fan is via the ignition switch, and the fan is switched on and off, according to coolant temperature, by a thermostatic switch mounted on the left hand side of the radiator.
2 If the fan does not appear to work, run the engine until normal operating temperature is reached, then allow it to idle. The fan should cut in within a few minutes (before the temperature gauge needle enters the red section, or before the coolant temperature warning light comes on). If not, switch off the ignition and disconnect the wiring plug from the cooling fan switch. Bridge the two contacts in the wiring plug using a length of

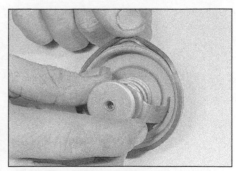

5.5b Remove the rubber seal from the thermostat

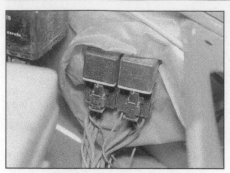

6.5 Cooling fan relays for twin fan system

spare wire, and switch on the ignition. If the fan now operates, the switch is probably faulty, and should be renewed.

3 If the fan still fails to operate, check that battery voltage is available at the feed wire to the switch; if not, then there is a fault in the feed wire (possibly due to a blown fuse). If there is no problem with the feed, check that there is continuity between the switch earth terminal and a good earth point on the body; if not, then the earth connection is faulty, and must be re-made.

4 If the switch and the wiring are in good condition, the fault must lie in the motor itself. The motor can be checked by disconnecting it from the wiring loom, and connecting a 12-volt supply directly to it.

Diesel models

5 Due to the larger engine size and the greater requirement for cooling, diesel models have dual fans, operating in series at low speed or in parallel at high speed. Current supply to the fans is via the ignition switch auxiliary relay. The fans are operated, according to coolant temperature, by a two-stage thermostatic switch mounted on the left hand side of the radiator. There are two relays located behind the battery on the left of the engine compartment, to modify the circuit and switch the fans between low speed and high speed **(see illustration)**. The circuit works as follows:
 a) At low temperature, the thermostatic switch is open and the fans do not operate.
 b) At intermediate temperature, the thermostatic switch closes at the first stage and the fans operate in series at low speed.

6.16 Disconnect the wiring (arrowed) and remove the fans

 c) At high temperature, the thermostatic switch closes at the second stage and the fans operate in parallel at high speed.

6 If the fans do not appear to work, run the engine until normal operating temperature is reached, then allow it to idle. The fans should cut in within a few minutes. If not, switch off the ignition and disconnect the wiring plug from the cooling fan switch. Connect the earth contact in the wiring plug to each of the two feed contacts in turn and switch on the ignition. Observe the effects as follows:
 a) One of the positions will close the circuit for first stage cooling. If the cooling fans begin to operate at low speed, the switch is probably faulty and should be renewed.
 b) The other position will change the circuit to second stage cooling. Both of the relays should operate when the ignition is switched on, and should make an audible click. If either of the relays is not working it should be renewed. **Note:** The fans will not actually operate at high speed unless both of the feed contacts are connected to the earth contact.

7 If the fans cannot be made to operate at low speed, check that battery voltage is available at the feed wires to the switch. If not, then there is a fault in the feed wires, possibly due to a blown fuse. If there is no problem with the feed, check that there is continuity between the switch earth terminal and a good earth point on the body; if not, then the earth connection is faulty, and must be re-made.

8 If the switch, wiring and relays are in good condition, the fault must lie in one or both of the motors. Check each motor in turn by disconnecting it from the wiring loom, and running it directly from a 12-volt supply.

Removal

9 Disconnect the battery negative lead.
10 Drain the cooling system as described in Chapter 1A or 1B.
11 On diesel models, remove the air cleaner and inlet duct as described in Chapter 4C.
12 Disconnect the hoses from the radiator. Disconnect the wiring from the thermal switch on the bottom left of the radiator, and from the level detector on the expansion tank, if fitted (see Section 4).
13 Remove the radiator grille and cross panel (see Chapter 11).
14 Remove the left headlamp (see Chapter 12).
15 Lift the radiator slightly so that it disengages from the lower studs, then disengage it from the upper plastic moulding on the fan assembly. Carefully lift the radiator from the engine compartment. Alternatively, leave the radiator in place for the time being so that the radiator and fans are removed as a complete assembly and then separated.
16 Disconnect the electrical connectors from the front of the fan. On a twin-fan system there are four wires, two for each fan. Carefully disengage the fan assembly from its locating lugs on the lower bodywork and lift it out of the vehicle **(see illustration)**.

Refitting

17 Refitting is the reverse of removal.

---

7 Cooling system electrical switches - testing, removal and refitting

Electric cooling fan thermostatic switch

Testing

1 Testing of the switch is described in Section 6 as part of the electric cooling fan test procedure.

Removal

Note: *Suitable sealing compound or a new sealing ring (as applicable) will be required on refitting.*

2 The switch is located in the left-hand side of the radiator above the drain bolt. The engine and radiator should be cold before removing the switch.
3 Disconnect the battery negative lead.
4 Drain the cooling system as described in Chapter 1A or 1B.
5 Disconnect the wiring plug from the switch.
6 Carefully unscrew the switch from the radiator, and recover the sealing ring (where applicable).

Refitting

7 Refitting is the reverse of removal, noting the following points.
8 If the switch was originally fitted using sealing compound, clean the switch threads thoroughly, and coat them with fresh sealing compound. If the switch was originally fitted using a sealing ring, use a new one.
9 Refill the cooling system as described in Chapter 1A or 1B.
10 On completion, start the engine and run it until it reaches normal operating temperature. Continue to run the engine, and check that the cooling fan cuts in and out correctly.

Coolant temperature warning light sender

Testing

11 On petrol engine models, the sender is located in the left-hand end of the cylinder head. On diesel engine models it is located in the thermostat housing **(see illustration)**.

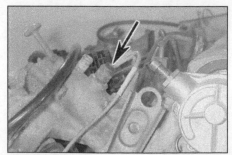

7.11 Temperature sender unit (arrowed) on thermostat housing - diesel model up to early 1993 (engine removed)

8.17a Unscrew the bolts . . .

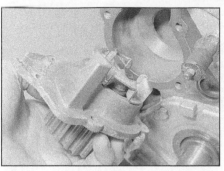

8.17.b . . . and withdraw the coolant pump

8.17c Coolant pump showing impeller vanes (diesel model)

12 The temperature warning light is fed with a voltage from the instrument panel. The warning light earth is controlled by the sender. The sender is effectively a switch that operates at a predetermined temperature to earth the light and complete the circuit.

13 If the sender is thought to be faulty, turn on the ignition and run the engine so that the instrument panel warning lights go out. Disconnect the wiring connector from the sender and earth it to the cylinder block. If the warning light comes on, the circuit is in good condition and the fault is in the sender unit.

Removal and refitting

14 The procedure is similar to that described previously in this Section for the electric cooling fan thermostatic switch. On some models, access to the sender is poor, and other components may need to be removed (or hoses, wiring, etc moved to one side) before the sender unit can be reached.

Fuel injection system coolant temperature sensor

Testing

15 No specific information is available about the location of the coolant temperature sensor on 1124cc petrol engines equipped with the Bosch A2.2 Monopoint fuel injection system. It is likely to be either in the thermostat housing or screwed into the cylinder head.

16 The sensor is a thermistor - an electronic component with a resistance that decreases as the temperature rises. The fuel injection ECU supplies the sensor with a set voltage, and by measuring the current flowing in the sensor circuit, it determines the engine temperature. This information is then used, in conjunction with other inputs, to control the injector opening time (pulse width) and idle speed.

17 If the sensor circuit should fail to provide adequate information, the ECU back-up facility will override the sensor signal. In this event, the ECU assumes a predetermined setting which will allow the fuel injection system to run, albeit at reduced efficiency. When this occurs, the engine warning light on the instrument panel will come on, and the advice of a Citroën dealer should be sought. The sensor itself can only be tested using special Citroën diagnostic equipment. *Do not* attempt to test the circuit

using any other equipment, as there is a high risk of damaging the ECU.

Removal and refitting

18 The procedure is similar to that described previously in this Section for the electric cooling fan thermostatic switch. If access to the sensor is poor, other components may need to be removed (or hoses, wiring, etc moved to one side) to reach the sensor.

Diesel engine post-heating system temperature sensor

Testing

19 On diesel models from 1996 onwards the glow plug post-heating is controlled by a coolant temperature sensor fitted to the thermostat housing (see Chapter 5C for details). No specific information is available about testing the sensor.

Removal and refitting

20 The procedure is similar to that described previously in this Section for the electric cooling fan thermostatic switch. If access to the sensor is poor, other components may need to be removed (or hoses, wiring, etc moved to one side) to reach the sensor.

8 Coolant pump - removal and refitting

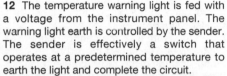

Petrol engines

Removal

1 The coolant pump is driven by the timing belt, and is located in a housing bolted to the rear of the cylinder block at the timing belt end of the engine.

2 Drain the cooling system as described in Chapter 1A.

3 Remove the timing belt as described in Chapter 2A.

4 Remove the securing bolts, and withdraw the coolant pump from the pump housing. Manipulate the pump past the engine mounting, and withdraw it from the engine compartment. Recover the O-ring.

5 If desired, the rear housing can be removed from the opposite end of the coolant pump

housing. Access is most easily obtained from underneath the vehicle (it may be necessary to remove the exhaust heat shield). Disconnect the coolant hoses from the impeller housing (be prepared for coolant spillage), then remove the securing bolts and withdraw the impeller housing. Again, recover the O-ring.

6 If it is required to remove the coolant pump rear housing from the rear of the cylinder block, it will be necessary to support the engine on a jack and piece of wood, and remove the right-hand engine mounting bracket (see Chapter 2A). Unbolt the housing from the cylinder block and recover the O-ring. The housing is located on the cylinder block by two dowels – make sure these are not displaced.

Refitting

7 Ensure that all mating faces are clean. Where applicable, refit the coolant pump main housing to the cylinder block, together with a new O-ring, and tighten the bolts to the specified torque. Make sure that the location dowels are in place.

8 Where applicable, refit the rear housing to the coolant pump housing, using a new O-ring, and tighten the bolts. Reconnect the coolant hoses.

9 Refit the coolant pump to the pump housing, using a new O-ring, and tighten the bolts to the specified torque.

10 Refit the timing belt as described in Chapter 2A.

11 Refill the cooling system as described in Chapter 1A.

Diesel engines

Removal

12 The coolant pump is located in the cylinder block and is driven by the timing belt.

13 Disconnect the battery negative lead.

14 Remove the timing belt as described in Chapter 2B.

15 Drain the cooling system as described in Chapter 1B.

16 To provide additional working room, loosen the clips and remove the bottom hose.

17 Unscrew the bolts and withdraw the pump from the cylinder block **(see illustrations)**. Remove the gasket.

Refitting

18 Clean the mating faces of the coolant pump and block.
19 Fit the coolant pump together with a new gasket. Insert the bolts and tighten them evenly to the specified torque.
20 Reconnect the bottom hose if removed.
21 Refit the timing belt as described in Chapter 2B.
22 Reconnect the battery negative lead.
23 Refill the cooling system as described in Chapter 1B.

9 Heating/ventilation system - general information

1 The heating/ventilation system consists of a two-speed blower motor (housed behind the facia), face level vents in the centre and at each end of the facia, and air ducts to the front footwells.
2 The control unit is located in the facia, and the controls operate flap valves to deflect and mix the air flowing through the various parts of the heating/ventilation system. The flap valves are contained in the air distribution housing, which acts as a central distribution unit, passing air to the various ducts and vents.

3 Cold air enters the system through the grille at the rear of the engine compartment. If required, the airflow is boosted by the blower, and then flows through the various ducts, according to the settings of the controls. If warm air is required, the cold air is passed over the heater matrix, which is heated by the engine coolant.

10 Heater control panel - removal and refitting

Removal

1 Disconnect the battery negative lead.
2 Remove the steering column lower shroud (see Chapter 11).
3 Remove the radio (see Chapter 12). Alternatively, leave the radio in place for the time being, so that the radio and control panel come out as an assembly.
4 Undo two screws at the top of the control panel, one on the left and one on the right **(see illustration)**.
5 Undo one screw at the left of the control panel **(see illustration)**. The corresponding screw at the right has already been removed with the lower shroud.

10.4 Undo the upper screws on the heater control panel

6 Undo the two lower screws, securing the control panel to the left and right parcel shelf lugs **(see illustration)**.
7 Carefully withdraw the control panel **(see illustrations)**. Disconnect the bulbholders and wiring, including the radio connectors and aerial if still in place. Note the location of all the wiring for refitting.
8 Separate the radio from the control panel, if still fitted.

Refitting

9 Refitting is the reverse of removal. Ensure that all connections are secure, and test all components on completion.

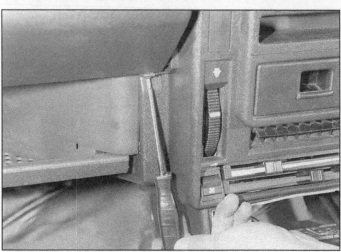

10.5 Undo one screw on the left

10.6 Undo the lower screws from the parcel shelf lugs

10.7a Carefully withdraw the control panel

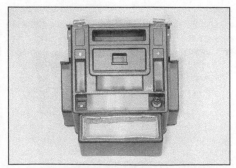

10.7b Front view of heater control panel . . .

10.7c . . . and rear view

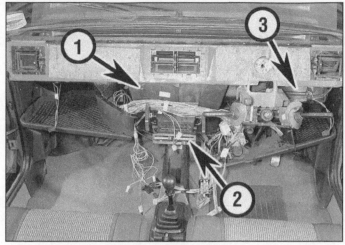

11.1 Heating and ventilation system components (facia removed)

11.2 Blower motor (facia removed)

1 Heating and ventilation system housing
2 Heating and ventilation cable controls
3 Blower motor

11 Heating/ventilation system components - removal and refitting

Access to components

1 Access to the heating and ventilation system components is available by removing the appropriate facia panels. In some cases it will be necessary to remove the complete facia. First remove the heater control panel (see Section 10), then continue to remove the facia panel assembly (see Chapter 11). This will give complete access to the components **(see illustration)**.

Heater blower motor

Removal

2 The blower motor is located under the facia, just forward of the right-hand air vent **(see illustration)**.
3 Disconnect the battery negative lead.
4 On right-hand drive models, remove the steering column lower shroud (see Chapter 11).
5 On left-hand drive models, remove the right-hand parcel shelf, if required for easier access (see Chapter 11).
6 If access to the motor still proves difficult, remove the heater control panel (see Section 10), then remove the complete facia panel assembly (see Chapter 11).

7 Undo the motor retaining nuts and recover any washers and spacers. Carefully pull the motor downwards and detach it from the inlet aperture, then detach it from the right-hand side of the heater assembly. Recover the seals from the inlet aperture or heater if they come off.
8 Detach the wiring and withdraw the motor from the vehicle.
9 To access the motor components, carefully undo the clips around the edge of the housing and separate the two halves.

Refitting

10 Refitting is the reverse of removal.

Heater assembly

Removal

11 Disconnect the battery negative lead.
12 Remove the heater control panel (see Section 10), then continue to remove the complete facia panel assembly (see Chapter 11).
13 Undo the blower motor retaining nuts and recover any washers and spacers. Carefully detach the ducting from the right-hand side of the heater assembly and recover the heater seal if it comes off. It is not necessary to remove the blower from the vehicle.
14 Drain the cooling system (see Chapter 1A or 1B).
15 The heater feed and return hoses pass through the bulkhead **(see illustration)** and connect to the left side of the heater

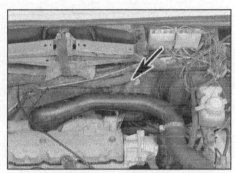

11.15 Heater hoses (arrowed) passing through bulkhead

assembly. Disconnect the hoses from their fittings on the heater.
16 Identify and remove the bolts attaching the heater assembly to the bulkhead and its mountings within the vehicle. Detach the assembly from the ventilation ducts and withdraw it from the vehicle, tilting it upwards slightly at the left to avoid spillage of coolant.
17 To access the heater matrix, undo the screws and clips and separate the left and right halves of the assembly.

Refitting

18 Refitting is the reverse of removal. Ensure that all connections are securely made, and refill the cooling system as described in Chapter 1A or 1B.

Chapter 4 Part A:
Fuel and exhaust systems - carburettor petrol engines

Contents

Degrees of difficulty

Easy, suitable for novice with little experience	**Fairly easy,** suitable for beginner with some experience	**Fairly difficult,** suitable for competent DIY mechanic	**Difficult,** suitable for experienced DIY mechanic	**Very difficult,** suitable for expert DIY or professional

Specifications

Fuel pump
Type . . . .   Mechanical, driven by eccentric on camshaft

Carburettor
Type:
  954 cc models . . . .   Solex 32 IBSN
  1124 cc models . . . .   Solex 32 PBISA
Designation:
  954 cc models . . . .   Solex 32 IBSN 16 Ref. 412
  1124 cc models . . . .   Solex 32 PBISA 16 Ref. 411
Choke type . . . .   Manual, cable-controlled

Solex 32 IBSN carburettor data - 954 cc (C1A) models
Venturi diameter . . . .   25 mm
Air correction jet . . . .   155 ± 10
Emulsion tube   31
Main jet . . . .   127.5 ± 2.5
Idling jet . . . .   47 ± 4
Accelerator pump jet . . . .   40 ± 5
Needle valve 1.6
Float weight . . . .   5.7 g
Float level . . . .   20 mm
Enrichment jet   50
Accelerator pump leakage   20
Throttle valve fast idle setting (throttle valve opening) . . . .   0.9 mm
Choke pull-down setting (choke valve opening) . . . .   4.5 mm (mechanical)
Idle speed . . . .   750 ± 50 rpm
Idle mixture CO content . . . .   0.8 to 1.2 %

Solex 32 PBISA carburettor data - 1124 cc (H1A) models

Venturi diameter	25 mm
Air correction jet	175 ± 10
Emulsion tube	EM
Main jet	127.5 ± 2.5
Idling jet	46 ± 4
Accelerator pump jet	40 ± 5
Needle valve	1.6
Float weight	5.7 g
Enrichment jet	45
Accelerator pump leakage	35
Float height setting	38 ± 1 mm
Throttle valve fast idle setting (throttle valve opening)	0.8 mm
Choke pull-down setting (choke valve opening)	3.0 mm (with 350 mb vacuum)
Idle speed	750 ± 50 rpm
Idle mixture CO content	0.8 to 1.2 %

Recommended fuel

Minimum octane rating	95 RON unleaded (Premium unleaded)

Note: *Leaded petrol may be used on carburettor models. For the correct grade, see the label on the filler neck, or consult your Citroën dealer.*

Torque wrench settings

	Nm	lbf ft
Fuel pump retaining bolts	15	11
Inlet manifold retaining nuts	8	6
Exhaust manifold retaining nuts	15	11
Exhaust system fasteners:		
Front downpipe-to-manifold nut	35	26
Clamping ring nut/bolts	15	11

1 General information and precautions

The carburettor petrol version of the Citroën C15 Van was launched in July 1988 and was later superseded by the monopoint petrol injection and diesel models described in Chapters 4B and 4C. Petrol models are relatively scarce in the UK, compared with the more popular diesel models. The 954 cc carburettor model was only sold in France, and there is little information available about the carburettor, although the other components of the fuel system are considered to be the same as the larger 1124 cc carburettor model.

The fuel system consists of a fuel tank mounted under the rear of the car, a mechanical fuel pump, and a carburettor. The fuel pump is operated by an eccentric on the camshaft, and is mounted on the rear of the cylinder head.

The air cleaner contains a disposable paper filter element. The air inlet ducting incorporates a flap valve to control the inlet air temperature, by mixing warm air from the exhaust manifold with cold air from the front of the vehicle in the correct proportions. The valve is either operated manually, for summer and winter conditions, or else automatically from a temperature switch, mounted on the air filter outlet ducting, which supplies vacuum from the inlet manifold to the valve.

The fuel pump lifts fuel from the fuel tank via a filter and sends it to the carburettor.

The 954 cc model is fitted with a Solex 32 IBSN single-choke carburettor. The 1124 cc model is fitted with a Solex 32 PBISA single-choke carburettor. On both types of carburettor, mixture enrichment for cold starting is by a cable-operated choke control. There is a choke pull-down which opens the flap slightly once the engine has started, to avoid flooding. On the IBSN carburettor, the choke pull-down is operated by the flow of inlet air, and on the PBISA carburettor it is operated by a vacuum diaphragm.

Refer to Section 18 for information on the exhaust system.

⚠ *Warning: Many of the procedures in this Chapter require the removal of fuel lines and connections, which may result in some fuel spillage. Before carrying out any operation on the fuel system, refer to the precautions given in "Safety first!" at the beginning of this manual, and follow them implicitly. Petrol is a highly-dangerous and volatile liquid, and the precautions necessary when handling it cannot be overstressed.*

2 Air cleaner assembly - removal and refitting

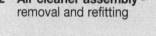

Removal

1 Slacken the retaining clips (where fitted) and disconnect the vacuum and breather hoses from the front of the air cleaner housing-to-carburettor duct **(see illustration)**. Where crimped-type hose clips or ties are fitted, cut and discard them; replace them with standard worm-drive hose clips or new cable ties on refitting.

2 Slacken the retaining clips, then lift the duct off the top of the carburettor and air cleaner housing **(see illustration)**. Where fitted, disconnect the air temperature control valve hose from the end of the duct, then remove the duct from the engine compartment. Recover the rubber sealing ring(s) from the top of the carburettor and/or air cleaner housing (as applicable).

2.1 Disconnect the vacuum and breather hoses (arrowed) from the front of the duct

2.2 Slacken the retaining clips (arrowed) and remove the duct

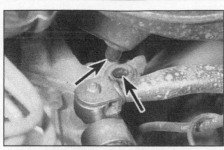

2.5 On refitting, ensure that the air cleaner housing peg is correctly located in its mounting rubber (arrowed)

3.2a Manual air temperature control set for summer conditions (above 20° C)

3.2b Manual air temperature control set for winter conditions (below 5° C)

3 Disconnect the intake duct from the front of the air cleaner housing, and remove the air cleaner housing from the engine compartment.
4 To remove the air intake duct assembly, release all clips and fasteners and disconnect the hoses, including the hot-air intake hose from the exhaust manifold shroud.

Refitting

5 Refitting is the reverse of removal, noting the following points:
  a) Examine the rubber sealing ring(s) for signs of damage or deterioration, and renew if necessary, including the O-ring which may be fitted to the carburettor seal.
  b) Ensure that the air cleaner housing locating peg is correctly engaged with its mounting on the top of the transmission (see Illustration).
  c) Before tightening the air cleaner-to-carburettor duct retaining clips, ensure that the duct is correctly seated on both the air cleaner housing and carburettor flanges.

3 Air temperature control - general information and component renewal

General information

1 The air inlet temperature is maintained by a flap control valve in the air inlet ducting, upstream of the air cleaner. The control valve mixes cold air from the front of the engine compartment with warm air from the exhaust shroud, in the correct proportions, to supply air to the carburettor at the correct

temperature. Operation of the control valve may be manual or automatic.

Manual air temperature control

2 On models with manual air temperature control, a lever on the control valve needs to be set to the correct position, according to ambient temperature, for summer conditions (above 20° C), spring and autumn conditions (5° C to 20° C) and winter conditions (below 5° C) **(see illustrations)**.

Automatic air temperature control

3 On models with automatic air temperature control, the control valve is operated by a vacuum diaphragm instead of a lever. A heat-sensitive bi-metal vacuum switch is mounted in the end of the air cleaner housing-to-carburettor duct. When the engine is started from cold, the switch is open, and allows vacuum from the inlet manifold to act on the diaphragm, operating the control valve flap so that the inlet air is from the exhaust manifold shroud.
4 As the temperature of the exhaust-warmed air in the air cleaner-to-carburettor duct rises, the bi-metal strip in the vacuum switch deforms and closes the vacuum supply to the diaphragm, so that the flap gradually closes against the hot-air intake and opens towards the cold air inlet from the front of the vehicle.
5 To check the system, run the engine and observe the position of the flap. When the engine is cold, the flap should close off the cold-air intake. As the engine warms up, the flap should gradually move so that it opens the cold-air intake and closes the hot-air intake.
6 To check the vacuum switch, disconnect the vacuum pipe from the control valve

diaphragm. Run the engine and place a finger over the pipe end. When the engine is cold, full inlet manifold vacuum should be present in the pipe, and when the engine is at normal operating temperature there should be no vacuum in the pipe.
7 To check the control valve, disconnect the vacuum pipe and observe that the cold-air intake is fully open. Then suck hard at the control valve stub and observe that the flap moves so that it opens the hot-air intake and closes the cold-air intake.
8 If either the vacuum switch or control valve is faulty, it must be renewed.

Air temperature control valve - renewal

9 On models with automatic air temperature control, disconnect the vacuum pipe from the control valve diaphragm.
10 Slacken the retaining clips and disconnect the ducting from the control valve. Undo any securing clips and remove the control valve from the vehicle.
11 Refitting is the reverse of removal. The control valve can only be renewed as a complete unit.

Automatic air temperature control vacuum switch - renewal

12 Remove the air cleaner housing-to-carburettor duct as described in Section 2.
13 Bend up the tags on the switch retaining clip, then remove the clip, along with its seal, and withdraw the switch from inside the duct **(see illustrations)**. Examine the seal for signs of damage or deterioration, and renew if necessary.

3.13a Remove the retaining clip . . .

3.13b . . . and seal . . .

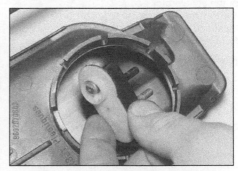

3.13c . . . then withdraw the vacuum switch from inside the duct

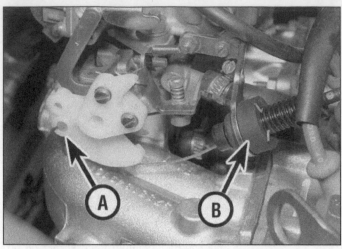

4.1 Accelerator cable removal from carburettor

4.6 Adjusting the accelerator cable

A Inner cable nipple   *B Rubber grommet*

14 When refitting, ensure that the switch and duct mating surfaces are clean and dry, and position the switch inside of the duct.

15 Fit the seal over the switch unions, and refit the retaining clip. Ensure that the switch is pressed firmly against the duct, and secure it in position by bending down the retaining clip tags.

16 Refit the duct as described in Section 2.

4 Accelerator cable - removal, refitting and adjustment

Removal

1 Free the accelerator inner cable from the carburettor throttle cam, then pull the outer cable out from its mounting bracket rubber grommet **(see illustration)**. Slide the flat washer off the end of the cable, and remove the spring clip.

2 Working back along the length of the cable, free it from any relevant retaining clips or ties, noting its correct routing.

3 Working inside the vehicle, remove the parcel shelf from below the steering column (see Chapter 11), then depress the retaining clips, and detach the inner cable from the top of the accelerator pedal.

4 Pull the spring shock absorber from the bulkhead and withdraw the accelerator cable from inside the engine compartment.

Refitting

5 Refitting is the reverse of removal, but adjust the cable as follows.

Adjustment

6 Remove the spring clip from the accelerator outer cable then, ensuring that the throttle cam is fully against its stop, gently pull the cable out of its grommet until all free play is removed from the inner cable **(see illustration)**.

7 With the cable held in this position, ensure that the flat washer is pressed securely against the grommet. Fit the spring clip to the last exposed outer cable groove in front of the rubber grommet and washer, so that when the outer cable is released, there is only a small amount of free play in the inner cable.

8 Have an assistant depress the accelerator pedal, and check that the throttle cam opens fully and returns smoothly to its stop.

5 Accelerator pedal - removal and refitting

Note: *On right-hand drive models with a brake linkage cross-tube, the accelerator pedal is integral with the brake pedal assembly (see Chapter 9). This section applies only to models with a separate accelerator pedal.*

Removal

1 Remove the parcel shelf from below the steering column (see Chapter 11).

2 Depress the retaining clips, and detach the inner cable from the top of the accelerator pedal.

3 Slacken and remove the two nuts securing the pedal mounting bracket to the bulkhead.

6.1 Release the choke inner cable from its linkage (1) then undo the clamp bolt (2) and release the outer cable

Slide off the outer part of the mounting clamp, then withdraw the pedal, and slide off the inner part of the clamp.

4 Examine the mounting bracket and pedal pivot points for signs of wear, and renew if necessary.

Refitting

5 Refitting is the reverse of removal. Apply a little multi-purpose grease to the pedal pivot point. On completion, adjust the accelerator cable as described in Section 4.

6 Choke cable - removal, refitting and adjustment

Removal

1 Free the choke inner cable from the carburettor linkage, then undo the retaining bolt and release the outer cable ferrule from its clamp **(see illustration)**.

2 Working back along the length of the cable, free it from any relevant retaining clips or ties, noting its correct routing. Tie a suitable length of string to the end of the choke inner cable.

3 Working from inside the vehicle, undo the screws and pull down the steering column lower shroud (see Chapter 11), then depress the two lugs and release the choke cable button from the shroud. Pull out the cable and untie the string, but leave the string in place for refitting.

Refitting

4 Refitting is the reverse of removal, using the string to draw the cable through the bulkhead. Adjust the cable as follows.

Adjustment

5 With the choke button fully pushed in and the carburettor linkage fully against its stop, fit the outer cable ferrule to its clamp so that there is only a small amount of free play in the inner cable, and tighten the clamp bolt.

6 Have an assistant operate the choke button, and check that the choke linkage closes fully and returns smoothly to its stop. If necessary, repeat the adjustment procedure.

7 Fuel pump - testing, removal and refitting

Note: *Refer to the warning note at the end of Section 1 before proceeding.*

Testing

1 To test the fuel pump without removing it, first disable the ignition system, then disconnect the outlet pipe which leads to the carburettor, and hold a wad of rag over the pump outlet while an assistant turns the engine on the starter motor. Regular spurts of fuel should be ejected as the engine turns.
Caution: Keep your hands away from the electric cooling fan.
2 The pump can also be tested by removing it. With the pump outlet pipe disconnected but the inlet pipe still connected, hold the wad of rag by the outlet and operate the pump lever by hand. If the pump is in a satisfactory condition, the lever should move and return smoothly, and a strong jet of fuel should be ejected.

Removal

3 Identify the pump inlet and outlet hoses, and slacken both retaining clips **(see illustration)**. Where crimped-type hose clips are fitted, cut the clips and discard them; replace them with standard worm-drive hose clips on refitting. Place wads of rag beneath the hose unions to catch any spilled fuel. Disconnect both hoses from the pump, and plug the hose ends to minimise fuel loss.
4 Slacken and remove the bolts securing the pump to the rear of the cylinder head. Remove the pump **(see illustration)**, and recover its gasket. Discard the gasket - a new one must be used on refitting.

Refitting

5 Ensure that the pump and cylinder head mating surfaces are clean and dry, then offer

7.4 Fuel pump removed from engine

7.3 Arrows on fuel pump unions indicate the direction of fuel flow

up the new gasket and refit the pump to the cylinder head. Tighten the pump retaining bolts to the specified torque.
6 Reconnect the inlet and outlet hoses to the relevant pump unions, and securely tighten their retaining clips.

8 Fuel gauge sender unit - removal and refitting

Note: *Refer to the warning note at the end of Section 1 before proceeding.*

Removal

1 Remove the fuel tank (see Section 11). There are two apertures at the top of the tank. The left-hand aperture (facing towards the front of the vehicle) is the fuel gauge sender unit.
2 Make a note of any alignment marks on the tank, sender unit and locking ring. If necessary, make your own alignment marks for correct refitting.
3 Unscrew the locking ring and remove it from the tank **(see illustration)**.

> **HAYNES HINT** *Use a screwdriver on the raised ribs of the locking ring. Carefully tap the screwdriver to turn the ring anti-clockwise until it can be unscrewed by hand.*

4 Carefully lift the sender unit from the top of the fuel tank, taking care not to bend the sender unit float arm or to spill fuel. Recover the rubber sealing ring and discard it - a new one must be used on refitting.

Refitting

5 Fit the new sealing ring to the top of the fuel tank.
6 Carefully manoeuvre the assembly into the fuel tank, and align the mark on the sender unit to its correct position.
7 Refit the locking ring, and tighten it until its mark is correctly aligned with the mark on the fuel tank.
8 Refit the fuel tank (see Section 11).

9 Fuel pick-up unit - removal and refitting

Note: *Refer to the warning note at the end of Section 1 before proceeding.*
1 Removing and refitting of the fuel pick-up unit, with the fuel tank removed, is the same as removing and refitting the fuel gauge sender unit, noting the following points.
 a) *The right-hand aperture at the top of the tank (facing towards the front of the vehicle) is the fuel pick-up unit.*
 b) *Take care not to damage the fuel filter when removing the pick-up unit from the tank. When refitting, make sure it is clean and free of debris.*
 c) *The fuel pick-up is only available as a complete assembly. No individual components are available separately.*

10 Fuel tank filler hose - removal and refitting

Note: *Refer to the warning note at the end of Section 1 before proceeding.*

Removal

1 The fuel filler hose consists of two sections, an upper section which is accessed from behind a trim panel within the load compartment, and a lower section that goes through the floor and connects to the tank. If the lower section is to be removed, the operation should be carried out with the fuel tank almost empty. There are also some vent hoses coming up through the floor, and a drain hose which prevents water from leaking in through the fuel filler orifice.
2 To access the upper section of the filler hose, pull out the studs and detach the trim panel from the bodywork. A vent chamber is riveted to the rear of the panel **(see illustrations)**.
3 Pull out the plastic tubes connecting the vent chamber to the pipework, then remove the trim panel completely.

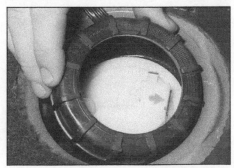

8.3 Unscrew the locking ring from the fuel gauge sender unit

10.2a Undo the clips and pull back the trim panel . . .

10.2b . . . to access the fuel filler hose. The vent chamber (arrowed) is riveted to the panel

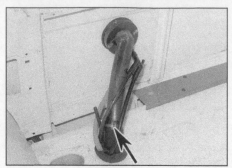

10.4 Slacken the clip (arrowed) connecting the upper and lower filler hoses

4 Slacken the clip connecting the upper and lower filler hose sections **(see illustration)**. Also slacken the upper clips at the filler orifice and disconnect the vent and drain pipes.

5 Working outside the vehicle, remove the filler cap and seal, then undo the screws connecting the filler orifice to the bodywork. The screws engage with metal tabs behind the orifice. Return to the load compartment and withdraw the filler hose and orifice. Recover the metal tabs if they have fallen off.

6 The lower section of the filler hose is connected to the side of the fuel tank. To remove it, syphon or hand-pump the remaining fuel from the tank, then chock the front wheels, jack up the rear of the vehicle and support it on axle stands (see *"Jacking and vehicle support"*). Undo the clip and disconnect the filler hose, then pull it out through the floor, from above or from underneath. If necessary, remove a section of the sealing ring from the floor. The sealing ring is made up of two sections. One section will remain in place because the vent hoses pass through it.

7 To remove the vent hoses, it may be necessary to lower the tank or remove it altogether as described in Section 11.

Refitting

8 Refitting is the reverse of removal, noting the following points:
 a) *Ensure that all pipes and hoses are correctly routed, and securely held in position with their retaining clips.*

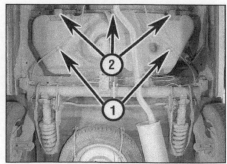

11.3 Fuel tank

1 Handbrake retaining clips
2 Fuel tank retaining nuts

 b) *On completion, refill the tank with fuel, and check for signs of leakage before taking the vehicle out on the road.*

11 Fuel tank - removal and refitting

Note: *Refer to the warning note at the end of Section 1 before proceeding.*

Removal

1 Before removing the fuel tank, all the fuel must be drained from the tank. Since a fuel tank drain plug is not provided, it is therefore preferable to carry out the removal operation when the tank is nearly empty. Before proceeding, disconnect the battery negative lead, and syphon or hand-pump the remaining fuel from the tank.

2 Remove the exhaust system and relevant heat shield(s) as described in Section 18.

3 Free both handbrake cables from their retaining clips on the base of the fuel tank **(see illustration)**.

4 Release the retaining clips, then disconnect the filler neck hose and vent pipe from the left side of the fuel tank. Also disconnect the vent pipe from the right side of the tank, above the front mounting, and any other relevant vent pipes or breather hoses. Some hoses are joined to the tank with quick-release fittings. To disconnect these fittings, slide the cover along the hose, then depress the centre ring and pull the hose out of its fitting. If any of the pipework is difficult to access, wait until the tank has been lowered slightly, then disconnect the pipe.

5 Place a trolley jack with an interposed block of wood beneath the tank, then raise the jack until it is supporting the weight of the tank.

6 Undo the three self-locking retaining nuts and recover the shims and rubbers, then lower the fuel tank slightly, without straining the upper fittings. Recover any other mounting rubbers from around the tank, noting their positions. Lower the tank sufficiently to access the upper fittings. The left-hand aperture (facing the front of the vehicle) is the fuel gauge sender unit. The right-hand aperture is the fuel pick-up unit.

7 Disconnect the wiring connector from the fuel gauge sender unit.

8 Slacken the retaining clip and disconnect the feed hose from the fuel pick-up unit, and plug the ends. If a crimped hose clip is fitted, cut the clip and discard it, and replace it with a worm-drive clip on refitting. On later models, the fuel hose may have a quick-release fitting which is released by depressing a metal collar with a small, flat-bladed screwdriver.

9 Disconnect any other relevant vent pipes as they become accessible, and remove the tank from underneath the vehicle.

10 If the tank is contaminated with sediment or water, remove the sender unit (Section 8) and swill the tank out with clean fuel. The tank is injection-moulded from a synthetic material and if damaged, it should be renewed. However, in certain cases, it may be possible to have small leaks or minor damage repaired. Seek the advice of a suitable specialist before attempting to repair the fuel tank.

Refitting

11 Refitting is the reverse of removal, noting the following points:
 a) *When lifting the tank back into position, make sure that the mounting rubbers are correctly positioned, and take care to ensure that none of the hoses become trapped between the tank and vehicle body.*
 b) *Use new self-locking nuts to refit the fuel tank.*
 c) *Ensure that all pipes and hoses are correctly routed, and securely held in position with their retaining clips. If crimped clips have been cut, use new worm-drive hose clips.*
 d) *On completion, refill the tank with fuel, and check for signs of leakage before taking the vehicle out on the road.*

12 Unleaded petrol - general information and usage

Note: *The information given in this Chapter is correct at the time of writing, and applies only to petrols currently available in the UK. If updated information is thought to be required, check with a Citroën dealer. If travelling abroad, consult one of the motoring organisations (or a similar authority) for advice on the petrols available and their suitability for your vehicle.*

1 The fuel recommended by Citroën is given in the Specifications at the beginning of this Chapter.

2 RON and MON are different testing standards; RON stands for Research Octane Number (also written as RM), whilst MON stands for Motor Octane Number (also written as MM).

13 Carburettor -
general information

Solex 32 PBISA carburettor - 1124 cc models

The Solex 32 PBISA carburettor is a downdraught single-venturi instrument, with a manually-controlled choke. The carburettor consists of three main components. These are the upper body, the main body and the throttle body (which contains the throttle valve assembly). An insulating block placed between the carburettor body and throttle body prevents excess heat transfer from the manifold to the main body.

The throttle body contains a drilling through which the engine coolant runs. The engine coolant warms the carburettor body quickly on cold starts, improving atomisation of the fuel/air mixture, and preventing carburettor icing during warm-up.

During slow running and at idle, fuel sourced from the float chamber passes into the idle channel through a metered idle jet. Here it is mixed with a small amount of air from a calibrated air bleed. The resulting mixture is drawn through a channel, to be discharged from the idle orifice under the throttle valve. A tapered mixture screw is used to vary the outlet, and this ensures fine control of the idle mixture.

A progression slot provides extra enrichment as it is uncovered by the opening of the throttle valve during initial acceleration.

Under normal operating conditions, fuel is drawn through a calibrated main jet, into the base of the auxiliary venturi. An emulsion tube is placed in the auxiliary venturi, capped with an air correction jet. The fuel is mixed with air, drawn in through the holes in the emulsion tube. The resulting mixture is discharged into the main airstream via four orifices, spaced at 90° apart, in the upper part of the auxiliary venturi.

The carburettor is also equipped with an accelerator pump, to provide an initial spurt of extra fuel during sudden acceleration. The accelerator pump is controlled by a diaphragm, and is mechanically operated by a lever and rod which is connected to the throttle linkage.

The idle speed and mixture are set by adjustable screws. The idle mixture adjustment requires the use of an exhaust gas analyser, and the mixture screw is sealed during production with a tamperproof plug, to prevent unnecessary adjustment.

The cold start enrichment device consists of a manually operated strangler-type choke.

When the choke control is pulled out, the choke flap is closed by a spring. When the engine starts, vacuum from the carburettor operates a diaphragm which overcomes the spring tension and opens the flap slightly (choke pull-down), causing the mixture to weaken and avoid flooding.

Solex 32 IBSN carburettor - 954 cc models

Apart from the Specifications given at the beginning of this Chapter, little information is available about the Solex 32 IBSN carburettor. It is thought to be similar to the PBISA carburettor fitted to the 1124 cc models, except that the choke pull-down is operated by the inlet air and there is no vacuum diaphragm. When the engine starts, the flow of air partially opens the choke flap against a spring.

14 Carburettor -
removal and refitting

Note: *Refer to the warning note at the end of Section 1 before proceeding.*

Removal

1 Disconnect the battery negative terminal.

2 Remove the air cleaner-to-carburettor duct as described in Section 2.

3 Slacken the retaining clips and disconnect the coolant hoses from the base of the carburettor (see illustration). Plug the hose ends to minimise coolant loss, and mop up any spilt coolant immediately.

4 Slacken the retaining clip and disconnect the fuel feed hose from the carburettor. Place wads of rag around the union to catch any spilled fuel. Plug the hose as soon as it is disconnected, to minimise fuel loss and prevent the entry of dirt into the system.

5 Free the accelerator inner cable from the throttle cam, then pull the outer cable out from its mounting bracket rubber grommet, along with its flat washer and spring clip.

6 Disconnect the choke inner cable from the carburettor linkage, then undo the retaining bolt and release the outer cable ferrule from its clamp. Position the cable clear of the carburettor.

7 Make a note of the correct fitted positions

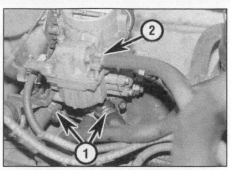

14.3 Coolant hoses (1) and fuel feed hose (2) - Solex 32 PBISA carburettor

of all the relevant vacuum pipes and breather hoses, to ensure that they are correctly positioned on refitting, then release the retaining clips (where fitted) and disconnect them from the carburettor.

8 Unscrew the nuts and washers securing the carburettor to the inlet manifold, and remove the carburettor assembly from the van **(see illustrations)**. Remove the insulating spacer and/or gasket(s), then discard the gasket(s); new ones must be used on refitting. Plug the inlet manifold aperture with a wad of clean cloth, to prevent the possible entry of foreign matter.

Refitting

9 Refitting is the reverse of removal, noting the following points:

a) *Ensure that the carburettor and inlet manifold sealing faces are clean and flat. Fit the insulating spacer and new gasket(s), as applicable, and securely tighten the carburettor retaining nuts.*

b) *Use the notes made on dismantling to ensure that all hoses are refitted to their original positions and, where necessary, are securely held by their retaining clips.*

c) *Where the original crimped-type hose clips were fitted, discard them and replace them with worm-drive hose clips.*

d) *Refit and adjust the accelerator and choke cables as described in Sections 4 and 6.*

e) *Refit the air cleaner duct as described in Section 2.*

f) *On completion check and, if necessary, adjust the idle speed and mixture settings as described in Chapter 1A.*

14.8a Carburettor retaining nut (arrowed) on left side . . .

14.8b . . . and on right side (arrowed) - Solex 32 PBISA carburettor

15 Solex 32 PBISA carburettor (1124 cc engine) - diagnosis, overhaul and adjustments

Note: *Apart from the Specifications given at the beginning of this Chapter, little information is available about the overhaul and adjustment of the Solex 32 IBSN carburettor fitted to 954 cc engines. If the carburettor is thought to be faulty, take it to your Citroën dealer.*

Diagnosis

1 If a carburettor fault is suspected, always check first that the ignition timing is accurate and the spark plugs are in good condition and correctly gapped, that the accelerator and choke cables are correctly adjusted, and that the air cleaner filter element is clean. See the relevant Sections of Chapter 1A or of this Chapter. If the engine is running very roughly, check the valve clearances and the compression pressures as described in Chapter 2A.

2 If careful checking of all the above produces no improvement, the carburettor must be removed for cleaning and overhaul.

3 Note that in the rare event of a complete carburettor overhaul being necessary, it may be more economical to renew the carburettor as a complete unit. Check the price and availability of a replacement carburettor and its component parts before starting work. Note that most sealing washers, screws and gaskets are available in kits, as are some of the major sub-assemblies. In most cases, it will be sufficient to dismantle the carburettor and to clean the jets and passages.

Overhaul

Note: *Refer to the warning note at the end of Section 1 before proceeding.*

4 Remove the carburettor from the vehicle as described in Section 14.

5 Disconnect the vacuum hose from the choke pull-down diaphragm.

6 Disconnect the choke spring, then undo the six screws and lift off the carburettor upper body.

7 Tap out the float pivot pin and remove the float assembly, needle valve and float chamber gasket. Check that the needle valve anti-vibration ball is free in the valve end, then examine the needle valve tip and seat for wear or damage. Examine the float assembly and pivot pin for signs of wear and damage. The float assembly must be renewed if it appears to be leaking.

8 Unscrew the fuel inlet union, and inspect the fuel filter. Clean the filter housing of debris and dirt, and renew the filter if it is blocked.

9 Undo the screws, then detach the accelerator pump cover and remove the pump diaphragm and spring, noting which way around they are fitted. Examine the diaphragm for signs of damage and deterioration, and renew if necessary.

10 Unscrew the idle jet from the main body.

11 Unscrew the main jet from the float chamber. Note that it may be necessary to remove a plug in the float chamber body to expose an opening through which the main jet can be withdrawn.

12 Remove the combined air correction jet and emulsion tube from the auxiliary venturi.

13 Remove the two screws, then separate the carburettor main body and throttle body assemblies, and recover the insulating spacer. Examine the throttle valve spindle and throttle bore for signs of wear or damage and, if necessary, renew the throttle body assembly.

14 Remove the idle mixture adjustment screw tamperproof cap. Turn the screw in until it seats lightly, counting the exact number of turns required, then unscrew and remove it. When refitting, turn the screw in until it seats lightly, then back the screw off by the number of turns noted on removal, to return the screw to its original location.

15 Clean the jets, carburettor body assemblies, float chamber and internal drillings. An air line may be used to clear the internal passages once the carburettor is fully dismantled.

 Aerosol cans of carburettor cleaner are widely available, and can be very useful in helping to clear internal passages of stubborn obstructions.

⚠ **Warning: If high pressure air is directed into drillings and passages were a diaphragm is fitted, the diaphragm is likely to be damaged.**

16 Use a straight edge to check all carburettor body assembly mating surfaces for distortion.

17 On reassembly, renew any worn components, and fit a complete set of new gaskets and seals. A jet kit and a gasket and seal kit are available from your Citroën dealer.

18 Reassembly is the reverse of dismantling. Ensure that all jets are securely locked in position, but take care not to overtighten them. Ensure that all mating surfaces are clean and dry, and that all body sections are correctly assembled with their fuel and air passages correctly aligned. Before refitting the carburettor to the vehicle, set the throttle valve fast idle and choke pull-down settings as described below.

Adjustments

Idle speed and mixture

19 Refer to Chapter 1A.

Float height setting

20 To accurately check the float height setting, a Solex PBISA float height checking gauge is required. Therefore, this task should be entrusted to a Citroën dealer. As a guide, with the carburettor body inverted, so that the float is at the top and the needle valve is depressed, the distance between the upper edge of the float and the sealing face of the upper body (with its gasket fitted) should be approximately according to the Specifications. To adjust the float height setting, carefully bend the pivot arm.

15.22 Throttle valve fast idle adjustment screw (arrowed) - Solex 32 PBISA carburettor

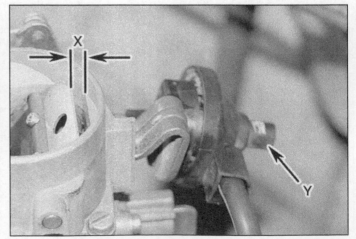

15.25 Choke pull-down setting - Solex 32 PBISA carburettor

Adjust screw Y until clearance X is as given in the Specifications

Note: *The Solex PBISA float height checking gauge is not suitable for use with Solex 32 IBSN carburettors fitted to 954 cc engines. Consult your Citroën dealer for adjustment of the float height.*

Throttle valve fast idle setting

21 Invert the carburettor, and operate the carburettor choke linkage to fully close the choke valve. The fast idle screw will butt against the fast idle cam, and force the throttle valve open slightly.

22 Using the shank of a suitable twist drill, measure the clearance between the edge of the throttle valve and bore, and compare this to the clearance given in the Specifications. If necessary, adjust by turning the fast idle adjustment screw in the appropriate direction until the specified clearance is obtained **(see illustration)**.

Choke pull-down setting

23 Operate the carburettor choke linkage to fully close the choke valve, and hold the linkage in this position.

24 Disconnect the vacuum hose from the choke pull-down diaphragm and attach a hand-held vacuum pump. Apply a vacuum to the diaphragm so that the diaphragm rod is pulled fully into the diaphragm body. In the absence of a vacuum pump, the rod can be pushed into the diaphragm using a small screwdriver.

25 With the rod fully retracted, use the shank of a suitable twist drill to measure the clearance between the edge of the choke valve and bore, and compare this to the clearance given in the Specifications. If necessary, remove the plug from the diaphragm cover, and adjust by turning the adjustment screw **(see illustration)**.

26 When the pull-down setting is correctly adjusted, refit the plug to the diaphragm cover. If a vacuum pump has been used, remove it and re-connect the vacuum hose to the diaphragm valve.

Note: *On the Solex 32 IBSN carburettor fitted to 954 cc engines, the choke pull-down is operated by the flow of inlet air and there is no vacuum diaphragm.*

16.5 Inlet manifold and upper retaining nuts (arrowed)

16 Inlet manifold - removal and refitting

Note: *Refer to the warning note at the end of Section 1 before proceeding.*

Removal

1 Remove the carburettor as described in Section 14.

2 Drain the cooling system as described in Chapter 1A, then slacken the retaining clip and disconnect the coolant hose from the manifold.

3 Slacken the retaining clip, and disconnect the vacuum servo unit hose from the left-hand side of the manifold.

4 Make a final check that all the necessary vacuum/breather hoses have been disconnected from the manifold.

5 Unscrew the six retaining nuts **(see illustration)**, then manoeuvre the manifold away from the head and out of the engine compartment. Note that there is no manifold gasket.

Refitting

6 Refitting is the reverse of removal, noting the following points:

a) *Ensure that the manifold and cylinder head mating surfaces are clean and dry, and apply a thin coating of suitable sealing compound to the manifold mating surface. Install the manifold, and tighten its retaining nuts to the specified torque.*

b) *Ensure that all relevant hoses are reconnected to their original positions, and are securely held (where necessary) by their retaining clips.*

c) *Refit the carburettor as described in Section 14.*

d) *On completion, refill the cooling system as described in Chapter 1A.*

17 Exhaust manifold - removal and refitting

Removal

1 Disconnect the hot-air intake hose from the manifold shroud, and remove it from the vehicle **(see illustration)**.

2 Slacken and remove the retaining bolts, and remove the shroud from the top of the exhaust manifold **(see illustration)**.

3 Firmly apply the handbrake, then jack up the front of the vehicle and support it on axle stands (see "*Jacking and vehicle support*").

4 Working underneath the vehicle, remove the bolt securing the exhaust front pipe to its mounting bracket. Then working above or below the vehicle, undo the nuts and disconnect the exhaust front pipe from the manifold, and recover the gasket.

5 Undo the eight retaining nuts securing the manifold to the cylinder head **(see illustration)**. Manoeuvre the manifold out of the engine compartment, and discard the manifold gaskets.

Refitting

6 Refitting is the reverse of removal, noting the following points:

a) *Examine all the exhaust manifold studs for signs of damage and corrosion. Remove all traces of corrosion, and repair or renew any damaged studs.*

b) *Ensure that the manifold and cylinder head sealing faces are clean and flat, and fit new manifold gaskets. Tighten the manifold retaining nuts to the specified torque.*

c) *Reconnect the front pipe to the manifold using the information given in Section 18.*

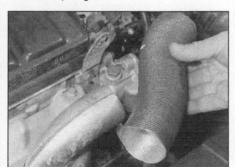

17.1 Remove the hot-air intake hose . . .

17.2 . . . then undo the three retaining bolts (arrowed) and remove the exhaust manifold shroud

17.5 The exhaust manifold is retained by eight nuts (upper four arrowed)

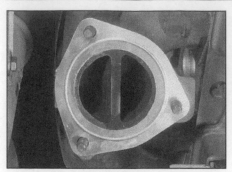

18.7 Exhaust manifold downpipe flange and gasket

18 Exhaust system - general information and component renewal

General information

1 The exhaust system consists of three sections; the front downpipe, the intermediate pipe and the tailpipe with silencer. On models with an extended cab there is an extension that fits onto the tailpipe.

2 The front downpipe is secured to the manifold flange by threaded spindles with nuts at both ends. The intermediate pipe connects to the front downpipe by a spring-loaded ball/ring joint, to allow movement of the exhaust system. The tailpipe connects to the intermediate pipe by a ring clamp consisting of two halves.

3 The system is suspended throughout its entire length by rubber mountings.

Removal

4 Each exhaust section can be removed individually, or the complete system can be removed as a unit.

5 To remove the system or part of the system, first jack up the front or rear of the van, and support it on axle stands (see *"Jacking and vehicle support"*). Alternatively, position the van over an inspection pit, or on car ramps.

⚠️ **Warning: When disconnecting the joints between sections of pipework, make sure the pipes on both sides of the joint are supported so that they do not drop down. Do not allow any disconnected pipework to hang down under its own weight, otherwise it will**

18.10 Ring clamp connecting intermediate exhaust pipe to tailpipe

place stress on the remaining joints. If necessary, tie the disconnected pipes to the bodywork with string.

Front downpipe

6 Undo the bolt securing the front pipe to its mounting bracket.

7 Undo the nuts securing the front pipe flange joint to the manifold, then separate the flange joint and recover the gasket **(see illustration)**.

8 Disconnect the front pipe from the intermediate pipe by removing the bolts, together with their springs and collars. Recover the sealing ring. Manoeuvre the front pipe out from underneath the vehicle.

Intermediate pipe

9 Undo the bolts securing the front pipe flange joint to the intermediate pipe. Recover the tension springs and sealing ring.

10 Slacken the bolts and disengage the ring clamp from the joint connecting the intermediate pipe to the tailpipe **(see illustration)**.

11 Free the intermediate pipe from its mounting rubbers, then withdraw it from underneath the vehicle.

Tailpipe

12 Slacken the bolts and disengage the ring clamp from the joint connecting the intermediate pipe to the tailpipe.

13 Unhook the tailpipe from its mounting rubbers, and remove it from the vehicle.

Tailpipe extension

14 The tailpipe extension is fitted to models with an extended cab, although no specific information is available about which models require it.

15 To remove the tailpipe extension, undo the ring clamp and pull it out.

Complete system

16 Undo the nuts and remove the exhaust manifold-to-downpipe spindles.

17 With the aid of an assistant, free the system from all its mounting rubbers and manoeuvre it out from underneath the vehicle.

Heat shield(s)

18 The heat shields are secured to the underside of the body by a mixture of screws and clips. They can be removed once the relevant exhaust section has been removed.

> **HAYNES HiNT** *If the heat shield is being removed to gain access to a component located behind it, it may be sufficient to simply remove the fasteners and lower the shield, without disturbing the exhaust system.*

Refitting

19 Each section is refitted by a reversal of the removal sequence, noting the following points:

a) *Ensure that all traces of corrosion have been removed from the flanges, and renew all necessary gaskets.*

b) *When fitting the front downpipe to the manifold, tighten the nuts to the specified torque.*

c) *When fitting the front downpipe to the intermediate pipe, fit a new sealing ring and apply a smear of high-temperature grease to the joint mating surfaces. Refit the bolts, with the springs and collars, and tighten the nuts progressively until approximately four threads are visible and the springs are compressed to 22.0 mm in length.*

d) *When fitting the intermediate pipe to the tailpipe, apply a smear of exhaust system jointing paste to the flange joint, to ensure a gastight seal. Tighten the clamping ring nuts evenly and progressively to the specified torque, so that the clearance between the clamp halves remains equal on either side.*

e) *Inspect all rubber mountings for signs of damage or deterioration, and renew as necessary. Ensure that the rubber mountings are correctly located, and that there is adequate clearance between the exhaust system and the vehicle underbody.*

Chapter 4 Part B:
Fuel and exhaust systems - single-point petrol injection engines

Contents

Degrees of difficulty

| Easy, suitable for novice with little experience | | Fairly easy, suitable for beginner with some experience | | Fairly difficult, suitable for competent DIY mechanic | | Difficult, suitable for experienced DIY mechanic | | Very difficult, suitable for expert DIY or professional | |

Specifications

System type
1124 cc (HDZ) models: . . . . Bosch Monopoint A2.2

Fuel system data
Fuel pump type . . . . Electric, immersed in tank
Fuel pump regulated constant pressure (approximate): . . . . 1.0 bar (approx)
Idle speed (not adjustable) . . . . 850 ± 50 rpm (controlled by ECU)
Idle mixture CO content (not adjustable) . . . . Less than 1.0 % (controlled by ECU)

Recommended fuel
Minimum octane rating . . . . 95 RON unleaded (Premium unleaded)
Note: *Do not use leaded fuel.*

Torque wrench settings

	Nm	lbf ft
Inlet manifold retaining nuts	8	6
Exhaust manifold retaining nuts	15	11
Exhaust system fasteners:		
Front downpipe-to-manifold nut	35	26
Clamping ring nuts	15	11

1 General information and precautions

1 Single-point petrol injection models, with the 1124 cc engine and the Bosch Monopoint A2.2 fuel injection system, were available in the UK from June 1994 until January 1996. They replaced the carburettor models which are documented in Chapter 4A, and were marketed alongside the diesel models which are documented in Chapter 4C.

2 The fuel system consists of a fuel tank mounted under the rear of the van with an electric fuel pump immersed in it, a fuel filter, fuel feed and return lines, the throttle body assembly (which incorporates the single fuel injector and the fuel pressure regulator), the Electronic Control Unit (ECU) and the various sensors, electrical components and related wiring.

3 The air cleaner contains a disposable paper filter element. It incorporates a flap valve air temperature control system, which allows cold air from the outside of the van and warm air from the exhaust manifold to enter the air cleaner in the correct proportions.

4 Refer to Section 7 for general information on the fuel injection system and Section 18 for information on the exhaust system.

⚠ **Warning: Many of the procedures in this Chapter require the removal of fuel lines and connections, which may result in some fuel spillage. Before carrying out any operation on the fuel system, refer to the precautions given in "Safety first!" at the beginning of this manual, and follow them implicitly. Petrol is a highly-dangerous and volatile liquid, and the precautions necessary when handling it cannot be overstressed.**

Note: *Residual pressure will remain in the fuel lines long after the vehicle was last used. Before disconnecting any fuel line, depressurise the fuel system as described in Section 8.*

4.3 To adjust the accelerator cable, fit the clip in the third groove

2 Air cleaner assembly - removal and refitting

1 The air cleaner assembly on monopoint injection models is essentially the same as on carburettor models. It connects to a throttle body instead of a carburettor, and the design of the ducting from the air cleaner housing to the throttle is slightly different, but removal and refitting is the same. To remove and refit the air cleaner assembly, follow the instructions for carburettor models given in Chapter 4A.

3 Air temperature control - general information

1 The air inlet temperature is maintained by a flap control valve in the air inlet ducting, upstream of the air cleaner. The control valve mixes cold air from the front of the engine compartment with warm air from the exhaust shroud, in the correct proportions, to supply air to the throttle at the correct temperature.

2 No specific information is available about manual or automatic control of the valve, but it should be self-evident.

3 The procedure for removing and refitting the control valve is the same as for carburettor models, described in Chapter 4A.

4 Accelerator cable - removal, refitting and adjustment

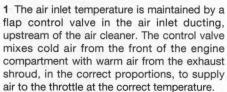

Removal and refitting

1 The procedure for removing and refitting the accelerator cable on monopoint injection models is the same as for carburettor models, described in Chapter 4A. The cable connects to a throttle body instead of a carburettor, but the cable end fittings are the same. Adjust the cable as described below.

Adjustment

2 Remove the spring clip from the accelerator outer cable. Ensure that the throttle cam is fully against its stop, and gently pull the cable out of its grommet until all free play is removed from the inner cable.

3 With the cable held in this position, ensure the flat washer is pressed securely against the grommet, then fit the spring clip to the third outer cable groove visible in front of the rubber grommet and washer **(see illustration)**. This will leave a fair amount of free play in the inner cable, which is necessary to ensure correct operation of the idle control stepper motor.

4 Have an assistant depress the accelerator pedal, and check that the throttle cam opens fully and returns smoothly to its stop.

5 Accelerator pedal - removal and refitting

1 The procedure for removing and refitting the accelerator cable on monopoint injection models is the same as for carburettor models, described in Chapter 4A. See also Chapter 9 for details of right-hand drive models with a brake linkage cross-tube, where the accelerator pedal is integral with the brake pedal assembly.

6 Unleaded petrol - general information and usage

Note: *The information given in this Chapter is correct at the time of writing, and applies only to petrols currently available in the UK. If updated information is thought to be required, check with a Citroën dealer. If travelling abroad, consult one of the motoring organisations (or a similar authority) for advice on the petrols available and their suitability for your vehicle.*

1 The fuel recommended by Citroën is given in the Specifications at the start of this Chapter, followed by the equivalent petrol currently on sale in the UK.

2 RON and MON are different testing standards; RON stands for Research Octane Number (also written as RM), while MON stands for Motor Octane Number (also written as MM).

3 All monopoint injection models are designed to run on fuel with a minimum octane rating of 95 (RON). They are equipped with catalytic converters, and therefore must be run on unleaded fuel only.

Caution: Under no circumstances should leaded fuel (UK "4-star") be used, or the catalytic converter may be damaged.

7 Fuel injection system - general information

1 The Bosch Monopoint A2.2 fuel injection system injects regular pulses of fuel into the throttle body, according to the engine load requirements calculated by an electronic control unit (ECU). The system includes a closed-loop catalytic converter and an evaporative emission control system, to comply with emission control standards.

2 The fuel system ECU has partial control over the breakerless electronic ignition system, so that the ignition timing is retarded at certain engine temperatures. Apart from the timing retard, the ignition system is independent of the fuel system and is described in Chapter 5B.

3 The fuel pump, immersed in the fuel tank, pumps fuel to a filter underneath the rear of

the vehicle and then to the injector mounted in the throttle body. Fuel supply pressure is controlled by the pressure regulator, also in the throttle body assembly, which lifts to allow excess fuel to return to the tank when the optimum operating pressure of the fuel system is exceeded.

4 The electrical control system consists of the ECU, along with the following sensors and idle control regulator:

a) *Throttle potentiometer - informs the ECU of the throttle valve position.*

b) *Coolant temperature sensor - informs the ECU of engine temperature.*

c) *Intake air temperature sensor - informs the ECU of the temperature of the air passing through the throttle body.*

d) *Lambda sensor - informs the ECU of the oxygen content of the exhaust gases (explained in Chapter 4D).*

e) *Idle control regulator - when the throttle is at the idle position, the ECU compares the engine idle speed with the idle speed stored in its memory. The idle control regulator is then activated by the ECU until the engine idle speed is correct.*

f) *Ignition coil - ECU monitors the pulses in the coil low tension (LT) circuit to determine the engine speed.*

5 All the above signals are compared by the ECU and, based on this information, the ECU selects the response appropriate to those values, and controls the fuel injector, varying its pulse width (the length of time the injector is held open), to provide a richer or weaker mixture, as appropriate. The mixture and idle speed are constantly varied by the ECU, to provide the best settings for cranking, starting (with either a hot or cold engine) and engine warm-up, idle, cruising and acceleration.

6 The ECU also has full control over the engine idle speed, via a stepper motor, which is fitted to the throttle body. The motor pushrod rests against a cam on the throttle valve spindle. When the throttle valve is closed (accelerator pedal released), the ECU uses the motor to vary the opening of the throttle valve, and so control the idle speed.

7 The ECU also controls the exhaust and evaporative emission control systems, which are described in detail in Part D of this Chapter.

8 If there is an abnormality in any of the readings obtained from either the coolant temperature sensor, the intake air temperature sensor or the lambda sensor, the ECU enters its back-up mode. If this happens, it ignores the sensor signal, and assumes a pre-programmed value which will allow the engine to continue running at reduced efficiency. If the ECU enters this back-up mode, the warning light on the instrument panel will come on, and the relevant fault code will be stored in the ECU memory.

9 If the warning light comes on, the vehicle should be taken to a Citroën dealer, or fuel injection specialist, at the earliest opportunity, for a complete test of the engine management

system, using a special electronic diagnostic test unit which is plugged into the system's diagnostic connector.

Note: *The injection system diagnostic socket is not the same as the TDC diagnostic socket which is fitted to all Citroën C15 models, including diesel models.*

8 Fuel system - depressurisation

Note: *Refer to the warning note at the end of Section 1 before proceeding.*

 Warning: The following procedure will merely relieve the pressure in the fuel system - remember that fuel will still be present in the system components, and take precautions accordingly before disconnecting any of them.

1 The fuel system referred to in this Section is defined as the tank-mounted fuel pump, the fuel filter, the fuel injector and the pressure regulator in the injector housing, and the metal pipes and flexible hoses connecting these components. All these contain fuel, which will be under pressure while the engine is running and/or while the ignition is switched on. The pressure will remain for some time after the ignition has been switched off, and must be relieved before any of these components are disturbed for servicing work.

2 Disconnect the battery negative terminal.

3 Place a suitable container beneath the relevant connection/union to be disconnected, and have a large rag ready to soak up any escaping fuel not being caught by the container.

4 Slowly loosen the connection or union nut (as applicable) to avoid a sudden release of pressure, and position the rag around the connection to catch any fuel spray which may be expelled. Once the pressure is released, disconnect the fuel line and insert plugs to minimise fuel loss and prevent the entry of dirt into the fuel system.

9 Fuel pump - removal and refitting

Note: *Refer to the warning notes at the end of Section 1 before proceeding.*

Removal

1 Remove the fuel tank (see Section 12). There are two apertures at the top of the tank. The right-hand aperture (facing towards the front of the vehicle) is the fuel pump.

2 Make a note of any alignment marks on the tank, pump cover and the locking ring. If necessary, make your own alignment marks for correct refitting.

3 Unscrew the locking ring and remove it from the tank.

 HAYNES HINT *Use a screwdriver on the raised ribs of the locking ring. Carefully tap the screwdriver to turn the ring anti-clockwise until it can be unscrewed by hand.*

4 Carefully lift the fuel pump assembly out of the fuel tank, taking care not to damage the filter or to spill fuel. Recover the rubber sealing ring and discard it - a new one must be used on refitting.

Refitting

5 Ensure that the fuel pump pick-up filter is clean and free of debris. Fit the new sealing ring to the top of the fuel tank.

6 Carefully manoeuvre the pump assembly into the fuel tank, and align the mark on the pump to its correct position.

7 Refit the locking ring, and tighten it until its mark is correctly aligned with the mark on the fuel tank.

8 Refit the fuel tank (see Section 12).

10 Fuel gauge sender unit - removal and refitting

The fuel gauge sender unit on monopoint injection models is the same as on carburettor models. See Chapter 4A for removal and refitting.

11 Fuel tank filler hose - removal and refitting

The fuel tank filler hose on monopoint injection models is the same as on carburettor models. See Chapter 4A for removal and refitting.

12 Fuel tank - removal and refitting

1 The procedure for removing and refitting the fuel tank on petrol injection models is the same as for carburettor models (see Chapter 4A) except for the following points.

a) *The fuel pump is mounted in the tank, and is integral with the fuel pick-up. There are fuel feed and return hoses which need to be marked for identification purposes, and it will be necessary to depressurise the fuel system before the hoses are disconnected from the fuel pump (see Section 8). It will also be necessary to disconnect the wiring connector from the fuel pump before lowering the tank out of position.*

b) *On models sold outside of the UK from 1996 onwards, the fuel return hose has to be disconnected from a one-way valve on the tank.*

13.3a Disconnect the wiring connectors from the throttle potentiometer ...

13.3b ... the injector cap and the stepper motor

13.4 Fuel feed and return hose connections (arrowed)

13 Throttle body - removal and refitting

Note: *Refer to the warning notes at the end of Section 1 before proceeding.*

Removal

1 Disconnect the battery negative terminal.
2 Remove the air cleaner housing-to-throttle body duct, referring to Section 2.
3 Depress the retaining clips and disconnect the wiring connectors from the throttle potentiometer, the idle control stepper motor, and the injector wiring loom connector which is situated on the side of the throttle body **(see illustrations)**.
4 Bearing in mind the information given in Section 8 about depressurising the fuel system, release the retaining clips and

13.5a Disconnect the accelerator inner cable from the throttle cam ...

disconnect the fuel feed and return hoses from the throttle body assembly **(see illustration)**. If the original crimped-type clips are still fitted, cut the clips and discard them; replace them with worm-drive hose clips on refitting.
5 Disconnect the accelerator inner cable from the throttle cam, then withdraw the outer cable from the mounting bracket along with its flat washer and spring clip **(see illustrations)**.
6 Where necessary, disconnect the distributor vacuum hose, and the idle control auxiliary air valve and/or purge valve hose from the throttle body (as applicable) **(see illustration)**.
7 Slacken and remove the bolts securing the throttle body assembly to the inlet manifold **(see illustration)**, then remove the assembly along with its gasket.
8 If necessary, the upper and lower sections of the throttle body can be separated after removing the retaining screws; note the insulating spacer and gasket which is fitted between the two.

Refitting

9 Refitting is the reverse of removal, bearing in mind the following points:
a) Where applicable, ensure that the mating surfaces of the upper and lower throttle body sections are clean and dry, then fit the insulating spacer and a new gasket and reassemble the two, tightening the retaining screws securely.
b) Ensure that the mating surfaces of the manifold and throttle body are clean and

dry, then fit a new gasket. Securely tighten the throttle body retaining bolts.
c) Ensure that all hoses are correctly reconnected and, where necessary, that their retaining clips are securely tightened.
d) On completion, adjust the accelerator cable using the information given in Section 4.

14 Fuel injection system - testing and adjustment

Testing

1 If a fault appears in the fuel injection system, first ensure that all the system wiring connectors are securely connected and free of corrosion. Then ensure that the fault is not due to poor maintenance - check that the air cleaner filter element is clean, the spark plugs are in good condition and correctly gapped, the valve clearances are correctly adjusted, the cylinder compression pressures are correct, the ignition timing is correct, and that the engine breather hoses are clear and undamaged, referring to the relevant Part of Chapters 1, 2 and 5 for further information.
2 If these checks fail to reveal the cause of the problem, the vehicle should be taken to a suitably-equipped Citroën dealer, or fuel injection specialist, for testing. A wiring block connector is incorporated in the engine management circuit, into which a special electronic diagnostic tester can be plugged.

13.5b ... then free the outer cable from its bracket, and recover the flat washer and spring clip (arrowed)

13.6 Disconnecting the purge valve hose from the throttle body

13.7 Throttle body retaining bolts (arrowed)

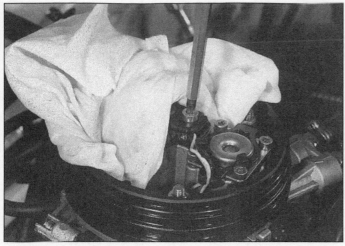

15.3a Undo the injector cap retaining screw, noting the use of a rag to catch any fuel spray . . .

15.3b . . . then lift off the cap and withdraw the injector

The tester will locate the fault quickly and simply, alleviating the need to test all the system components individually, which is a time-consuming operation that carries a high risk of damaging the ECU.

Adjustment

3 Experienced home mechanics with a considerable amount of skill and equipment (including a good-quality tachometer and a good-quality, carefully-calibrated exhaust gas analyser) may be able to *check* the exhaust CO level and the idle speed. No adjustments are possible, and if either of the values are found to be incorrect, there is a fault in the fuel injection system.

Note: *There are two diagnostic sockets on monopoint injection models. The injection system diagnostic socket is for use with special test equipment. The TDC diagnostic socket is for use with a tachometer, and is fitted to all Citroën C15 models, including diesel models.*

15 Bosch Monopoint system components - removal and refitting

Fuel injector

Note: *Refer to the warning notes at the end of Section 1 before proceeding. If a faulty injector is suspected, before condemning the injector, it is worth trying the effect of one of the proprietary injector-cleaning treatments.*
Note: *The fuel injector and its seals may not be available separately. If the injector is faulty, the complete throttle body assembly must be renewed. Refer to your Citroën dealer for the latest information on parts availability. Although the unit can be dismantled for cleaning, it should not be disturbed unless absolutely necessary.*
1 Disconnect the battery negative terminal.

2 Remove the air cleaner-to-throttle body duct.
3 Undo the injector cap retaining screw, then lift off the cap and withdraw the injector from the housing, noting the sealing ring locations. As the cap screw is slackened and the injector is withdrawn, place a rag over the injector to catch any fuel spray which may be released **(see illustrations)**.
4 Refitting is the reverse of removal, ensuring that the injector sealing ring(s) and injector cap O ring are in good condition. When refitting the injector cap, ensure that the injector pins are correctly aligned with the cap terminals; the terminals are marked "+" and "-" for identification **(see illustration)**.

Fuel pressure regulator

Note: *Refer to the warning notes at the end of Section 1 before proceeding.*
Note: *The fuel pressure regulator assembly may not be available separately. If it is faulty, the complete throttle body assembly may have to be renewed. Refer to a Citroën dealer for the latest information on parts availability. Although the unit can be dismantled for cleaning, it should not be disturbed unless absolutely necessary.*
5 Disconnect the battery negative terminal.

6 Remove the air cleaner-to-throttle body duct.
7 Using a suitable marker pen, make alignment marks between the regulator cover and throttle body, then slacken and remove the cover retaining screws **(see illustration)**. As the screws are slackened, place a rag over the cover to catch any fuel spray which may be released.
8 Lift off the cover, then remove the spring and withdraw the diaphragm, noting its correct fitted orientation. Remove all traces of dirt, and examine the diaphragm for signs of splitting. If damage is found, renewal will be necessary.
9 Refitting is the reverse of removal, ensuring that the diaphragm and cover are fitted the correct way round, and the retaining screws are securely tightened.

Idle control stepper motor

Note: *The idle control stepper motor may not be available separately. If the motor is faulty, the complete throttle body assembly may have to be renewed. Refer to your Citroën dealer for the latest information on parts availability.*
10 Disconnect the battery negative terminal.

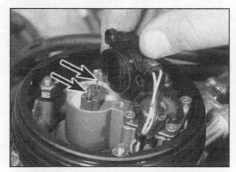

15.4 On refitting, ensure that the cap terminals are correctly aligned with the injector pins (arrowed)

15.7 Fuel pressure regulator retaining screws (arrowed)

15.12 Idle control stepper motor retaining screws (arrowed)

15.15 The intake air temperature sensor (arrowed) is an integral part of the injector cap

11 Depress the retaining clip, and disconnect the wiring connector from the idle control stepper motor.

12 Undo the retaining screws, and remove the motor from the throttle body (see illustration).

13 Refitting is the reverse of removal, ensuring that the motor retaining screws are securely tightened.

15.16a Undo the three retaining screws (arrowed) . . .

Throttle potentiometer

14 The throttle potentiometer is a sealed unit, and must not be disturbed under any circumstances. On some models it is secured to the throttle body assembly by tamperproof screws. If the throttle potentiometer is faulty, the complete throttle body assembly must be renewed. Refer to your Citroën dealer for further information.

Intake air temperature sensor

Note: *Refer to the warning notes at the end of Section 1 before proceeding.*

Note: *The intake air temperature sensor may not be available separately. If the sensor is faulty, the complete throttle body assembly may have to be renewed. Refer to your Citroën dealer for the latest information on parts availability.*

15 The intake air temperature sensor is an integral part of the throttle body injector cap (see illustration). To remove the cap, first disconnect the battery negative terminal, then remove the air cleaner-to-throttle body duct using the information given in Section 2.

16 Undo the three retaining screws, and remove the circular plastic ring from the top of the throttle body. Recover its sealing ring (see illustrations).

17 Depress the retaining clip, and disconnect the injector wiring connector (see illustration).

18 Undo the injector cap retaining screw, then lift off the cap and recover the gasket and/or sealing ring (as applicable). As the cap screw is slackened, place a rag over the injector to catch any fuel spray which may be released. Release the injector cap connector from the throttle body as the cap is removed (see illustration).

19 Refitting is the reverse of removal, ensuring that the injector cap gasket and/or O-ring is in good condition. Take care to ensure that the cap terminals are correctly aligned with the injector pin, and securely tighten the cap retaining screw.

15.16b . . . then lift off the plastic ring. Recover the sealing ring

15.17 Disconnecting the injector wiring connector. Injector retaining screw is arrowed

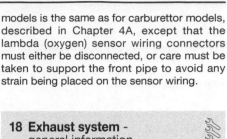

15.18 Injector cap wiring connector is a push fit in the throttle body

Coolant temperature sensor

20 The coolant temperature sensor informs the ECU of the engine temperature. For removal and refitting, see Chapter 3.

Ignition LT coil

21 The ECU monitors the pulses in the coil low tension (LT) circuit to determine the engine speed. The LT coil drives the HT coil and is housed within the same unit. For removal and refitting of the ignition HT coil, see Chapter 5B.

Lambda Sensor

22 The lambda sensor is fitted to the exhaust front downpipe, just below the exhaust manifold joint. It measures the oxygen content of the exhaust gases and sends a signal to the ECU, to adjust the mixture strength and provide the optimum conditions for the catalytic converter to operate. See Section 18 for details of removal and refitting.

Inertia switch

23 On models from 1996 onwards an inertia switch is fitted to the electrical supply circuit between the ECU and the fuel injection system relay unit, to shut off the fuel supply in the event of an impact. The switch is mounted in the engine compartment in front of the right-hand suspension strut. It contains a ball in a conical housing, held in place by a magnet. The ball is released from the housing when there is rapid deceleration exceeding 8 times the force of gravity, equivalent to a wall impact of about 16 mph.
24 To remove the unit, disconnect the battery negative lead, then disconnect the multi-plug from the inertia switch.
25 Undo the nuts or bolts as required, and remove the switch from its support.
26 Refitting is the reverse of removal.

ECU, fuel injection system relay unit and injector resistor

27 These components interact with each other as follows. The fuel injection system relay unit is a double relay supplied as a single component. One relay is energised by the

ignition and switches on the ECU. The other relay is energised by the ECU and switches on the fuel pump, injector, and lambda sensor heater, using power from the battery. The injector resistor is in the supply circuit to the injector, providing the correct current during the injector pulses.
28 No specific instructions are available about removing and refitting these components, except that they are close to each other on a mounting plate. In general, the procedure is as follows.
29 Disconnect the battery negative terminal.
30 To remove the ECU or the relay unit, disconnect the appropriate wiring connector, then undo the bolts or clips as required and remove the unit from the mounting plate. Refitting is the reverse of removal.
31 To remove the injector resistor, undo the resistor retaining nut, then disconnect the wiring connector and remove the resistor from the vehicle. If the resistor is riveted in position, drill out the rivets or cut them to release the resistor from the mounting plate. Refitting is the reverse of removal. Where the resistor was riveted in position, secure it with either a suitable nut and bolt or new pop-rivets.

16 Inlet manifold - removal and refitting

1 The procedure for removing and refitting the inlet manifold on monopoint injection models is the same as for carburettor models, described in Chapter 4A, except for the following:
a) There is a throttle body instead of a carburettor. For removal and refitting, see Section 13.
b) There may be more than one coolant hose to be disconnected from the inlet manifold.

17 Exhaust manifold - removal and refitting

1 The procedure for removing and refitting the exhaust manifold on monopoint injection

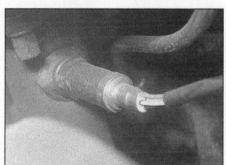

18.7 Lambda sensor screwed into top of exhaust front downpipe

models is the same as for carburettor models, described in Chapter 4A, except that the lambda (oxygen) sensor wiring connectors must either be disconnected, or care must be taken to support the front pipe to avoid any strain being placed on the sensor wiring.

18 Exhaust system - general information and component renewal

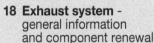

General information

1 The exhaust system consists of three sections; the front downpipe with lambda sensor, the intermediate pipe with catalytic converter, and the tailpipe with silencer. On models with an extended cab there is an extension that fits onto the tailpipe.
2 The front downpipe is secured to the manifold flange by threaded spindles with nuts at both ends. The intermediate pipe connects to the front downpipe by a spring-loaded ball/ring joint, to allow movement of the exhaust system. The tailpipe connects to the intermediate pipe by a ring clamp consisting of two halves.
3 The system is suspended throughout its entire length by rubber mountings.

Exhaust system - removal and refitting

4 Each exhaust section can be removed individually, or the complete system can be removed as a unit.
5 The catalytic converter is integral with the intermediate section. If the catalytic converter needs to be renewed, it is necessary to renew the complete intermediate section.
6 Removal and refitting of the exhaust sections is the same as for carburettor models (see Chapter 4A) except that when removing the front downpipe it is necessary to disconnect the lambda sensor wiring. Always disconnect the wiring first, before disconnecting the manifold joint, to avoid placing the wiring under stress. When renewing the front downpipe, the lambda sensor needs to be transferred to the new pipe as follows.

Lambda sensor

Note: *The lambda sensor is DELICATE. It will not work if it is dropped or knocked, if its power supply is disrupted, or if any cleaning materials are used on it.*

Removal

7 The lambda sensor is fitted to the exhaust front downpipe, just below the exhaust manifold joint **(see illustration)**. It measures the oxygen content of the exhaust gases and sends a signal to the ECU, to adjust the mixture strength and provide the optimum conditions for the catalytic converter to operate. For further details see Chapter 4D.

8 Trace the wiring back from the lambda sensor and disconnect the wiring connectors. Free the wiring from any retaining clips or ties.
9 Unscrew the sensor from the exhaust system front downpipe, and remove it along with its sealing washer.

Refitting

10 Refitting is the reverse of removal, using a new sealing washer. Before installing the sensor, apply a smear of high-temperature grease to the sensor threads. Ensure that the sensor is securely tightened. Also, the wiring must be correctly routed, and in no danger of contacting either the exhaust system or the engine.

Chapter 4 Part C:
Fuel and exhaust systems - diesel engines

Contents

Degrees of difficulty

Easy, suitable for novice with little experience	Fairly easy, suitable for beginner with some experience 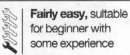	Fairly difficult, suitable for competent DIY mechanic 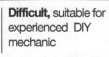	Difficult, suitable for experienced DIY mechanic	Very difficult, suitable for expert DIY or professional

Specifications

General

System type	Rear-mounted fuel tank, injection pump with integral transfer pump, indirect injection
Firing order	1-3-4-2 (No 1 at flywheel end)
Fuel:	
Type	Commercial diesel fuel for road vehicles (DERV)
Tank capacity	47 litres

Injection pump types

Lucas CAV/Roto-diesel:	
From February 1993	047 - R 8443B 930 A
Later models, about 1997*	052 - 8443B 931 A
Bosch:	
Up to February 1993	VER 171-1
From February 1993	523 R171-3

The pump type was available at this date but there is no specific information about it when it was introduced.

Injection pump (Lucas)

Direction of rotation	Clockwise, viewed from sprocket end
Dynamic timing	14° ± 1° BTDC @ idle speed
Static timing	dimension marked on pump
Fast idle speed	950 ± 50 rpm
Anti-stall speed	See Chapter 1B

	047 - R 8443B 930 A	052 - 8443B 931 A
Idle speed	800 ± 50 rpm	775 ± 25 rpm
Maximum engine speed	5150 ± 125 rpm	5110 ± 125 rpm
Injector opening pressure:		
No colour code	(not applicable)	138 to 143 bars
Green collar	138 to 143 bars	142 to 147 bars
Green collar and green spot	142 to 147 bars	(not applicable)

Injection pump (Bosch)

Direction of rotation	Clockwise, viewed from sprocket end
Dynamic timing	14° ± 1° BTDC @ idle speed
Static timing (pump ABDC)	0.89 to 0.90 mm
Fast idle speed	950 ± 50 rpm
Anti-stall speed	See Chapter 1B
Idle speed	800 ± 50 rpm
Maximum engine speed	5130 ± 125 rpm
Injector opening pressure (colour code)	130 to 135 bars (mauve)

Injector

Type	Pintle
Injector opening pressure	According to colour code. Must be correct for pump type.

Heater plug

Type	Champion CH 68 or equivalent

Torque wrench settings

	Nm	lbf ft
Heater plug	22	16
Injection pump front mounting nuts	18	13
Injection pump rear mounting nut	22	16
Injection pump (Bosch) blanking plug	20	15
Injection pump sprocket nut	50	37
Injector	90	66
Injector pipe union nuts	25	19

---

1 General information and precautions

General

1 The fuel system consists of a rear-mounted fuel tank, a fuel filter with integral water separator, a fuel injection pump, injectors and associated components. The exhaust system is conventional and there is no catalytic converter fitted to UK models.

2 Fuel is drawn from the tank by a vane-type transfer pump incorporated in the delivery head of the injection pump. Before reaching the pump, the fuel passes through a fuel filter where foreign matter and water are removed. Any excess fuel lubricates the moving components of the pump, and is then returned to the tank.

3 The fuel is heated by the engine coolant, to reduce the chance of wax crystals forming in the fuel in cold weather, which would cause fuel starvation. There are two types of fuel heater:

a) On models up to early 1993, the fuel heater is mounted behind the cylinder block, integral with the coolant pump inlet manifold. Fuel from the tank is heated as it passes to the fuel filter, mounted on the right-hand inner wing.

b) From early 1993 onwards, the fuel heater is integral with the fuel filter, mounted on top of the thermostat housing in front of the cylinder head.

4 The injection pump is driven at half crankshaft speed by the timing belt. The high pressure required to inject the fuel into the compressed air in the swirl chambers is achieved by a cam plate acting on a single piston on the Bosch pump, or by two opposed pistons forced together by rollers running in a cam ring on the Lucas/CAV pump. The fuel passes through a central rotor with a single outlet drilling which aligns with ports leading to the injector pipes.

5 Fuel metering is controlled by a centrifugal governor that reacts to accelerator pedal position and engine speed. The governor is linked to a metering valve, which increases or decreases the amount of fuel delivered at each pumping stroke.

6 Basic injection timing is determined when the pump is fitted. When the engine is running, it is varied automatically to suit the prevailing engine speed, by a mechanism which turns the cam plate or ring.

7 There are four precision-made injectors that produce a homogeneous spray of fuel into the swirl chambers located in the cylinder head. The injectors are calibrated to open and close at critical pressures to provide efficient and even combustion. Each injector needle is lubricated by fuel, which accumulates in the spring chamber and is channelled to the injection pump return hose by leak-off pipes.

8 Bosch or Lucas CAV/Roto-diesel fuel system components may be fitted, depending on model. Lucas components are marked either "CAV", "Roto-diesel", or "Con-Diesel", depending on their date and place of manufacture. Replacement components must be of the same make as those originally fitted, with the exception of the fuel filter assembly.

9 Cold starting is assisted by pre-heater or "glow" plugs fitted to each swirl chamber (see Chapter 5C for further details).

10 On Lucas pumps, the fast idle system is operated by a thermostatic valve which is screwed into the fuel filter/thermostat housing. The valve operates a fast idle lever on the injection pump, via a cable, to increase the idle speed when the engine is cold. Bosch pumps have the same system, but additionally, the fast idle system incorporates an electrically-heated device which advances the pump timing when the engine is cold.

11 A stop solenoid cuts the fuel supply to the injection pump rotor when the ignition is switched off. There is also a hand-operated stop lever for use in an emergency (see illustrations).

1.11a Hand-operated stop lever (arrowed) - Roto-diesel pump

1.11b Hand-operated stop lever (arrowed) - Bosch pump

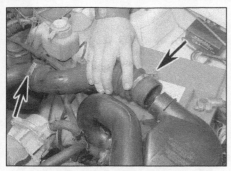

2.1 Undo the clips (arrowed) and remove the air cleaner outlet ducting

2.2 Undo the clip and remove the air cleaner inlet ducting

2.3a Release the clips on the air cleaner cover . . .

12 Provided that the specified maintenance is carried out, the fuel injection equipment will give long and trouble-free service. The injection pump itself may well outlast the engine. The main potential cause of damage to the injection pump and injectors is dirt or water in the fuel.

13 Servicing of the injection pump and injectors is very limited for the home mechanic, and any dismantling or adjustment other than that described in this Chapter must be entrusted to a Citroën dealer or fuel injection specialist.

14 Following the introduction of new EEC emission standards, all engines from February 1993 are equipped with modified injection pumps and cylinder heads as follows:

a) *On Lucas pumps, the advance curve is modified and reduced volume injectors are used. The new injectors enable the volume of unburnt fuel in the exhaust to be reduced.*

b) *On Bosch pumps, a device known as NLK is used, to adjust the advance point by means of a hydraulic piston, making it quicker and more accurate.*

c) *Valve stem oil seals are fitted, to reduce hydrocarbon emissions, by preventing oil from entering the combustion chambers.*

Precautions

⚠️ *Warning: It is necessary to take certain precautions when working on the fuel system components, particularly the fuel injectors. Before*

carrying out any operations on the fuel system, refer to the precautions given in 'Safety first!' at the beginning of this manual, and to any additional warning notes at the start of the relevant Sections.

2 Air cleaner assembly - removal and refitting

Removal

1 Undo the clips and remove the air cleaner outlet ducting **(see illustration)**.

2 Undo the clip and disconnect the inlet ducting from the air cleaner housing, then pull the ducting away from the front grille and withdraw it from the vehicle **(see illustration)**.

3 To remove the air filter element, release the clips and open the cover, then withdraw the element **(see illustrations)**.

4 Undo the clip and disconnect the breather hose from the oil filler **(see illustration)**. Alternatively, disconnect the other end of the hose from the air cleaner housing.

5 Undo the nut connecting the air cleaner housing to the battery support plate **(see illustration)**.

6 Pull the locating lugs from their grommets on the front grille, and withdraw the air cleaner housing from the vehicle **(see illustration)**.

Refitting

7 Refitting is the reverse of removal. Lightly oil the air filter element when fitting a new one.

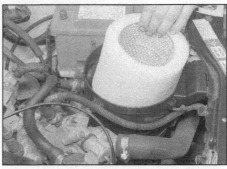

2.3b . . . then open the cover and remove the air filter element

3 Accelerator cable - removal, refitting and adjustment

1 The accelerator cable connects to the accelerator lever on the fuel pump, and the adjustable ferrule on the outer sleeve connects to a mounting bracket **(see illustration)**. The limits of travel are between the anti-stall adjustment screw and the maximum speed adjustment screw.

Caution: The maximum speed adjuster on Roto-diesel pumps is sealed under warranty.

2.4 Undo the clip and disconnect the breather hose from the oil filler

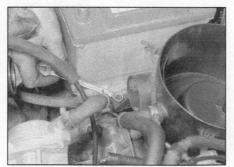

2.5 Undo the nut connecting the air cleaner housing to the battery support plate

2.6 Pull the locating lugs (arrowed) from their grommets on the front grille, and withdraw the air cleaner housing

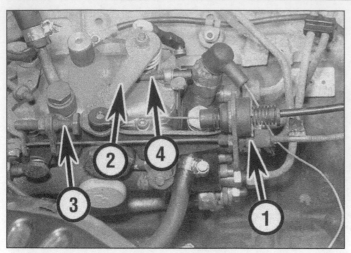

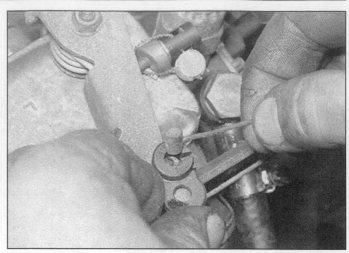

3.1 Accelerator cable and lever assembly - Roto-diesel pump

1 *Accelerator cable adjustment ferrule*
2 *Accelerator lever*
3 *Anti-stall adjuster*
4 *Maximum speed adjuster*

3.2a Disconnect the inner cable by passing it through the special slot - Roto-diesel pump

Removal

2 Open the accelerator lever on the injection pump and disconnect the inner cable by passing it through the special slot **(see illustrations)**.

3 Withdraw the outer cable and adjustment ferrule from the grommet on the mounting bracket, and recover the washer and spring clip.

4 Working inside the vehicle, remove the parcel shelf from below the steering column (see Chapter 11), then depress the retaining clips, and detach the inner cable from the top of the accelerator pedal.

5 Pull the spring shock absorber from the bulkhead and withdraw the accelerator cable from inside the engine compartment.

Refitting

6 Refitting is the reverse of removal, but adjust the cable as follows.

Adjustment

7 The outer cable adjustment ferrule is held in position against the mounting grommet by a washer and spring clip **(see illustrations)**.

8 Remove the spring clip from the ferrule then, ensuring that the pump accelerator lever is fully against the anti-stall screw, gently pull the cable out of its grommet until all free play is removed from the inner cable.

9 With the cable held in this position, ensure that the flat washer is pressed securely against the grommet, then fit the spring clip to the first outer cable groove visible in front of the rubber grommet and washer.

10 Have an assistant depress the accelerator pedal, and check that the accelerator lever opens fully, so that it contacts the maximum speed screw. Check also that it returns smoothly to its stop against the anti-stall screw.

4 Accelerator pedal - removal and refitting

1 On right-hand drive models with a brake linkage cross-tube, the accelerator pedal is integral with the brake pedal assembly (see Chapter 9).

2 For models with a separate accelerator pedal, see Chapter 4A.

5 Fuel gauge sender unit - removal and refitting

The fuel gauge sender unit is located in the fuel tank, and is the same as the unit fitted to carburettor petrol models. Removal and refitting of the fuel tank and fuel gauge sender unit is described in Chapter 4A.

6 Fuel pick-up unit - removal and refitting

The fuel pick-up unit is located in the fuel tank, and is similar to the unit fitted to carburettor petrol models. Removal and refitting of the fuel tank and pick-up unit is described in Chapter 4A, but there is a fuel return hose that needs to be disconnected. Mark the fuel feed and return hoses for identification purposes so that they are not mixed up.

3.2b Accelerator cable attachment - Bosch pump

3.7a Accelerator adjustment ferrule with washer and spring clip (arrowed) - Roto-diesel pump

3.7b The same adjustment ferrule is used on the Bosch pump

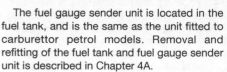

7 Fuel tank filler hose - removal and refitting

The fuel tank filler hose is the same as the unit fitted to carburettor petrol models. Removal and refitting is described in Chapter 4A.

8 Fuel tank - removal and refitting

The fuel tank and fittings are similar to the units fitted to carburettor petrol models. Removal and refitting is described in Chapter 4A, but there is a fuel return hose that needs to be disconnected. Mark the fuel feed and return hoses for identification purposes so that they are not mixed up.

9 Fuel system - priming and bleeding

⚠️ **Warning: Diesel injection pumps supply fuel at very high pressure. Extreme care must be taken when working on the fuel injectors and fuel pipes. It is advisable to place an absorbent cloth around the union before slackening a fuel pipe. Never expose the hands, face or any other part of the body to injector spray; the high working pressure can penetrate the skin, with potentially-fatal results. Refer also to "Safety first!" at the start of this manual.**

1 After disconnecting part of the fuel supply system or running out of fuel, it is necessary to prime the system and bleed off any air which may have entered the system components.
2 All models are fitted with a hand-operated priming pump. On models up to February 1993, the priming pump is integral with the fuel filter, mounted on the right-hand inner wing, and may be the Roto-diesel or Purflux type. When the plunger is operated, fuel is forced into the fuel pump and any excess is expelled through the fuel pump outlet and returned to the tank. Later models, from February 1993 onwards, are fitted with a separate pump consisting of a rubber priming bulb, and the fuel filter is mounted on the thermostat housing. There is an automatic bleed valve which collects air from various parts of the low-pressure circuit and returns it to the tank without having to pass through the fuel pump **(see illustrations).**

Bleeding air from the priming pump

3 Some early models have a bleed screw on the priming pump, so that air can be expelled from the pump and surrounding pipework before priming. Open the bleed screw and

9.2a Roto-diesel fuel filter and priming pump - up to February 1993 (the bolt is for removing the filter element)

place a rag around it, then operate the plunger until fuel, free from air, is ejected. Then tighten the bleed screw and mop up any spilt fuel. Prime the low-pressure circuit as described below.

Priming the low-pressure circuit

4 Switch on the ignition to position "M" so that the stop solenoid is energised and the pre-heater warning light comes on. If the vehicle has run out of fuel, pour at least two litres of fuel into the tank, with the van on level ground.
5 Operate the pump plunger or squeeze the priming bulb until appreciable resistance is felt.
6 Make sure the pre-heater warning light has gone out.
7 Depress the accelerator pedal by three-quarters of its travel and start the engine. If the engine does not start at the first attempt, wait 15 seconds and try again. If it still doesn't start, operate the plunger or priming bulb again, then make another attempt to start the engine. When the engine starts, accelerate slightly to finish priming the fuel circuit.

Bleeding the high-pressure circuit

8 If the engine still fails to start after thoroughly priming the low-pressure circuit, it is likely that air has entered the high-pressure circuit downstream of the fuel pump. To bleed the injector pipes, place wads of rag around the pipe unions at the injectors, to absorb spilt fuel, then slacken the unions. Refer to the

9.2c Priming bulb - February 1993 onwards

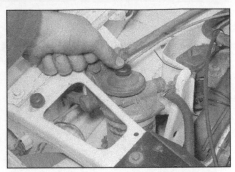

9.2b Purflux fuel filter and priming pump - up to February 1993

warning at the start of this Section. Crank the engine on the starter motor until fuel emerges from the unions, then stop cranking the engine and retighten the unions. Mop up any spilt fuel.
9 Start the engine with the accelerator pedal fully depressed. Additional cranking may be necessary to finally bleed the system before the engine starts.

10 Maximum engine speed - checking and adjustment

Caution: On Roto-diesel injection pumps the maximum speed setting is sealed by the manufacturers at the factory using locking wire and a lead seal. Do not disturb the wire if the vehicle is still within the warranty period otherwise the warranty will be invalidated.
Note: *This adjustment requires the use of a tachometer. The usual type of tachometer that works from ignition system pulses cannot be used on diesel engines. A TDC diagnostic socket is provided for use with Citroën test equipment which will not normally be available to the home mechanic. If you cannot obtain the correct equipment, take the vehicle to a Citroën dealer or other specialist.*

Checking

1 Run the engine to normal operating temperature.

9.2d Automatic bleed valve - February 1993 onwards

10.2 Connect a tachometer to the TDC sensor diagnostic socket (arrowed)

10.4 Maximum engine speed adjustment screw (arrowed) on the Roto-diesel injection pump - sealed under warranty

10.5 Maximum engine speed adjustment screw (arrowed) on the Bosch injection pump

2 Connect a tachometer to the TDC sensor diagnostic socket **(see illustration)**.
3 Have an assistant fully depress the accelerator pedal and check that the maximum engine speed is as given in the Specifications. Do not keep the engine at maximum speed for more than two or three seconds.

Adjustment

Roto-diesel injection pump

4 If adjustment is necessary on a Roto-diesel pump, take the vehicle to a Citroën dealer or suitable diesel specialist. The adjustment screw is sealed and should not be altered by the home mechanic **(see illustration)**.

Bosch injection pump

5 If adjustment is necessary on a Bosch pump, stop the engine and loosen the locknut, then turn the maximum engine speed adjustment screw as necessary, and retighten the locknut **(see illustration)**.
6 Repeat the procedure in paragraph 3 to check the adjustment.
7 Switch off the ignition.

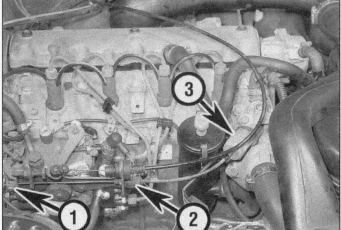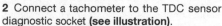

◀ **11.1a Fast idle thermostatic valve and cable connected to a Roto-diesel injection pump**

1 *Fast idle lever cable clamp*
2 *Fast idle cable adjustment ferrule*
3 *Thermostatic valve in thermostat housing - up to February 1993*

11 Fast idle thermostatic valve and cable - removal, refitting and adjustment

1 The fast idle thermostatic valve is screwed into the thermostat housing and operates a cable which connects to the fast idle lever on the injection pump, to increase the idling speed when the engine is cold **(see illustrations)**.

Removal

2 Loosen the clamp screw or nut and remove the end fitting from the inner cable **(see illustration)**.
3 Unscrew the locknut and remove the adjustment ferrule and outer cable from the bracket on the injection pump **(see illustration)**.
4 Partially drain the cooling system as described in Chapter 1B.
5 Unscrew the thermostatic valve from the thermostat housing cover and recover the sealing washer.

Refitting

6 Refit the thermostatic valve to the thermostat housing, using a new sealing washer, and tighten it securely.
7 Insert the cable and ferrule in the bracket and screw on the locknut finger tight.
8 Slide the end fitting onto the inner cable and lightly tighten the clamp screw or nut.
9 Refill the cooling system as described in Chapter 1B, then adjust the cable as follows.

11.1b Fast idle thermostatic valve (arrowed) - February 1993 onwards

11.2 Fast idle inner cable and end fitting (arrowed) on the Bosch injection pump

11.3 Fast idle cable adjustment ferrule (arrowed) on the Roto-diesel injection pump

12.5a Disconnect the stop solenoid wire - Roto-diesel pump

12.5b Disconnect the stop solenoid wire - Bosch pump

Refitting

4 Refitting is the reverse of removal.

14 Fuel injection pump - removal and refitting

⚠️ **Warning: Exercise extreme caution when working on the high-pressure pipework from the pump to the fuel injectors. The fuel may be under pressure for some time after the engine has been switched off. Never expose the hands or any part of the body to fuel under pressure because it can penetrate the skin, with possibly fatal results. Place rags around the pipework before you undo the unions.**

Adjustment

10 With the engine cold, push the fast idle lever fully towards the flywheel end of the engine. Hold it in this position and slide the cable end fitting along the cable so that it touches the fast idle lever, then tighten the clamp screw or nut.

11 Adjust the ferrule to ensure that the fast idle lever is touching its stop, then tighten the locknut.

12 Measure the exposed length of the inner cable between the ferrule and end fitting.

13 Run the engine to normal operating temperature, then check that the length of the inner cable has increased by at least 6.0Êmm indicating that the thermostatic sensor is functioning correctly.

14 Switch off the engine.

12 Stop solenoid - general, removal and refitting

General

1 The stop solenoid is located on top of the fuel injection pump near the injector pipes. Its purpose is to cut the fuel supply when the ignition is switched off. If an open-circuit occurs in the solenoid or its electrical supply, or if the solenoid jams shut, it will be impossible to start the engine because the fuel will not be allowed to pass through the injection pump. If the solenoid jams open, the engine will not stop when the ignition is switched off.

2 If the solenoid has failed and the engine will not run, a temporary repair may be made by removing the solenoid as described in the following paragraphs. Refit the solenoid body without the plunger and spring, and tape up its feed wire so that it cannot touch earth. The engine can then be started as usual, but it will be necessary to use the manual stop lever on the injection pump, or to stall the engine in gear, to stop it.

Removal

3 Disconnect the battery negative lead.

4 Clean the area around the solenoid, to prevent dust and dirt entering the fuel system.

5 Remove the rubber cover, where fitted, then slacken and remove the retaining nut and washer, and disconnect the solenoid wiring connector (see illustrations).

6 Unscrew the solenoid valve from the pump, and recover the plunger, spring and sealing ring.

Refitting

7 Refitting is the reverse of removal, using a new sealing ring.

13 Inertia switch - general, removal and refitting

General

1 On models from 1996 onwards an inertia switch is fitted to the electrical supply circuit to the stop solenoid, to shut off the fuel supply in the event of an impact. The switch is mounted in the engine compartment in front of the right-hand suspension strut. It contains a ball in a conical housing, held in place by a magnet. The ball is released from the housing when there is rapid deceleration exceeding 8 times the force of gravity, equivalent to a wall impact of about 16 mph.

Removal

2 Disconnect the battery negative lead, then disconnect the multi-plug from the inertia switch.

3 Undo the nuts or bolts as required, and remove the switch from its support.

Removal

1 Disconnect the battery negative lead.

2 Cover the alternator with a plastic bag as a precaution against spillage of diesel fuel.

3 Apply the handbrake, then jack up the front right-hand corner of the vehicle until the wheel is just clear of the ground, and support the vehicle on an axle stand (see "Jacking and vehicle support"). Engage 4th or 5th gear, to enable the engine to be turned easily by turning the right-hand wheel.

4 Pull up the special clip, release the spring clips, and withdraw the two timing cover sections.

5 Open the accelerator lever on the injection pump and disconnect the cable by passing it through the special slot. Disconnect the cable adjustment ferrule from the bracket.

6 Note the position of the end fitting on the fast idle cable, then loosen the screw and disconnect the inner cable. Unscrew the adjustment locknut and remove the cable and ferrule from the bracket.

7 Loosen the clip and disconnect the fuel supply hose.

8 Disconnect the main fuel return pipe and the injector leak off return pipe from the union tube (see illustration).

9 Disconnect the wire from the stop solenoid.

10 Place rags around the unions connecting the injection pump to the injector pipes, to absorb any fuel that comes out under pressure, then unscrew the union nuts (see illustration).

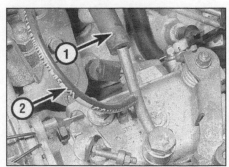

14.8 Main fuel return pipe (1) and injector leak off return pipe (2) - Roto-diesel pump

14.10 Injector pipe union nuts on the Roto-diesel injection pump

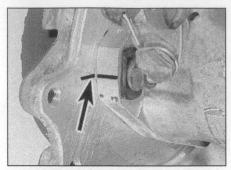

14.13 Mark the injection pump in relation to the mounting bracket (arrowed)

14.14a Injection pump mounting nut and plate (arrowed)

Counterhold the unions on the pump while unscrewing the nuts. Cover the open unions to keep dirt out, using small plastic bags secured with rubber bands.

11 Turn the engine by means of the front right-hand wheel or crankshaft pulley bolt until the two bolt holes in the injection pump sprocket are aligned with the corresponding holes in the engine front plate.

12 Insert two M8 bolts through the holes and hand tighten them. The bolts must retain the sprocket while the injection pump is removed thereby making it unnecessary to remove the timing belt.

13 Mark the injection pump in relation to the mounting bracket using a scriber or felt tip pen **(see illustration)**. This will ensure the correct timing when refitting. If a new pump is being fitted transfer the mark from the old pump to give an approximate setting.

14 Unscrew the three mounting nuts and remove the plates or washers. Unscrew and remove the rear mounting nut and bolt, noting the locations of the washers, and support the injection pump on a block of wood **(see illustrations)**.

15 Unscrew the sprocket nut until the taper is released from the sprocket. The nut acts as a puller, together with the plate bolted to the sprocket. From late 1992, the fuel injection pump sprocket bolt no longer incorporates a puller. To free the sprocket from the taper on the injection pump shaft, an improvised puller consisting of a plate and two bolts must be fitted to the sprocket before unscrewing the nut. Ideally, a plate should be removed from an old sprocket and used as a puller.

16 Continue to unscrew the sprocket nut and withdraw the injection pump from the mounting bracket **(see illustration)**. Recover the Woodruff key from the shaft groove if it is loose.

Refitting

17 Unbolt the improvised or built-in puller plate from the injection pump sprocket.

18 Refit the Woodruff key to the injection pump shaft groove (if removed), then insert the injection pump from behind the sprocket, making sure that the Woodruff key enters the groove in the sprocket. Screw on the nut and hand tighten it.

19 Fit the front mounting nuts, together with their plates or washers, and hand tighten the nuts.

20 Tighten the sprocket nut to the specified torque.

21 On models with a built-in puller, refit the puller plate and tighten the bolts.

22 Unscrew and remove the two M8 bolts from the injection pump sprocket.

23 If the original injection pump is being refitted, align the scribed marks and tighten the mounting nuts. If fitting a new pump, transfer the mark from the old pump to give an approximate setting of the injection timing, then set the timing as described in Section 16.

24 Refit the rear mounting bolt and special nut, tightening the nut slowly to allow the bush to align itself.

25 Refit the injector pipes to the injection pump and tighten the union nuts.

26 Reconnect the wire to the stop solenoid.

27 Refit the fuel supply and return pipes.

28 Refit the accelerator cable and fast idle cable and adjust them, referring to Sections 3 and 11.

29 Refit the two timing cover sections and secure them with the spring clips.

30 Lower the vehicle to the ground.

31 Remove the plastic bag from the alternator and reconnect the battery negative lead.

32 Prime the fuel circuit, then start the engine (see Section 9).

33 Adjust the idling speed, referring to Chapter 1B.

15 Fuel injection timing - checking methods

1 Checking the injection timing is not a routine operation. It should only be necessary after the injection pump has been disturbed. There are two methods of timing the injection pump, known as dynamic and static timing. Dynamic timing should only be used within the limitations of the checking equipment. If the timing requires adjustment, the static timing method must be used.

2 Dynamic timing equipment is unlikely to be available to the home mechanic. The equipment works by converting pressure pulses in an injector pipe into electrical signals. If such equipment is available, use it in accordance with its maker's instructions. Note that most dynamic checking testers are only accurate to approximately ±2°. The engine should be at normal operating temperature, to ensure that it is running at the correct idle speed. On the Bosch injection pump, running at normal operating temperature will also ensure that the cold advance system is not functioning.

3 Static timing, as described in the next Section, gives good results if carried out carefully. A dial test indicator will be needed, with probes and adapters appropriate to the type of injection pump. Read through the procedures before starting work, to find out what is involved.

16 Fuel injection static timing - checking and adjustment

⚠ *Warning: Exercise extreme caution when working on the high-pressure pipework from the pump to the fuel injectors. The fuel may be under pressure for some time after the engine has been switched off. Never expose the hands or any part of the body to fuel under pressure because it can penetrate the skin, with possibly fatal results. Place rags around the pipework before you undo the unions.*

Caution: The maximum engine speed and transfer pressure settings, together with

14.14b Injection pump mounting bolt (arrowed)

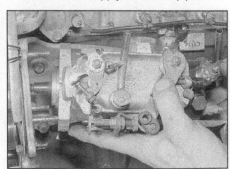

14.16 Remove the injection pump from its mounting bracket

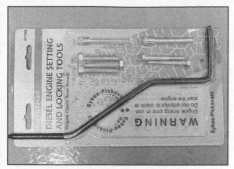

16.5a TDC setting and locking tools for Citroën diesel engines

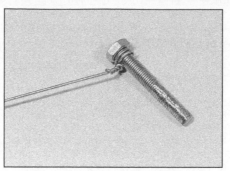

16.5b Home-made TDC setting tool

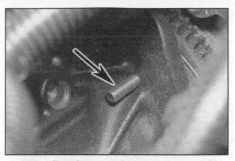

16.5c Rod (arrowed) inserted through cylinder block into TDC hole in flywheel with starter motor removed

timing access plugs, are sealed by the manufacturers at the factory using locking wire and lead seals. On the Roto-diesel pump, the maximum engine speed adjuster is also sealed under warranty. Do not disturb any locking wires or seals if the vehicle is still within the warranty period, otherwise the warranty will be invalidated. Also do not attempt the timing procedure unless accurate instrumentation is available.

Note: A dial test indicator will be required for this job, together with a special probe, available from your Citroën dealer, for use with the type of pump being measured. The probe used with the Bosch pump comes with a special adapter to screw into the pump.

1 Disconnect the battery negative lead.
2 Cover the alternator with a plastic bag as a precaution against spillage of diesel fuel.

16.6a Remove the inspection plug . . .

3 Apply the handbrake, then jack up the front right-hand corner of the vehicle until the wheel is just clear of the ground, and support the vehicle on an axle stand (see "Jacking and vehicle support"). Engage 4th or 5th gear, to enable the engine to be turned easily by turning the right-hand wheel.
4 Turn the engine to bring No 4 cylinder (timing belt end) to TDC on compression. To establish which cylinder is on compression, either remove No 4 cylinder heater plug and feel for pressure, or remove the valve cover and observe when No 1 cylinder valves are "rocking" (inlet opening and exhaust closing).
5 Insert an 8.0 mm diameter rod or drill through the hole in the left hand flange of the cylinder block next to the starter motor, so that it enters the TDC setting hole in the flywheel. The hole in the cylinder block is awkwardly placed but can be accessed using locking tools, available from your Citroën dealer. You can make up an improvised tool from an M8 bolt with the threads filed away, attached to a piece of welding rod. Alternatively, remove the starter motor and insert a rod or twist drill (see illustrations). Turn the engine back and forward slightly until the tool enters the hole in the flywheel, then leave the tool in position.

Roto-diesel pumps

6 Remove the inspection plug from the top of the pump and position a dial test indicator on the pump so that it can read the movement of a probe inserted into the hole (see illustrations).

Note: A gauge mounted anywhere other than on the pump will also measure the pump movement.
7 Insert a probe into the inspection hole so that the tip of the probe rests on the rotor timing piece. Position the dial test indicator so that it reads the movement of the probe.
8 Remove the TDC setting tool. Turn the engine approximately a quarter-turn backwards. Zero the dial test indicator.
9 Turn the engine forwards slowly until the TDC setting tool can be re-inserted. Read the dial test indicator; the reading should correspond to the value marked on the pump, to within ± 0.03 mm. The marking may be on a plastic disc on the front of the pump, or on a label on the top of the pump or on a tag on the accelerator lever (see illustrations).

Bosch pumps

10 Place rags around the union nuts to catch any fuel that comes out under pressure, and disconnect the injector pipes for cylinders 1 and 2 from the injection pump. Counterhold the unions on the pump while unscrewing the nuts. Cover the open unions to keep dirt out, using small plastic bags secured with rubber bands.
11 Unscrew the blanking plug from the end of the injection pump between the injector pipe connections. Be prepared for the loss of some fuel.
12 Insert the probe and screw adapter, and connect it to the dial test indicator positioned directly over the hole (see illustration).

16.6b . . . and position a dial test indicator to measure the injection timing - Roto-diesel pump

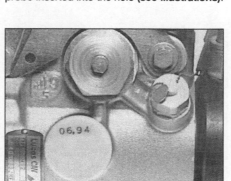

16.9a Pump timing value marked on plastic disc - Roto-diesel pump

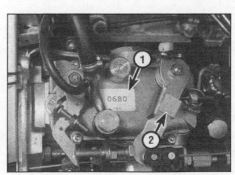

16.9b Pump timing values marked on label (1) and tag (2) - Roto-diesel pump

16.12 Dial test indicator positioned to read injection pump timing - Bosch pump

13 Remove the TDC setting tool. Turn the engine approximately a quarter-turn back-wards. Zero the dial test indicator.

14 Turn the engine forwards slowly until the TDC setting tool can be re-inserted. Read the dial test indicator; the value should correspond to that given in the Specifications.

All pumps

15 If the reading is not as specified, continue as follows.

16 Place rags around the union nuts to catch any fuel that comes out under pressure, and disconnect the injector pipes from the pump. Counterhold the unions on the pump while unscrewing the nuts. Cover the open unions to keep dirt out, using small plastic bags secured with rubber bands.

17 Slacken the pump mounting nuts and bolts, and swing the pump away from the engine. Zero the dial test indicator.

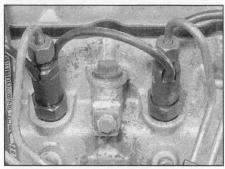

17.4 Leak off pipe connected between two injectors

17.7a Removing an injector

18 With the engine still at TDC, slowly swing the pump back towards the engine until the dial test indicator displays the correct value. In this position, tighten the pump mountings, then remove the TDC setting tool and recheck the timing as described above.

19 When the timing is correct, remove the dial test indicator. Refit the inspection plug on Roto-diesel pumps. Reconnect the injector pipes, and remove the TDC setting tool.

20 Refit any other disturbed components, remove the plastic bag from the alternator, and lower the vehicle to the ground.

17 Fuel injectors - testing, removal and refitting

⚠️ **Warning: Exercise extreme caution when working on the fuel injectors. Never expose the hands or any part of the body to injector spray, as the high working pressure can cause the fuel to penetrate the skin, with possibly fatal results. You are strongly advised to have any work that involves testing the injectors under pressure, carried out by a dealer or fuel injection specialist.**
Caution: Take great care not to allow dirt into the injectors or fuel pipes during this procedure. Cover all open unions to keep dirt out, using suitable plugs, or small plastic bags secured with rubber bands. Do not drop the injectors or allow the needles at their tips to become damaged. The injectors are precision-made to fine limits, and must

17.6 Disconnect the injector pipes

17.7b An injector

not be handled roughly. In particular, do not mount them in a bench vice.

Testing

1 Injectors deteriorate with prolonged use, and it is reasonable to expect them to need reconditioning or renewal after 60 000 miles (100 000 km) or so. Accurate testing, overhaul and calibration of the injectors must be left to a specialist. A defective injector which is causing knocking or smoking can be located without dismantling as follows.

2 Run the engine at fast idle. Slacken each injector union in turn, placing rag around the union to catch spilt fuel, and being careful not to expose the skin to any spray. When the union on the defective injector is slackened, the knocking or smoking will stop.

Removal

3 Clean around the injectors and injector pipe union nuts. Cover the alternator with a clean cloth or plastic bag to protect it from spilt fuel.

4 Pull the leak off pipes from the injectors **(see illustration)**.

5 Loosen the injector pipe union nuts at the injection pump. Counterhold the unions on the pump while unscrewing the nuts.

6 Unscrew the union nuts and disconnect the pipes from the injectors **(see illustration)**. If required, the injector pipes may be completely removed. Cover the ends of the injectors to prevent the entry of dirt.

7 Unscrew the injectors using a deep socket or box spanner, and remove them from the cylinder head **(see illustrations)**.

8 Recover the copper washers and fire-seal washers from the cylinder head. Also remove the sleeves if they are loose or damaged **(see illustrations)**.

> **HAYNES HiNT**
> *If an injector sleeve is tight in the cylinder head, it can be removed as follows. First block the injector sleeve hole with grease, to prevent debris entering the combustion chamber. Cut a thread in the sleeve using a tap, then screw in a stud or bolt, which should have a thread on its entire length. Using a thick washer in contact with the cylinder head, tighten a nut onto the washer, and pull out the sleeve.*

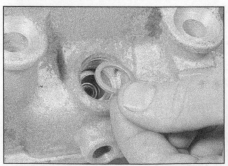

17.8a Remove the injector copper washer . . .

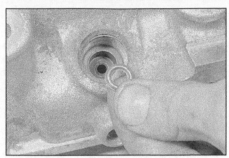

17.8b . . . fire-seal washer . . .

17.8c . . . and sleeve

17.11 Tighten the injector

Refitting

9 Obtain new copper washers and fire-seal washers. Also renew the sleeves, if they are damaged.

10 Insert the sleeves (if removed) into the cylinder head, followed by the fire-seal washers (convex face uppermost), and copper washers.

 HAYNES HiNT *The new injector sleeve may be inserted in the cylinder head by using an old injector as a drift, but do not fit the sealing washer or fire ring while using this method.*

11 Insert the injectors and tighten them to the specified torque **(see illustration)**.
12 Refit the injector pipes and tighten the union nuts to the specified torque.
13 Reconnect the leak off pipes.

18 Inlet manifold - removal and refitting

Removal

1 Remove the air inlet duct from the inlet manifold **(see illustration)**.
2 Using a hexagon key, undo the bolts and remove the inlet manifold from the cylinder head **(see illustrations)**. There are no gaskets.

Refitting

3 Refitting is the reverse of removal, but ensure that the inlet manifold and cylinder head sealing faces are completely clean and flat. If in doubt, apply a thin layer of a non-hardening sealing compound. Tighten the bolts evenly.

19 Exhaust manifold - removal and refitting

Removal

1 If necessary, to enable the exhaust manifold and resonator to be removed from above the engine, remove the inlet manifold as described in Section 18.
2 Apply the handbrake, then jack up the front of the van and support on axle stands (see "*Jacking and vehicle support*").
3 Unscrew and remove the exhaust manifold-to-downpipe bolts, together with their springs and collars **(see illustration)**. Recover the sealing ring from the downpipe or manifold, as applicable. Tie the downpipe to one side.
4 Undo the nuts and withdraw the exhaust manifold and resonator from the studs in the cylinder head. Recover the gaskets **(see illustrations)**.

18.1 Remove the air inlet duct from the inlet manifold

18.2a Undo the bolts (arrowed) . . .

18.2b . . . and remove the inlet manifold

19.3 Undo the exhaust manifold-to-downpipe bolts (arrowed)

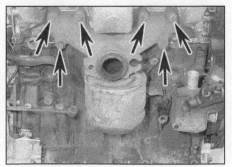

19.4a Undo the nuts (arrowed) and remove the exhaust manifold

19.4b Exhaust manifold gasket

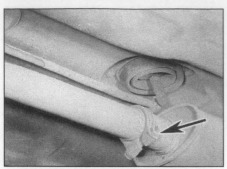

20.6 Slacken the ring clamp (arrowed) connecting the exhaust front downpipe to the intermediate pipe

Refitting

5 Refitting is the reverse of removal, noting the following points:

a) *Refit any studs that have fallen out, and renew any studs that are damaged. Studs can be removed and refitted using two manifold nuts locked together.*

b) *Clean the mating faces between the manifold and cylinder head and fit new gaskets.*

c) *Tighten the manifold nuts evenly.*

d) *When refitting the exhaust downpipe to the manifold, remove all traces of corrosion from the flanges. Use a new sealing ring and apply a smear of high-temperature grease to the joint mating surfaces. Refit the bolts, with the springs and collars, and tighten the nuts progressively until approximately four threads are visible and the springs are compressed to 22.0 mm in length.*

---

20 Exhaust system - general information and component renewal

General information

1 The exhaust system consists of three sections; the front downpipe, the intermediate pipe and silencer box, and the tailpipe and main silencer box. The sections are joined together by ring clamps. The front section connects to the exhaust manifold by a spring-loaded ball/ring joint, to allow movement of the exhaust system.

2 The system is suspended throughout its length by clamps and rubber mountings.

Removal

3 Each exhaust section can be removed individually, or the complete system can be removed as a unit.

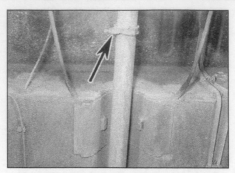

20.10 Slacken the ring clamp (arrowed) connecting the exhaust intermediate pipe to the tailpipe

4 To remove the system or part of the system, first jack up the front or rear of the van, and support it on axle stands (see "*Jacking and vehicle support*"). Alternatively, position the van over an inspection pit, or on car ramps.

⚠ *Warning: When disconnecting the joints between sections of pipework, make sure the pipes on both sides of the joint are supported so that they do not drop down. Do not allow any disconnected pipework to hang down under its own weight, otherwise it will place stress on the remaining joints. If necessary, tie the disconnected pipes to the bodywork with string.*

Front downpipe

5 Unscrew and remove the exhaust manifold-to-downpipe bolts, together with their springs and collars, referring to Section 19. Recover the sealing ring from the downpipe or manifold, as applicable.

6 Slacken the ring clamp and separate the front downpipe from the intermediate pipe **(see illustration)**.

7 Undo any support clamps and remove the front exhaust pipe from the vehicle.

Intermediate pipe

8 The intermediate pipe is difficult to remove by itself because it is held by ring clamps and needs to be pulled out from both the front and rear sections. It may be easier to remove the complete exhaust system. To remove the intermediate pipe, do either of the following:

a) *Remove the front downpipe (see above), then slacken the ring clamp and separate the intermediate pipe from the tailpipe.*

b) *Remove the tailpipe (see below), then slacken the ring clamp and separate the intermediate pipe from the front downpipe.*

9 Undo any support clamps and remove the intermediate exhaust pipe from the vehicle.

Tailpipe

10 Slacken the ring clamp and separate the intermediate pipe from the tailpipe **(see illustration)**.

11 Undo the support clamps and remove the tailpipe from the vehicle.

Complete system

12 Unscrew and remove the exhaust manifold-to-downpipe bolts, together with their springs and collars, referring to Section 19. Recover the sealing ring from the downpipe or manifold, as applicable.

13 With the aid of an assistant, free the system from all its mountings and manoeuvre it out from underneath the vehicle.

Heat shield(s)

14 The heat shields are secured to the underside of the body by a mixture of screws and clips. They can be removed once the relevant exhaust section has been removed.

> **HAYNES HINT**
>
> *If the heat shield is being removed to gain access to a component located behind it, it may be sufficient to simply remove the fasteners and lower the shield, without disturbing the exhaust system.*

Refitting

15 Each section is refitted by a reversal of the removal sequence, noting the following points:

a) *Ensure that all traces of corrosion have been removed from the joints.*

b) *When fitting the front downpipe to the manifold, use a new sealing ring and apply a smear of high-temperature grease to the joint mating surfaces. Refit the bolts, with the springs and collars, and tighten the nuts progressively until approximately four threads are visible and the springs are compressed to 22.0 mm in length.*

c) *On joints which are secured by ring clamps, apply a smear of exhaust system jointing paste to the joint mating surfaces, to ensure a gastight seal.*

d) *Inspect all rubber mountings for signs of damage or deterioration, and renew as necessary. Ensure that the rubber mountings are correctly located, and that there is adequate clearance between the exhaust system and the vehicle underbody.*

Chapter 4 Part D:
Emission control systems

Contents

Degrees of difficulty

Easy, suitable for novice with little experience	**Fairly easy,** suitable for beginner with some experience	**Fairly difficult,** suitable for competent DIY mechanic	**Difficult,** suitable for experienced DIY mechanic	**Very difficult,** suitable for expert DIY or professional 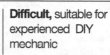

1 General information

1 All C15 petrol engine models can use unleaded petrol. All petrol injection models have a catalytic converter, so that unleaded petrol becomes a requirement.

2 On carburettor petrol models, emissions are controlled mainly by the efficient adjustment of the carburettor, although other emission control systems may be fitted.

3 All petrol injection models are fitted with a carbon canister and purge valve, for evaporative emission control. Together with the closed loop catalytic converter, this became mandatory in 1993 to meet emission control standards. Petrol injection models are also fitted with an ignition timing retard system, to reduce the nitrous oxide (NOx) content of the exhaust gases (see Chapter 5B).

4 Diesel engine models are designed to meet strict emission requirements. In 1993, all diesel models were modified as follows to meet new standards:

a) *On models fitted with Bosch systems, the injection pump was fitted with a device known as NLK, giving quicker and more accurate adjustment of the ignition advance point, using a hydraulic piston.*

b) *On models fitted with Lucas Roto-diesel systems, the ignition advance curve was modified. Reduced volume injectors were also fitted to reduce the amount of unburnt fuel in the exhaust.*

c) *Valve stem oil seals were fitted in January 1992 to both the inlet and exhaust valves to prevent oil from entering the combustion chambers.*

5 Both petrol and diesel models may be fitted with crankcase emission control systems or exhaust gas recirculation systems, if they are required to meet local environmental standards. Unregulated catalytic converters were fitted to the exhaust systems of diesel engines meeting the emission standard W3 (non-UK).

6 The emission control systems are as follows.

Crankcase emission control

7 To reduce emissions of unburned hydrocarbons from the crankcase into the atmosphere, the engine is sealed. The blow-by gases and oil vapour are drawn from inside the crankcase, through a wire-mesh oil separator, into the inlet tract, to be burned by the engine during normal combustion.

8 On petrol engines, under conditions of high manifold depression (idling, deceleration) the gases are sucked positively out of the crankcase. Under conditions of low manifold depression (acceleration, full-throttle running) the gases are forced out of the crankcase by the relatively higher crankcase pressure. If the engine is worn, the raised crankcase pressure, due to increased blow-by, causes some of the flow to return under all manifold conditions. The system operates similarly on diesel engines, except that the manifold depression remains relatively constant, since the engine is not throttled.

Exhaust emission control - petrol models

9 To minimise the amount of pollutants which escape into the atmosphere, all petrol injection models are fitted with a catalytic converter in the exhaust system. The catalytic converter system is the closed-loop type, in which a lambda sensor in the exhaust system provides the fuel-injection/ignition system ECU with constant feedback on the oxygen content of the exhaust gases. This enables the ECU to adjust the mixture to provide the best possible conditions for the converter to operate.

10 The lambda sensor has a built-in heating element, controlled by the ECU through the injection system relay, to quickly bring the sensor's tip to an efficient operating temperature. The sensor's tip is sensitive to oxygen, and sends the ECU a varying voltage depending on the amount of oxygen in the exhaust gases. If the intake air/fuel mixture is too rich, the sensor sends a high-voltage signal. The voltage falls as the mixture weakens. Peak conversion efficiency of all major pollutants occurs if the intake air/fuel mixture is maintained at the chemically-correct ratio for the complete combustion of petrol - 14.7 parts by weight of air to 1 part of fuel, known as the 'stoichiometric' ratio. The sensor output voltage alters in a large step at this point, the ECU using the signal change as a reference point, and correcting the intake air/fuel mixture accordingly by altering the fuel injector pulse width (injector opening time).

Exhaust emission control - diesel models

11 An unregulated catalytic converter consists of a canister containing a fine mesh impregnated with a catalyst material, over which the hot exhaust gases pass. The catalyst speeds up the oxidation of harmful carbon monoxide, unburnt hydrocarbons and soot, effectively reducing the quantity of harmful products released into the atmosphere via the exhaust gases.

Exhaust gas recirculation system

12 This system circulates a proportion of the inert exhaust gas back to the combustion chambers during certain operating conditions, reducing the peak temperatures so that harmful exhaust gas emissions are reduced.

Evaporative emission control - petrol models

13 To minimise the escape into the atmosphere of unburned hydrocarbons, an evaporative emission control system is fitted to petrol injection models. The fuel tank filler cap is sealed, and a charcoal canister collects the petrol vapours generated in the tank when the van is parked. The vapours are stored until they can be cleared from the canister, under the control of the fuel system ECU, via the purge valve(s), into the inlet tract, to be burned by the engine during normal combustion.

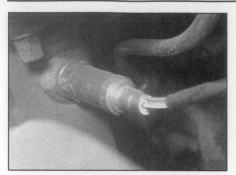

2.7 The lambda sensor is screwed into the top of the exhaust front pipe

14 To ensure that the engine runs correctly when it is cold and/or idling, and to protect the catalytic converter from the effects of an over-rich mixture, the purge control valve(s) are not opened by the ECU until the engine has warmed up, and the engine is under load. The valve solenoid is then modulated on and off to allow the stored vapour to pass into the inlet tract.

2 Emission control systems - testing and component renewal

Crankcase emission control

1 The components of this system require no attention, other than to check at regular intervals that the hose(s) are clear and undamaged.

Exhaust emission control

Testing

2 The performance of the catalytic converter can be checked only by measuring the exhaust gases using a good-quality, carefully-calibrated exhaust gas analyser as described in Chapter 1A or 1B.

3 On petrol injection models, if the CO level at the tailpipe is too high, the vehicle should be taken to a Citroën dealer so that the fuel injection and ignition systems, including the lambda sensor, can be thoroughly checked using special diagnostic equipment. Once these have been checked and are known to be free from faults, the fault must be in the catalytic converter, which must be renewed.

4 On diesel engine models, if the CO level is too high, it is worth checking that the problem is not due to faulty injector(s), particularly before condemning the catalytic converter (if fitted). Refer to your Citroën dealer for further information.

Catalytic converter - renewal

5 The catalytic converter on petrol injection models is integral with the intermediate exhaust section, so that the complete section needs to be renewed. Removal and refitting is the same as for carburettor petrol models (see Chapter 4A), but take care not to damage the lambda sensor in the front downpipe or stress the wires.

6 Catalytic converters fitted to diesel models are also integral with the intermediate exhaust section (see Chapter 4C).

Lambda sensor (petrol models with catalytic converter) - renewal

Note: *The lambda sensor is DELICATE. It will not work if it is dropped or knocked, if its power supply is disrupted, or if any cleaning materials are used on it.*

7 Trace the wiring back from the lambda sensor, which is screwed into the top of the exhaust front pipe, and disconnect it. Free the wiring from any relevant retaining clips or ties **(see illustration)**.

8 Unscrew the sensor from the exhaust system front pipe, and remove it along with its sealing washer.

9 Refitting is the reverse of removal, using a new sealing washer. Before installing the sensor, apply a smear of high-temperature grease to the sensor threads. Ensure that the sensor is securely tightened. Also, the wiring must be correctly routed, and in no danger of contacting either the exhaust system or the engine.

Exhaust gas recirculation (EGR) system

Testing

10 Testing of the system should be entrusted to a Citroën dealer.

Component renewal

11 No specific information is available about removal and refitting the system components.

Evaporative emission control - petrol models

Testing

12 If the system is thought to be faulty, disconnect the hoses from the charcoal canister and purge control valve(s), and check that they are clear by blowing through them. If the purge control valve(s) or charcoal canister are thought to be faulty, they must be renewed.

Component renewal

13 No specific information is available about removal and refitting the system components.

3 Catalytic converter - general information and precautions

1 The catalytic converter is a simple, reliable device which needs no maintenance in itself, but there are some facts which an owner should be aware of, if the converter is to function properly for its full service life.

Petrol models

a) DO NOT use leaded petrol in a van equipped with a catalytic converter - the lead will coat the precious metals, reducing their converting efficiency, and will eventually destroy the converter.

b) Always keep the ignition and fuel systems well-maintained in accordance with the manufacturer's schedule (see Chapter 1A).

c) If the engine develops a misfire, do not drive the van at all (or at least as little as possible) until the fault is cured.

d) DO NOT push or tow-start the van - this will soak the catalytic converter in unburned fuel, causing it to overheat when the engine starts.

e) DO NOT switch off the ignition at high engine speeds - i.e., do not "blip" the throttle immediately before switching off.

f) DO NOT use fuel or engine oil additives - these may contain substances harmful to the catalytic converter.

g) DO NOT continue to use the van if the engine burns oil to the extent of leaving a visible trail of blue smoke.

h) Remember that the catalytic converter operates at very high temperatures. DO NOT, therefore, park the van in dry undergrowth, over long grass, or over piles of dead leaves, after a long run.

i) Remember that the catalytic converter is FRAGILE - do not strike it with tools during servicing work.

j) In some cases, a sulphurous smell (like that of rotten eggs) may be noticed from the exhaust. This is common to many catalytic converter-equipped vans. Once the van has covered a few thousand miles, the problem should disappear. In the meantime, try changing the brand of petrol used.

k) The catalytic converter, used on a well-maintained and well-driven van, should last for between 50,000 and 100,000 miles. If the converter is no longer effective, it must be renewed.

Diesel models

The advice given in paragraphs d), f), g), h), i) and k) above applies equally to the converter on diesel models, if fitted.

Chapter 5 Part A:
Starting and charging systems

Contents

Degrees of difficulty

Easy, suitable for novice with little experience	**Fairly easy,** suitable for beginner with some experience	**Fairly difficult,** suitable for competent DIY mechanic	**Difficult,** suitable for experienced DIY mechanic	**Very difficult,** suitable for expert DIY or professional

Specifications

System type .... 12-volt, negative earth, with alternator and pre-engaged starter motor.

Battery

Type .... Lead-acid, low-maintenance or "maintenance-free" (sealed for life).
Battery capacity:
  Petrol models .... 35 amp hr
  Diesel models .... 55 amp hr
Charge condition:
  Poor .... 12.5 volts
  Normal .... 12.6 volts
  Good .... 12.7 volts

Alternator

Type .... Paris-Rhone, Valeo, Bosch or Mitsubishi (depending on model)

Starter motor

Type .... Ducellier, Valeo, Bosch, Mitsubishi or Iskra (depending on model)

Torque wrench settings	**Nm**	**lbf ft**
Oil pressure switch:		
Petrol models	28	21
Diesel models	20	15
Starter motor bolts (diesel models)	34	25

1 General information and precautions

General information

Because of their engine-related functions, the components of the starting and charging systems are covered separately from the body electrical devices such as the lights and instruments, which are covered in Chapter 12. On petrol models, refer to Part B of this Chapter for information on the ignition system. On diesel models, refer to Part C for information on the preheating system.

The electrical system is the 12-volt negative earth type.

The battery is the low-maintenance or "maintenance-free" (sealed for life) type and is charged by the alternator, which is belt-driven from the crankshaft pulley.

The starter motor is the pre-engaged type, incorporating an integral solenoid. On starting, the solenoid moves the drive pinion into engagement with the flywheel ring gear before the starter motor is energised. Once the engine has started, a one-way clutch prevents the motor armature being driven by the engine until the pinion disengages from the flywheel.

Precautions

Further details of the various systems are given in the relevant Sections of this Chapter. While some repair procedures are given, the usual course of action is to renew the component concerned. The owner whose interest extends beyond mere component renewal should obtain a copy of the "Automobile Electrical & Electronic Systems Manual", available from the publishers of this manual.

It is necessary to take extra care when working on the electrical system, to avoid damage to semi-conductor devices (diodes and transistors), and to avoid the risk of personal injury. In addition to the precautions given in "Safety first!" at the beginning of this manual, observe the following when working on the system:

Always remove rings, watches, etc before working on the electrical system. Even with the battery disconnected, capacitive discharge could occur if a component's live terminal is earthed through a metal object. This could cause a shock or nasty burn.

Do not reverse the battery connections. Components such as the alternator, electronic control units, or any other components having semi-conductor circuitry could be irreparably damaged.

If the engine is being started using jump leads and a slave battery, connect the batteries *positive-to-positive* and *negative-to-negative* (see "Jump starting"). This also applies when connecting a battery charger.

Never disconnect the battery terminals, the alternator, any electrical wiring or any test instruments when the engine is running.

Do not allow the engine to turn the alternator when the alternator is not connected.

Never "test" for alternator output by "flashing" the output lead to earth.

Never use an ohmmeter of the type incorporating a hand-cranked generator for circuit or continuity testing.

Always ensure that the battery negative lead is disconnected when working on the electrical system.

Before using electric-arc welding equipment on the car, disconnect the battery, alternator and components such as the fuel injection/ignition electronic control unit, to protect them from the risk of damage.

Some radio/cassette units have a built-in security code, to deter thieves. If the power source to the unit is cut, the anti-theft system will activate. Even if the power source is immediately reconnected, the radio/cassette unit will not function until the correct security code has been entered. Therefore, if you do not know the correct security code for the radio/cassette unit, do not disconnect the battery negative terminal, or remove the radio/cassette unit from the vehicle. Refer to *"Radio/cassette unit anti-theft system - precaution"* for further information.

2 Electrical fault-finding - general information

Refer to Chapter 12.

3 Battery - testing and charging

Standard and low-maintenance battery - testing

1 If the vehicle covers a small annual mileage, it is worthwhile checking the specific gravity of the electrolyte every three months, to determine the state of charge of the battery. Use a hydrometer to make the check, and compare the results with the following table. Note that the specific gravity readings assume an electrolyte temperature of 15°C (60°F); for every 10°C (18°F) below 15°C (60°F), subtract 0.007. For every 10°C (18°F) above 15°C (60°F), add 0.007. However, for convenience, the temperatures quoted in the table are ambient (outdoor air) temperatures, above or below 25°C (77°F):

	Above 25°C (77°F)	Below 25°C (77°F)
Fully-charged	1.210 to 1.230	1.270 to 1.290
70% charged	1.170 to 1.190	1.230 to 1.250
Discharged	1.050 to 1.070	1.110 to 1.130

2 If the battery condition is suspect, first check the specific gravity of electrolyte in each cell. A variation of 0.040 or more between any cells indicates loss of electrolyte or deterioration of the internal plates.

3 If the specific gravity variation is 0.040 or more, the battery should be renewed. If the cell variation is satisfactory but the battery is discharged, it should be charged as described later in this Section.

Maintenance-free battery - testing

4 In cases where a "sealed for life" maintenance-free battery is fitted, topping-up and testing of the electrolyte in each cell is not possible. The condition of the battery can therefore only be tested using a battery condition indicator or a voltmeter.

5 Certain models may be fitted with a battery with a built-in charge condition indicator. The indicator is located in the top of the battery casing, and indicates the condition of the battery from its colour. If the indicator shows green, then the battery is in a good state of charge. If the indicator turns darker, eventually to black, then the battery requires charging, as described later in this Section. If the indicator shows clear/yellow, then the electrolyte level in the battery is too low to allow further use, and the battery should be renewed. Do not attempt to charge, load or jump-start a battery when the indicator shows clear/yellow.

6 If testing the battery using a voltmeter, connect the voltmeter across the battery, and compare the result with those given in the Specifications under "charge condition". The test is only accurate if the battery has not been subjected to any kind of charge for the previous six hours, including charging by the alternator. If this is not the case, switch on the headlights for 30 seconds, then wait four to five minutes after switching off the headlights before testing the battery. All other electrical circuits must be switched off, so check, for example, that the doors are fully shut when making the test.

7 If the voltage reading is less than 12.2 volts, then the battery is discharged. A reading of 12.2 to 12.4 volts indicates a partially-discharged condition.

8 If the battery is to be charged, remove it from the vehicle (see Section 4) and charge it as follows.

Standard and low-maintenance battery - charging

Note: *The following is intended as a guide only. Always refer to the manufacturer's recommendations (often printed on a label attached to the battery) before charging a battery.*

9 Charge the battery at a rate of 3.5 to 4 amps, and continue to charge the battery at this rate until no further rise in specific gravity is noted over a four-hour period.

10 Alternatively, a trickle charger at the rate of 1.5 amps can safely be used overnight.

11 Specially rapid "boost" charges which are claimed to restore the power of the battery in 1 to 2 hours are not recommended, as they can cause serious damage to the battery plates through overheating.

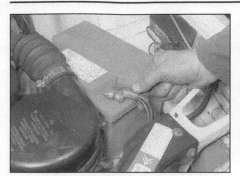

4.2 Disconnect the battery leads

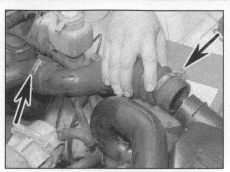

4.3 Undo the clips (arrowed) and remove the air cleaner outlet ducting

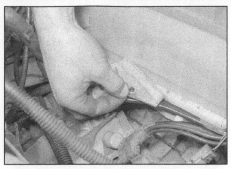

4.4 Remove the battery retaining clamp (air cleaner removed)

12 While charging the battery, note that the temperature of the electrolyte should never exceed 37.8°C (100°F).

Maintenance-free battery - charging

Note: *The following is intended as a guide only. Always refer to the manufacturer's recommendations (often printed on a label attached to the battery) before charging a battery.*

13 This battery type takes longer to fully recharge than the standard type, the time taken being dependent on the extent of discharge, but it can take anything up to three days.

14 A constant-voltage type charger is required, to be set, where possible, to 13.9 to 14.9 volts with a charger current below 25 amps. Using this method, the battery should be usable within three hours, giving a voltage reading of 12.5 volts, but this is for a partially-discharged battery and, as mentioned, full charging can take considerably longer.

4 Battery - removal and refitting

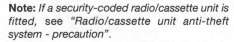

Note: *If a security-coded radio/cassette unit is fitted, see "Radio/cassette unit anti-theft system - precaution".*

Removal

1 The battery is located on the left-hand side of the engine compartment.

2 Remove the terminal insulation covers (where fitted), then loosen the clamp nuts and disconnect the leads from the battery terminals **(see illustration)**. Always disconnect the negative (earth) terminal first, then the positive terminal.

3 On diesel models, undo the clips and remove the air cleaner outlet ducting to gain access to the battery retaining clamp **(see illustration)**. If required, for improved access, remove the complete air cleaner assembly (see Chapter 4C).

4 Undo the nut and remove the battery retaining clamp **(see illustration)**.

5 Lift the battery out of the engine compartment , taking care not to tilt it excessively.

Refitting

6 Refitting is the reverse of removal. Smear petroleum jelly on the terminals after reconnecting the leads, and always reconnect the positive lead first, and the negative lead last.

5 Charging system - testing

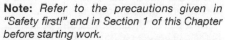

Note: *Refer to the precautions given in "Safety first!" and in Section 1 of this Chapter before starting work.*

1 If the ignition/no-charge warning light does not come on when the ignition is switched on, first check the alternator wiring connections for security. If satisfactory, check that the warning light bulb has not blown, and that the bulbholder is secure in its location in the instrument panel. If the light still fails to illuminate, check the continuity of the warning light feed wire from the alternator to the bulbholder. If all is satisfactory, the alternator is at fault, and should be renewed or taken to an auto-electrician for testing and repair.

2 If the ignition warning light comes on when the engine is running, stop the engine and check that the drivebelt is correctly tensioned (see Chapter 1A or 1B) and that the alternator connections are secure. If all is so far satisfactory, check the alternator brushes and slip rings as described in Section 8. If the fault persists, the alternator should be renewed, or taken to an auto-electrician for testing and repair.

3 If the alternator output is suspect even though the warning light functions correctly, the regulated voltage may be checked as follows.

4 Connect a voltmeter across the battery terminals and start the engine.

5 Increase the engine speed until the voltmeter reading remains steady. The reading should be approximately 12 to 13 volts, and no more than 14 volts.

6 Switch on as many electrical accessories as possible (for example the headlights, heated rear window and heater blower) and check that the alternator maintains the regulated voltage at around 13 to 14 volts.

7 If the regulated voltage is not as stated, the fault may be due to worn brushes, weak brush springs, a faulty voltage regulator, a faulty diode, a severed phase winding, or worn or damaged slip rings. The brushes and slip rings may be checked (see Section 8), but if the fault persists, the alternator should be renewed or taken to an auto-electrician for testing and repair.

6 Alternator drivebelt - removal, refitting and tensioning

Refer to the procedure given for the auxiliary drivebelt in Chapter 1A or 1B.

7 Alternator - removal and refitting

Removal

1 Disconnect the battery negative lead.

2 Slacken the auxiliary drivebelt as described in Chapter 1A or 1B, and disengage it from the alternator pulley.

3 Disconnect the wiring from the rear of the alternator **(see illustration)**.

4 On petrol models, unscrew the alternator upper and lower mounting bolts, then manoeuvre the alternator away from its mounting brackets and out of position.

5 On diesel models, remove the adjuster locknut and swivel the alternator outwards,

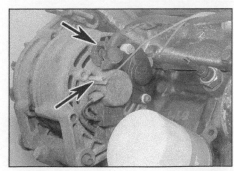

7.3 Disconnect the wiring (arrowed) from the rear of the alternator - diesel model

8.3a On the Valeo alternator, undo the retaining nuts (arrowed) . . .

8.3b . . . and lift off the rear cover

8.5 Pull the plastic cover from the armature shaft

then lift it from the engine. The alternator is slotted so that it can be removed without removing the lower pivot bolt.

Refitting

6 Refitting is the reverse of removal. Tension the auxiliary drivebelt as described in Chapter 1A or 1B, then tighten the pivot bolt and adjuster locknut securely.

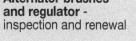

8 Alternator brushes and regulator - inspection and renewal

1 Remove the alternator as described in Section 7. Proceed as described below, under the relevant sub-heading.

Valeo alternator

2 Where applicable, scrape the sealing compound from the rear plastic cover to expose the three rear cover retaining nuts.
3 Undo the retaining nuts and remove the rear cover (see illustrations).
4 If necessary, scrape the sealing compound from the rear of the alternator to expose the regulator/brush holder assembly fixings. The assembly is retained by two nuts and a single screw.
5 Pull the plastic cover from the rear of the armature shaft (see illustration).
6 Undo the retaining nuts and screw, and withdraw the regulator/brush holder assembly from the rear of the alternator (see illustration).
7 Measure the protrusion of each brush from

the its holder. No minimum dimension is specified by the manufacturers, but excessive wear should be self-evident (see illustration). If either brush requires renewal, the complete regulator/brush holder assembly must be renewed. It is not possible to renew the brushes separately.
8 If the brushes are still serviceable, clean them with a petrol-moistened cloth. Check that the brush spring tension is equal for both brushes and provides a reasonable pressure. The brushes must move freely in their holders.
9 Clean the alternator slip rings with a petrol-moistened cloth. Check for signs of scoring, burning or severe pitting on the surface of the slip rings. It may be possible to have the slip rings renovated by an electrical specialist (see illustration).
10 Refit the regulator/brush holder assembly using a reverse of the removal procedure.
11 Refit the alternator as described in Section 7.

Bosch alternator

12 Unclip the cover from the rear of the alternator.
13 If necessary, scrape the sealing compound from the rear of the alternator to expose the regulator/brush holder assembly retaining screws. Slacken and remove the two retaining screws, and remove the regulator/brush holder from the rear of the alternator.
14 Examine the alternator components as described above in paragraphs 7 to 9.
15 Refit the regulator/brush holder assembly,

and securely tighten its retaining screws.
16 Clip the rear cover onto the alternator, and refit the alternator as described in Section 7.

Paris-Rhone alternator

17 Unclip the cover from the rear of the alternator.
18 Disconnect the brush holder/regulator wire, then unscrew and remove the two retaining screws and remove the unit from the rear of the alternator.
19 Examine the alternator components as described above in paragraphs 7 to 9.
20 Refit the regulator/brush holder assembly, and securely tighten its retaining screws.
21 Refit the cover to the rear of the alternator.

Mitsubishi alternator

22 At the time of writing, no information on the Mitsubishi alternator was available. Although the components differ in detail, the same basic principles outlined previously for the Valeo alternator are applicable.

9 Starting system - testing

Note: *Refer to the precautions given in "Safety first!" and in Section 1 of this Chapter before starting work.*
1 If the starter motor fails to operate when the ignition key is turned to the appropriate position, the causes may be as follows:
 a) *The battery is faulty.*

8.6 Regulator/brush holder retaining nuts and screw (arrowed)

8.7 Check the protrusion of the brushes (arrowed) from the holder

8.9 Clean the armature slip rings, and examine them for signs of damage

b) The electrical connections between the switch, solenoid, battery and starter motor are somewhere failing to pass the necessary current from the battery through the starter to earth.

c) The solenoid is faulty.

d) The starter motor is mechanically or electrically defective.

2 To check the battery, switch on the headlights. If they dim after a few seconds, this indicates that the battery is discharged. Recharge the battery (see Section 3) or renew it. If the headlights glow brightly, operate the ignition switch and observe the lights. If they dim, this indicates that current is reaching the starter motor, therefore the fault must be in the starter motor. If the lights continue to glow brightly (and no clicking sound can be heard from the starter motor solenoid), this indicates that there is a fault in the circuit or solenoid - see the following paragraphs. If the starter motor turns slowly when operated, but the battery is in good condition, this indicates that either the starter motor is faulty, or there is considerable resistance somewhere in the circuit.

3 If a fault in the circuit is suspected, disconnect the battery leads (including the earth connection to the body), the starter/solenoid wiring and the engine/transmission earth strap. Thoroughly clean the connections, and reconnect the leads and wiring, then use a voltmeter or test lamp to check that full battery voltage is available at the battery positive connection to the solenoid, and that the earth is sound.

 HAYNES HINT *Smear petroleum jelly around the battery terminals to prevent corrosion - corroded connections are amongst the most frequent causes of electrical system faults.*

4 If the battery and all connections are in good condition, check the circuit by disconnecting the wire from the solenoid blade terminal. Connect a voltmeter or test lamp between the wire end and a good earth (such as the battery negative terminal), and check that the wire is live when the ignition switch is turned to the "start" position. If it is, then the circuit is sound - if not, the circuit wiring can be checked as described in Chapter 12.

5 The solenoid contacts can be checked by connecting a voltmeter or test lamp between the battery positive feed connection on the starter side of the solenoid, and earth. When the ignition switch is turned to the "start" position, there should be a reading or lighted bulb, as applicable. If there is no reading or lighted bulb, the solenoid is faulty and should be renewed.

6 If the circuit and solenoid are proved sound, the fault must lie in the starter motor. In this event, it may be possible to have the

10.6 Remove the starter motor - petrol model

starter motor overhauled by a specialist, but check on the cost of repairs before proceeding as it may be more economical to obtain a new or exchange motor.

10 Starter motor - removal and refitting

1 The starter motor is bolted to the transmission casing, at the rear of the engine on petrol models, and at the front of the engine on diesel models.

Removal

2 Disconnect the battery negative lead.

Petrol models

3 So that access to the motor can be gained both from above and below, firmly apply the handbrake, then jack up the front of the vehicle and support it on axle stands (see "Jacking and Vehicle Support").

4 Disconnect the wiring from the starter motor solenoid. Undo the nuts, as required, and recover the washers.

5 Support the starter motor and undo the three mounting bolts. Two of the bolts are behind the starter motor support bracket, and one is at the top of the transmission housing. Recover the washers, and note the locations of any wiring or support brackets that are secured by the bolts.

6 Withdraw the starter motor and manoeuvre it out from underneath the engine **(see illustration)**. Recover the locating dowel(s) from the motor or transmission.

Diesel models

7 Remove the air cleaner and ducting (see Chapter 4C).

8 Disconnect the wiring from the starter motor solenoid. Undo the nuts, as required, and recover the washers **(see illustration)**.

9 Support the starter motor and undo the three mounting bolts from the transmission using a hexagon key. Recover the washers, and note the locations of any wiring or support brackets that are secured by the bolts **(see illustration)**.

10 Withdraw the starter motor from the transmission.

10.8 Disconnect the starter motor wiring (arrowed) - diesel model

Refitting

11 Refitting is the reverse of removal. Make sure that any wiring or support brackets are in place under the bolt heads, as noted during removal. Tighten the bolts to the specified torque, where given. On petrol models, ensure that the locating dowel(s) are correctly positioned.

11 Starter motor overhaul - general

If the starter motor is thought to be suspect, it should be removed from the vehicle and taken to an auto-electrician for testing. Most auto-electricians will be able to supply and fit brushes at a reasonable cost. However, check on the cost of repairs before proceeding as it may be more economical to obtain a new or exchange motor.

12 Ignition switch - removal and refitting

The ignition switch is integral with the steering column lock, and can be removed as described in Chapter 10.

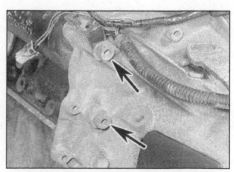

10.9 Undo the starter motor mounting bolts (arrowed) from the transmission - diesel model

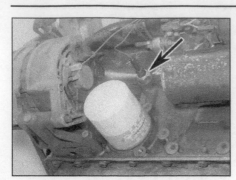

13.1 Oil pressure switch (arrowed) - diesel model

13 Oil pressure warning light switch - removal and refitting

Removal

Note: *If the switch was originally fitted with a sealing ring, a new sealing ring should be used on refitting.*

1 The switch is located at the front of the cylinder block, above the oil filter, on both petrol and diesel models **(see illustration)**.

2 Disconnect the battery negative lead.

3 Remove the protective sleeve from the wiring plug (where applicable), then disconnect the wiring from the switch.

4 Unscrew the switch from the cylinder block, and recover the sealing ring, where applicable.

Be prepared for oil spillage. If the switch is to be left removed from the engine for any length of time, plug the hole in the cylinder block.

Refitting

5 Refitting is the reverse of removal, bearing in mind the following points.

6 If the switch was originally fitted using a sealing compound, clean the switch threads thoroughly, then coat them with fresh compound.

7 If the switch was originally fitted with a sealing ring, use a new sealing ring on refitting.

8 Tighten the switch to the specified torque.

9 When finished, check and, if necessary, top-up the engine oil as described in "*Weekly checks*".

Chapter 5 Part B:
Ignition system - petrol engine models

Contents

Degrees of difficulty

Easy, suitable for novice with little experience	**Fairly easy,** suitable for beginner with some experience	**Fairly difficult,** suitable for competent DIY mechanic	**Difficult,** suitable for experienced DIY mechanic	**Very difficult,** suitable for expert DIY or professional

Specifications

System type
954 cc and 1124 cc carburettor models .... Breakerless electronic ignition system
1124 cc monopoint injection models .... Breakerless electronic ignition system with ignition timing retard system.

Firing order .... 1-3-4-2 (No 1 cylinder at transmission end)

Ignition timing
954 cc and 1124 cc carburettor models .... 8° BTDC @ 750 rpm
1124 cc monopoint injection models* .... 8° BTDC @ idle speed
Idle speed controlled by ECU, not adjustable.

Ignition HT coil
Ignition HT coil resistances:*
  Primary windings .... 0.8 ohms
  Secondary windings .... 6.5 k ohms
The above results are approximate values, and are accurate only when the coil is at 20°C. See text for further information.

Torque wrench setting	Nm	lbf ft
Distributor mounting nuts	8	5

1 General information

All models

A breakerless electronic ignition system is used on all petrol engine models. The system comprises solely of the HT ignition coil and the distributor, both of which are mounted on the left-hand end of the cylinder head, the distributor being driven from the end of the camshaft.

The distributor contains a reluctor mounted onto its shaft and a magnet and stator fixed to its body. The ignition amplifier unit is also mounted onto the side of the distributor body. The system operates as follows.

When the ignition is switched on but the engine is stationary, the transistors in the amplifier unit prevent current flowing through the ignition system primary (LT) circuit.

As the crankshaft rotates, the reluctor moves through the magnetic field created by the stator. When the reluctor teeth are in alignment with the stator projections, a small AC voltage is created. The amplifier unit uses this voltage to switch the transistors in the unit and complete the ignition system primary (LT) circuit.

As the reluctor teeth move out of alignment with the stator projections, the AC voltage changes and the transistors in the amplifier unit are switched again to interrupt the primary (LT) circuit. This causes a high voltage to be induced in the coil secondary (HT) windings, which then travels down the HT lead to the distributor and onto the relevant spark plug.

A TDC sensor is fitted to the rear of the flywheel, but the sensor is not part of the ignition system. It is there to be used for diagnostic purposes only.

1124 cc Monopoint fuel-injected models

On 1124 cc models with the Bosch Monopoint A2.2 fuel injection system, there is an ignition timing retard system, to reduce the nitrous oxide (NOx) content of the exhaust gases. This is achieved by reducing the temperature at the end of the combustion by reducing the ignition advance at certain engine temperatures. This system is controlled by the fuel injection ECU. The ECU has control over an electrically-operated solenoid valve, fitted in the vacuum pipe linking the distributor vacuum diaphragm unit to the throttle body or inlet manifold. At certain engine temperatures, the ECU switches off the solenoid valve, which cuts off the vacuum supply to the distributor vacuum diaphragm, reducing the ignition advance.

2 Ignition system - testing

Warning: Voltages produced by an electronic ignition system are considerably higher than those produced by conventional ignition systems. Extreme care must be taken when working on the system with the ignition switched on. Persons with surgically-implanted cardiac pacemaker devices should keep well clear of the ignition circuits, components and test equipment.

Note: *Refer to the precautions in Chapter 5A, Section 1, before starting work. Always switch off the ignition before disconnecting or connecting any component, and when using a multi-meter to check resistances.*

General

1 The components of electronic ignition systems are normally very reliable. Most faults are far more likely to be due to loose or dirty connections or to "tracking" of HT voltage due to dirt, dampness or damaged insulation than to the failure of any of the system's components. Always check all wiring thoroughly before condemning an electrical component, and work methodically to eliminate all other possibilities before deciding that a particular component is faulty.

2 The old practice of checking for a spark by holding the live end of an HT lead a short distance away from the engine is **not** recommended; not only is there a high risk of a powerful electric shock, but the HT coil or amplifier unit will be damaged. Similarly, **never** try to "diagnose" misfires by pulling off one HT lead at a time.

Engine will not start

3 If the engine either will not turn over at all, or only turns very slowly, check the battery and starter motor. Connect a voltmeter across the battery terminals (meter positive probe to battery positive terminal). Disconnect the ignition coil HT lead from the distributor cap, and earth it. Note the voltage reading obtained while turning over the engine on the

starter for (no more than) ten seconds. If the reading obtained is less than approximately 9.5 volts, first check the battery, starter motor and charging system as described in the relevant Sections of Chapter 5A.

4 If the engine turns over at normal speed but will not start, check the HT circuit by connecting a timing light (following its manufacturer's instructions) and turning the engine over on the starter motor. If the light flashes, voltage is reaching the spark plugs, so these should be checked first. If the light does not flash, check the HT leads themselves, followed by the distributor cap, carbon brush and rotor arm, using the information given in Chapter 1A.

5 If there is a spark, check the fuel system for faults, referring to the relevant Part of Chapter 4 for further information.

6 If there is still no spark, check the voltage at the ignition HT coil "+" terminal; it should be the same as the battery voltage (i.e., at least 11.7 volts). If the voltage at the coil is more than 1 volt less than at the battery, check the feed back through the fusebox and ignition switch to the battery and its earth until the fault is found.

7 If the feed to the HT coil is sound, check the coil's primary and secondary winding resistance as described in Section 3. Renew the coil if faulty, but be careful to check carefully the condition of the LT connections themselves before doing so, to ensure that the fault is not due to dirty or poorly-fastened connectors.

8 If the HT coil is in good condition, the fault is probably within the amplifier unit or distributor stator assembly. Testing of these components should be entrusted to a Citroën dealer.

Engine misfires

9 An irregular misfire suggests either a loose connection or intermittent fault on the primary circuit, or an HT fault on the coil side of the rotor arm.

10 With the ignition switched off, check carefully through the system, ensuring that all connections are clean and securely fastened. If the equipment is available, check the LT circuit as described above.

11 Check that the HT coil, the distributor cap

and the HT leads are clean and dry. Check the leads themselves and the spark plugs (by substitution, if necessary), then check the distributor cap, carbon brush and rotor arm as described in Chapter 1A.

12 Regular misfiring is almost certainly due to a fault in the distributor cap, HT leads or spark plugs. Use a timing light (paragraph 4 above) to check whether HT voltage is present at all leads.

13 If HT voltage is not present on any particular lead, the fault will be in that lead or in the distributor cap. If HT is present on all leads, the fault will be in the spark plugs; check and renew them if there is any doubt about their condition.

14 If no HT is present, check the HT coil; its secondary windings may be breaking down under load.

3 Ignition HT coil - removal, testing and refitting

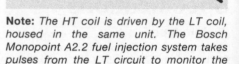

Note: *The HT coil is driven by the LT coil, housed in the same unit. The Bosch Monopoint A2.2 fuel injection system takes pulses from the LT circuit to monitor the engine speed. See Chapter 4B.*

Removal

1 Disconnect the battery negative terminal.

2 Disconnect the hot-air intake hose from the exhaust manifold shroud and air temperature control valve, and remove it from the engine. Release the intake duct fastener, and position the duct clear of the coil.

3 Disconnect the wiring connector from the capacitor mounted on the coil mounting bracket, and release the TDC sensor wiring connector from the front of the bracket **(see illustration)**.

4 Disconnect the HT lead from the coil, then depress the retaining clip and disconnect the coil wiring connector **(see illustrations)**.

5 Slacken and remove the two retaining bolts, and remove the coil and mounting bracket from the cylinder head. Where necessary, slacken and remove the four screws and nuts, and separate the HT coil and mounting bracket **(see illustrations)**.

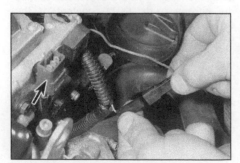

3.3 Disconnect the capacitor wiring connector and release the TDC sensor connector (arrowed) . . .

3.4a . . . then disconnect the HT lead . . .

3.4b . . . and wiring connector (arrowed) from the ignition HT coil

Testing

6 Testing of the coil is carried out using a multi-meter set to its resistance function, to check the primary (LT "+" to "-" terminals) and secondary (LT "+" to HT lead terminal) windings for continuity. Compare the results obtained to those given in the Specifications at the start of this Chapter. The resistance of the coil windings will vary slightly according to the coil temperature - the results in the Specifications are approximate values for when the coil is at 20°C.

7 Check that there is no continuity between the HT lead terminal and the coil body/mounting bracket.

8 If the coil is thought to be faulty, have your findings confirmed by a Citroën dealer before renewing the coil.

Refitting

9 Refitting is the reverse of removal. Ensure that the wiring connectors and the HT lead are securely reconnected.

4 Distributor - removal and refitting

Removal

1 Disconnect the battery negative terminal. If necessary, to improve access to the distributor, remove the ignition HT coil as described in Section 3, and the intake duct as described in the relevant Part of Chapter 4.

3.5a Undo the two retaining bolts (arrowed) . . .

2 Peel back the waterproof cover, then slacken and remove the distributor cap retaining screws. Remove the cap, and place it clear of the distributor body **(see illustrations)**. Recover the seal from the cap.

3 Depress the retaining clip, and disconnect the wiring connector from the distributor. Disconnect the hose from the vacuum diaphragm unit **(see illustrations)**.

4 Check the cylinder head and distributor flange for signs of alignment marks. If no marks are visible, using a scriber or suitable marker pen, mark the relationship of the distributor body to the cylinder head. Slacken and remove the two mounting nuts **(see illustration)**, and withdraw the distributor from the cylinder head. Remove the O-ring from the end of the distributor body, and discard it; a new one must be used on refitting.

3.5b . . . and remove the coil and mounting bracket from the engine

Refitting

5 Lubricate the new O-ring with a smear of engine oil, and fit it to the groove in the distributor body. Examine the distributor cap seal for wear or damage, and renew if necessary.

6 Align the distributor rotor shaft drive coupling key with the slots in the camshaft end, noting that the slots are offset to ensure that the distributor can only be fitted in one position. Carefully insert the distributor into the cylinder head, rotating the rotor arm slightly to ensure that the coupling is correctly engaged.

7 Align the marks noted or made (as applicable) on removal. Install the distributor retaining nuts, tightening them only lightly at this stage.

8 Ensure that the seal is correctly located in its groove, then refit the cap assembly to the

4.2a Peel back the waterproof cover . . .

4.2b . . . then undo the retaining screws . . .

4.2c . . . and remove the cap from the end of the distributor

4.3a Disconnect the distributor wiring connector . . .

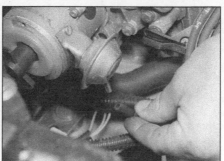

4.3b . . . and the vacuum diaphragm hose . . .

4.4 . . . then undo the retaining nuts and remove the distributor

5.2 The ignition amplifier unit is secured to the side of the distributor by two screws

distributor, and tighten its retaining screws securely. Fold the waterproof cover back over the distributor cap, ensuring that it is correctly located.

9 Reconnect the vacuum hose to the diaphragm unit, and the distributor wiring connector. Where necessary, refit the ignition HT coil as described in Section 3, and the intake duct as described in the relevant Part of Chapter 4.

10 Check and, if necessary, adjust the ignition timing as described in Section 7, then tighten the distributor mounting nuts to the specified torque.

5 Ignition system amplifier unit - removal and refitting

Removal

1 Disconnect the battery negative terminal.

2 The amplifier unit is mounted onto the side of the distributor body **(see illustration)**. To improve access to the unit, disengage the hot-air intake hose from the control valve and the exhaust manifold shroud, and remove it from the vehicle.

3 Disconnect the wiring connector, then undo the two retaining screws and remove the amplifier unit.

Refitting

4 Refitting is the reverse of removal.

6 Ignition timing retard solenoid valve (injection models) - general

1 The solenoid valve is part of the ignition timing retard system, fitted to 1124 cc models with the Bosch Monopoint A2.2 fuel injection system. The valve is fitted to the vacuum pipe linking the distributor vacuum diaphragm unit to the throttle body or inlet manifold.

2 No specific instructions are available about testing, removing or refitting the solenoid valve. If the valve is thought to be faulty, see your Citroën dealer.

7 Ignition timing - checking and adjustment

Checking

1 To check the ignition timing, a stroboscopic timing light will be required. It is also recommended that the flywheel timing mark is highlighted as follows.

2 Remove the plug from the aperture on the front of the clutch bellhousing on the transmission. Using a socket and suitable extension bar on the crankshaft pulley bolt, slowly turn the engine over until the timing mark (a straight line) scribed on the edge of the flywheel appears in the aperture. (Turning the engine will be much easier if the spark plugs are removed first - see Chapter 1A.) **(see illustrations)**.

 HAYNES HINT *Highlight the timing mark with quick-drying white paint - typist's correction fluid is ideal.*

3 Start the engine, allow it to warm up to normal operating temperature, then switch it off.

4 Disconnect the vacuum hose from the distributor diaphragm, and plug the hose end.

5 Connect the timing light to No 1 cylinder plug lead (nearest the transmission), as described in the timing light manufacturer's instructions.

6 Start the engine, allowing it to idle at the specified speed, and point the timing light at the transmission housing aperture. The flywheel timing mark should be aligned with the relevant notch on the timing plate (refer to the Specifications for the correct timing setting); the numbers on the plate indicate degrees Before Top Dead Centre (BTDC).

Adjustment

7 If adjustment is necessary, slacken the two distributor mounting nuts, then slowly rotate the distributor body as required until the flywheel mark and relevant timing plate notch are brought into alignment. Once the marks are correctly aligned, hold the distributor stationary, and tighten its mounting nuts to the specified torque. Recheck that the timing marks are still correctly aligned and, if necessary, repeat the adjustment procedure.

8 When the timing is correctly set, increase the engine speed and check that the pulley mark advances to beyond the beginning of the timing plate reference marks, returning to close to the specified mark when the engine is allowed to idle. This check shows that the centrifugal advance mechanism is at least functioning - a detailed check must be left to a Citroën dealer. Reconnect the vacuum hose to the distributor, and repeat the check. The rate of advance should significantly increase if the vacuum diaphragm is functioning correctly, but again, a detailed check must be left to a Citroën dealer.

9 When the ignition timing is correct, stop the engine and disconnect the timing light.

7.2a Remove the plug from the transmission bellhousing . . .

7.2b . . . to reveal the timing plate and flywheel timing mark (arrowed)

Chapter 5 Part C:
Preheating system - diesel engine models

Contents

Degrees of difficulty

Easy, suitable for novice with little experience	**Fairly easy,** suitable for beginner with some experience	**Fairly difficult,** suitable for competent DIY mechanic 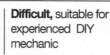	**Difficult,** suitable for experienced DIY mechanic	**Very difficult,** suitable for expert DIY or professional

Specifications

Glow plugs

Type	Champion CH68 or equivalent
Resistance	0.4 ohms to 0.6 ohms
Current draw:	
Up to 1996	12 amp after 20 sec
From 1996	9 amp after 20 sec

Heating times*

Preheating time until warning light goes out	5 sec @ 0°C
	4 sec @ 20°C
Post-heating time after cold start:	
Up to 1996	50 sec (typically)
From 1996	60 sec @ -30°C
	26 sec @ 0°C
Coolant temperature cut off	70°C

Heating times depend on the underbonnet air temperature measured by the heater control unit. From 1996 onwards a coolant temperature sensor, fitted to the thermostat housing, interrupts the post-heating when the coolant temperature reaches the cut-off value.

Torque wrench setting

	Nm	lbf ft
Glow plugs	22	16

---

1 Preheating system - description and testing

Description

1 Each swirl chamber has a heater plug (commonly called a glow plug) screwed into it. The plugs are electrically operated by the preheating control unit before, during and immediately after starting a cold engine. Heating is not applied if the engine has already been warmed up.

2 A warning light on the instrument panel tells the driver that preheating is taking place. When the light goes out, the engine is ready to be started. If no attempt is made to start, the timer then cuts off the supply, in order to avoid draining the battery and overheating the glow plugs.

3 After a cold engine has been started, the glow plugs continue to operate for a short time known as the "post-heating" period.

4 Heating times depend on the underbonnet air temperature measured by the heater control unit. On models from 1996 onwards, a temperature sensor is fitted to the thermostat housing so that post-heating is discontinued when the coolant has warmed up to the specified temperature.

Testing

5 If the system malfunctions, testing is ultimately by substitution of known good units, but some preliminary checks may be made as follows.

6 Disconnect the main supply cable from the No 1 cylinder glow plug (counting from the flywheel).

7 Connect a voltmeter or 12-volt test lamp between the supply cable and earth, making sure that the cable is kept clear of the engine and bodywork. Have an assistant switch on the ignition and check that there is a voltage supply for several seconds. Note the time for which the preheating warning light is lit, and the total time for which voltage is supplied

before the system cuts out, then switch off the ignition.

8 At an under-bonnet temperature of 20°C, typical times noted should be about 4 seconds for warning light operation, followed by a further 10 seconds supply after the light goes out. Warning light time will increase with lower temperatures and decrease with higher temperatures.

9 If there is no supply, the relay or associated wiring is at fault.

10 To locate a defective glow plug, disconnect the main supply cable and interconnecting wire from the top of the glow plugs. Be careful not to drop the nuts and washers.

11 Use a continuity tester, or a 12-volt test light connected to the battery positive terminal, to check for continuity between each glow plug terminal and earth. The resistance of a glow plug in good condition is very low (see Specifications), so if the test light does not come on or the continuity tester shows a high resistance, the glow plug is certainly defective.

2.2 Undo the nut and disconnect the wiring from the glow plug. Note the interconnecting cable (arrowed) between all four plugs.

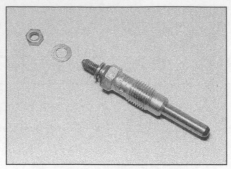

2.4 Glow plug and terminal nut

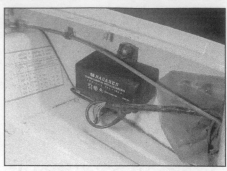

3.1 Preheating system control unit

12 If an ammeter is available, the current draw of each glow plug can be checked and should be according to the Specifications, after an initial surge of about 50% higher. Any plug which draws much more or less than this is probably defective.

13 As a final check, the glow plugs can be removed and inspected (see Section 2).

2 Glow plugs - removal, inspection and refitting

Removal

Caution: If the preheating system has just been energised, or if the engine has been running, the glow plugs will be very hot.

1 Disconnect the battery negative lead.

2 Unscrew the nuts from the relevant glow plug terminals, recover the washers, and disconnect the wiring. The main supply cable is connected to No 1 cylinder glow plug (counting from the flywheel), and an interconnecting cable is fitted between the four plugs **(see illustration)**.

3 Where applicable, carefully move any obstructing pipes or wires to one side, to gain access to the relevant glow plugs.

4 Unscrew the glow plugs and remove them from the cylinder head **(see illustration)**.

Inspection

5 Inspect each glow plug for physical damage. Burnt or eroded glow plug tips can be caused by a bad injector spray pattern. Have the injectors checked if this sort of damage is found.

6 If the glow plugs are in good physical condition, check them electrically using a 12-volt test light or continuity tester, as described in Section 1.

7 The glow plugs can be energised by applying 12 volts to them, to verify that they heat up evenly and in the required time. Observe the following precautions:

  a) Support the glow plug by clamping it carefully in a vice or in self-locking pliers. Remember - it will become red-hot.
  b) Make sure that the power supply or test lead incorporates a fuse or overload trip, to protect against damage from a short-circuit.
  c) After testing, allow the glow plug to cool for several minutes before attempting to handle it.

8 A glow plug in good condition will start to glow red at the tip after drawing current for about 5 seconds. Any plug which takes much longer to start glowing, or which starts glowing in the middle instead of at the tip, is defective.

Refitting

9 Refitting is the reverse of removal. Apply a smear of copper-based anti-seize compound to the plug threads, and tighten the glow plugs to the specified torque. Do not overtighten, as this can damage the glow plug element.

3 Preheating system control unit - removal and refitting

Removal

1 The unit is located on the left-hand side of the engine compartment, on the inner wing panel **(see illustration)**.

2 Disconnect the battery negative lead.

3 Unscrew the retaining bolt securing the unit to the inner wing panel.

4 Note the location of the wiring, then disconnect it from the base of the unit. Remove the unit from the engine compartment.

Refitting

5 Refitting is the reverse of removal, ensuring that all wiring is securely connected to its original location, as noted on removal.

Chapter 6
Clutch

Contents

Degrees of difficulty

| Easy, suitable for novice with little experience | 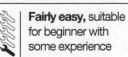 | Fairly easy, suitable for beginner with some experience | | Fairly difficult, suitable for competent DIY mechanic | | Difficult, suitable for experienced DIY mechanic | | Very difficult, suitable for expert DIY or professional | |

Specifications

Type .... Single dry plate with diaphragm spring. Cable-operated release mechanism.

Clutch pedal travel .... 130 mm

Friction plate diameter
Petrol models* .... 180 mm or 181.5 mm
Diesel models .... 200 mm
*Take the old friction plate to your Citroën dealer for an exact replacement part.

Pressure plate and release bearing (BE3 transmissions)
Pressure plate thickness from flywheel to ends of diaphragm spring fingers:
    Up to 1998 .... 34.0 mm
    From 1998 .... 36.0 mm
Release bearing thickness in direction of input shaft:
    Up to 1998 .... 20.5 mm
    From 1998 .... 18.5 mm

Torque wrench settings	Nm	lbf ft
Pressure plate retaining bolts:		
Petrol models	15	11
Diesel models	22	16

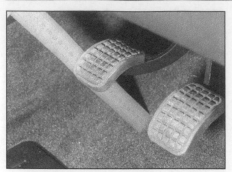

2. 1 Check the clutch cable adjustment by measuring the pedal travel

1 General information

1 The clutch consists of a friction plate, a pressure plate assembly, a release bearing, and the release mechanism. All of these components are contained in the large cast aluminium alloy bellhousing, sandwiched between the engine and the transmission. The release mechanism is mechanical, being operated by a cable.
2 The friction plate is fitted between the engine flywheel and the clutch pressure plate, and is allowed to slide on the transmission input shaft splines. It consists of two circular facings of friction material riveted in position to provide the clutch friction surface, and a spring-cushioned hub to damp out transmission shocks.
3 The pressure plate assembly is bolted to the flywheel, and is located by three dowel pins. When the engine is running, drive is transmitted from the crankshaft via the flywheel to the friction plate (these components being clamped securely together by the pressure plate assembly) and from the friction plate to the transmission input shaft.
4 The clutch is disengaged by a sealed release bearing fitted concentrically around the transmission input shaft. When the driver depresses the clutch pedal, the release bearing is pressed against the fingers at the centre of the diaphragm spring. Since the spring is held by rivets between two annular

fulcrum rings, the pressure at its centre causes it to deform, so that it releases the clamping force from the pressure plate.
5 Depressing the clutch pedal pulls the control cable, and this in turn pulls the lever which rotates the release fork. The release fork forces the release bearing against the diaphragm spring.
6 As the friction plate facings wear, the pressure plate moves towards the flywheel. This causes the diaphragm spring fingers to push against the release bearing, thus changing the position of the clutch pedal. To ensure correct operation, the clutch cable must be regularly adjusted.
7 The clutch components are essentially the same for all models, except for the following:
   a) *Diesel models up to April 1989 with BE1 transmissions have a clutch release fork which is separate from the release shaft and pivots on a ball-stud. The clutch pressure plate has a different design but is functionally the same as on later models.*
   b) *The clutch pressure plate and release bearing were redesigned in 1998 with different thickness dimensions. When renewing, make sure you get the identical replacement parts from your Citroën dealer.*

2 Clutch - adjustment

1 The clutch adjustment is checked by measuring the clutch pedal travel. Ensure that there are no obstructions beneath the pedal. Depress the pedal fully to the floor, measuring the distance that the centre of the pedal pad travels, from the at-rest position to the floor **(see illustration)**. If this is not according to the Specifications, adjust the clutch as follows.

> **HAYNES HINT** *Using a ruler for this check might prove awkward - as an alternative, mark the at-rest and fully-depressed positions on a strip of wood, and measure the distance (the pedal travel) between them.*

2.2 Clutch cable adjuster nut and locknut (arrowed) - air cleaner, ducting, battery and tray removed for clarity

2 The clutch cable adjuster nut and locknut are on the transmission end of the cable **(see illustration)**. To access the adjuster, it may be necessary to remove the air cleaner and/or ducting (see the relevant part of Chapter 4).
3 Slacken the locknut and adjust the position of the adjuster nut, then re-measure the clutch pedal travel. Repeat this procedure until the clutch pedal travel is as specified.
4 Once the adjuster nut is correctly positioned and the pedal travel is correctly set, securely tighten the locknut. Refit the air cleaner and ducting, if removed.

3 Clutch cable - removal and refitting

Removal

1 To improve access to the transmission end of the clutch cable, remove the air cleaner and/or ducting (see the relevant part of Chapter 4).
2 The inner clutch cable and outer sleeve are held to the clutch release lever and mounting bracket by plastic collars. Slacken the locknut and adjuster nut at the end of the clutch cable, then release the inner cable from the release lever by pulling out the collar. Then pull out the retaining clip and release the outer cable collar from the mounting bracket **(see illustrations)**. **Note:** *The cable end fittings might be slightly different on petrol models, but the method of removal should be self-evident.*
3 Working inside the vehicle, remove the parcel shelf from below the steering column (see Chapter 11), then unhook the clutch inner cable from the top of the clutch pedal.
4 Working within the engine compartment, release the outer cable from the engine compartment bulkhead, then withdraw the cable through the bulkhead, noting its correct routing, and remove it from the vehicle.
5 Examine the cable, looking for worn end fittings or a damaged outer sleeve, and for signs of fraying of the inner wire. Check the cable's operation - the inner wire should move smoothly and easily through the outer sleeve. Remember, however, that a cable that

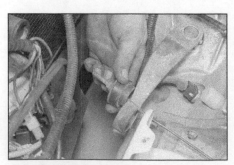

3.2a Slacken the locknut and adjuster nut and release the inner cable from the clutch release lever . . .

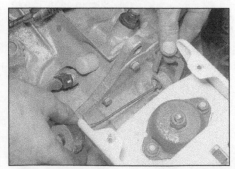

3.2b . . . then pull out the clip and release the outer cable from the mounting bracket

appears serviceable when tested off the van may well be much heavier in operation, when compressed into its working position. Renew the cable if it shows any signs of excessive wear or damage.

Refitting

6 Apply a thin smear of multi-purpose grease to the cable end fittings, then pass the cable through the engine compartment bulkhead. Clip the cable into position, making sure that it is routed correctly.

7 From inside the vehicle, hook the inner cable over the clutch pedal end, and check that it is securely retained. Refit the parcel shelf.

8 Refit the inner cable to the release lever, ensuring that the plastic collar is securely fitted to the lever. Then refit the outer cable plastic collar to the mounting bracket and refit the clip.

9 Refit the flat washer, adjusting nut and locknut (if removed) and adjust the clutch cable as described in Section 2.

4 Clutch pedal - general, removal and refitting

General

1 The clutch pedal is integral with either the brake pedal assembly or the brake linkage cross-tube, depending on the relative positions of the brake pedal and brake vacuum servo unit.

a) On carburettor petrol models and on all left-hand drive models, the brake servo unit is directly in front of the brake pedal. The clutch pedal and brake pedal are supported by the same mounting bracket.

b) On diesel models and monopoint petrol injection models, the brake servo unit is always on the left hand side of the bulkhead, regardless of whether the vehicle is left-hand or right-hand drive. On right-hand drive models, a cross-tube connects the brake pedal to the servo unit pushrod. The clutch pedal is integral with the cross-tube.

Models with cross-tube

2 Remove and refit the cross-tube as described in Chapter 9. No specific instructions are available about separation of the clutch pedal from the cross-tube, although it should become self-evident. Consult your Citroën dealer about the availability of parts.

Models without cross-tube

Removal

3 The clutch pedal is integral with the brake pedal mounting bracket.

4 Remove the parcel shelf from below the steering column (see Chapter 11).

5 Disconnect the battery negative lead, then disconnect the cable from the stop lamp switch, mounted on the pedal bracket.

6 Remove the clevis pin and disconnect the servo unit pushrod from the brake pedal.

7 Disconnect the cable from the clutch pedal.

8 Undo the nuts and detach the pedal bracket from the bulkhead.

9 No specific instructions are available about separation of the pedal from the bracket, although it should become self-evident. Consult your Citroën dealer about the availability of parts.

Refitting

10 Refitting is the reverse of removal. Adjust the clutch cable as described in Section 2.

5 Clutch assembly - removal, inspection and refitting

Warning: Dust created by clutch wear and deposited on the clutch components may contain asbestos, which is a health hazard. DO NOT blow it out with compressed air, and don't inhale any of it. DO NOT use petrol or petroleum-based solvents to clean off the dust. Brake system cleaner or methylated spirit should be used to flush the dust into a suitable receptacle. After the clutch components are wiped clean with rags, dispose of the contaminated rags and cleaner in a sealed, marked container. Although some friction materials may no longer contain asbestos, it is safest to assume that they DO, and to take precautions accordingly.

Removal

1 Unless the engine/transmission is to be removed from the van and separated for major overhaul (Chapter 2C), the clutch can be reached by removing the transmission alone, as described in Chapter 7.

2 Before disturbing the clutch, use chalk or a marker pen to mark the relationship of the pressure plate assembly to the flywheel.

3 Working in a diagonal sequence, slacken the pressure plate bolts by half a turn at a time, until the spring pressure is released and the bolts can be unscrewed by hand.

4 Prise the pressure plate assembly off its locating dowels, and collect the friction plate, noting which way round the friction plate is fitted **(see illustration)**. Note: Models with BE1 transmissions up to April 1989 have a pressure plate design which is different from later models, although the method of removal is the same **(see illustration)**.

Inspection

Note: Due to the amount of work necessary to gain access to the clutch components, it is usually considered good practice to renew the clutch friction plate, pressure plate assembly and release bearing as a matched set, even if only one of these is actually worn enough to require renewal.

5.4a Remove the pressure plate and friction plate, noting which way around the friction plate is fitted

5.4b Removing the pressure plate - BE1 transmissions up to April 1989

5.16a Insert the aligning tool into the end of the crankshaft . . .

5.16b . . . and engage it with the friction plate hub

5 When cleaning clutch components, first read the warning at the beginning of this Section. Remove any dust using a clean, dry cloth, and working in a well-ventilated atmosphere.

6 Check the friction plate facings for signs of wear, damage or oil contamination. If the friction material is cracked, burnt, scored or damaged, or if it is contaminated with oil or grease (shown by shiny black patches), the friction plate must be renewed.

7 If the friction material is still serviceable, check that the centre boss splines are unworn, that the torsion springs are in good condition and securely fastened, and that all the rivets are tightly fastened. If any wear or damage is found, the friction plate must be renewed.

8 If the friction material is fouled with oil, this must be due to an oil leak from the crankshaft left-hand oil seal, from the sump-to-cylinder block joint, or from the transmission input shaft. If a leak is evident, renew the seal or repair the joint (as appropriate) as described in the relevant part of Chapter 2 or 7, before installing the new friction plate.

9 Check the pressure plate assembly for obvious signs of wear or damage. Shake it to check for loose rivets or damaged fulcrum rings. Check that the drive straps securing the pressure plate to the cover do not show signs

5.17 Once the friction plate is centralised, tighten the pressure plate retaining bolts to the specified torque

of overheating (such as a deep yellow or blue discolouration). If the diaphragm spring is worn or damaged, or if its pressure is in any way suspect, the pressure plate assembly should be renewed.

10 Examine the machined bearing surfaces of the pressure plate and the flywheel. They should be clean, completely flat and free from scratches or scoring (minor damage of this nature can sometimes be polished away using emery paper). If either is discoloured from excessive heat, or shows signs of cracking, it should be renewed.

11 Check that the release bearing contact surface rotates smoothly and easily, with no sign of noise or roughness, and that the surface itself is smooth and unworn, with no signs of cracks, pitting or scoring. If there is any doubt about its condition, the bearing must be renewed. It is generally considered good practice to renew the bearing regardless of its apparent condition, whenever a new clutch plate is fitted (see note above).

Refitting

12 On reassembly, ensure that the bearing surfaces of the flywheel and pressure plate are completely clean, smooth, and free from oil or grease. Use solvent to remove any protective grease from new components.

13 Fit the friction plate so that its spring hub assembly faces away from the flywheel; there may also be a marking showing which way round the plate is to be refitted.

14 Refit the pressure plate assembly, aligning the marks made on dismantling (if the original pressure plate is re-used) and locating the pressure plate on its three locating dowels. Fit the pressure plate bolts, but tighten them only finger-tight, so that the friction plate can still be moved.

15 The friction plate must now be centralised, so that when the transmission is refitted, its input shaft will pass through the splines at the centre of the friction plate.

16 Centralisation can be achieved by passing

a screwdriver or other long bar through the friction plate and into the hole in the crankshaft. The friction plate can then be moved around until it is centred on the crankshaft hole. Alternatively, a clutch-aligning tool can be used to eliminate the guesswork. These can be obtained from most accessory shops **(see illustrations)**.

TOOL TiP *A home-made aligning tool can be made up from a length of metal rod or wooden dowel which fits closely inside the crankshaft hole, and has insulating tape wound around it to match the diameter of the friction plate splined hole.*

17 When the friction plate is centralised, tighten the pressure plate bolts evenly and in a diagonal sequence to the specified torque **(see illustration)**.

18 Apply a thin smear of high-melting point grease to the splines of the friction plate and the transmission input shaft, and also to the release bearing bore and release fork shaft. Do not apply too much grease, or it may find its way onto the friction surfaces of the new clutch plate.

19 Refit the transmission as described in Chapter 7.

---

6   Clutch release mechanism -
removal, inspection
and refitting

Note: *Refer to the warning concerning the dangers of asbestos dust at the beginning of Section 5.*

MA4, MA5 and BE3 transmissions

Note: *The release fork and pivot bushes are slightly different for MA and BE3 transmissions, but removal and refitting is the same.*

6.2a Clutch release bearing location

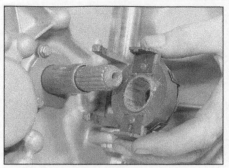

6.2b Clutch release bearing removal - MA transmission

6.3 Drive out the roll pin (arrowed) and remove the clutch release lever - BE3 transmission

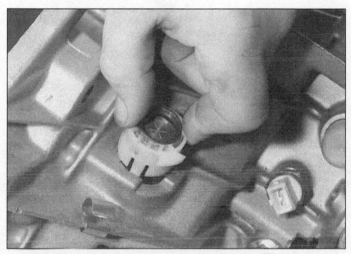

6.4a Release the upper pivot bush and slide it off the shaft . . .

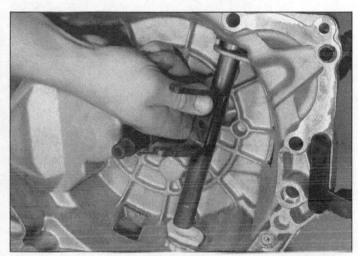

6.4b . . . then manoeuvre the shaft out from the transmission . . .

Removal

1 Unless the engine/transmission is to be removed from the van and separated for major overhaul (Chapter 2C), the clutch release mechanism can be reached by removing the transmission alone, as described in Chapter 7.
2 Unhook the release bearing from the fork, and slide it off the input shaft **(see illustrations)**.
3 If required, the release lever and shaft can be removed as follows. Drive out the roll pin, then remove the release lever from the top of the release fork shaft **(see illustration)**.
4 Depress the retaining tabs, then slide the upper bush off the end of the release fork shaft. Disengage the shaft from its lower bush, and manoeuvre it out from the transmission **(see illustrations)**. Depress the retaining tabs, and remove the lower pivot bush from the transmission housing.

Inspection

5 Check the release mechanism, renewing any components which are worn or damaged. Carefully check all bearing surfaces and points of contact.

6 Check the release bearing itself, noting that it is generally considered worthwhile to renew it as a matter of course, having got this far. Check that the contact surface rotates smoothly and easily, with no sign of noise or roughness, and that the surface itself is smooth and unworn, with no signs of cracks, pitting or scoring. If there is any doubt about its condition, the bearing must be renewed.

Refitting

7 Apply a smear of high-melting point grease to the shaft pivot bushes and the contact surfaces of the release fork.
8 Locate the lower pivot bush in the transmission, ensuring that it is securely retained by its locating tangs, and refit the release fork. Slide the upper bush down the shaft, and clip it into position in the transmission housing.
9 Refit the release lever to the shaft. Align the lever with the shaft hole, and secure it in position by tapping in a new roll pin. Slide the release bearing onto the input shaft, and engage it with the release fork.
10 Refit the transmission as described in Chapter 7.

Clutch release fork - BE1 transmissions

11 Diesel models were fitted with BE1 transmissions until April 1989 when the BE3 transmission was introduced. The clutch release mechanism for the BE1 transmission is similar to the MA4, MA5 and BE3 transmissions except that it has a release fork which is separate from the shaft, and pivots on a ball-stud. To remove and refit the release bearing and fork, proceed as follows:

6.4c . . . then remove the lower bush - BE3 transmission

6.12 Remove the clutch release bearing from the release fork - BE1 transmission

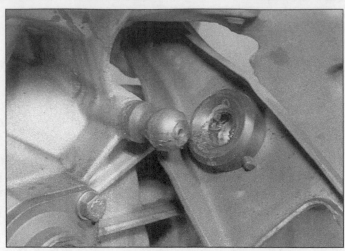

6.13 Detach the release fork from the ball stud - BE1 transmission

Removal

12 With the transmission separated from the engine, release the spring clips and slide the release bearing from the guide sleeve **(see illustration)**.

13 Detach the release fork from the ball stud **(see illustration)**.

14 If required, the pivot bush can be removed from the fork by drilling out the rivets, and the ball stud can be removed from the transmission casing using a slide hammer with a suitable claw.

Refitting

15 If the pivot bush has been removed from the fork, fit a new one by passing its rivets through the holes. Heat the rivets with a cigarette lighter or a low powered blowlamp, then flatten or bend over their heads while they are still hot.

16 If the ball stud has been removed, support the transmission casing and drive in a new stud using a soft metal drift.

17 Lightly coat the bush with high-melting point grease, then refit the release fork. Press it onto the ball-stud until it snaps home.

18 Refit the remaining components using a reversal of the removal procedure.

Chapter 7
Manual transmission

Contents

Degrees of difficulty

| Easy, suitable for novice with little experience | Fairly easy, suitable for beginner with some experience | Fairly difficult, suitable for competent DIY mechanic 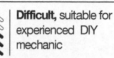 | Difficult, suitable for experienced DIY mechanic | Very difficult, suitable for expert DIY or professional |

Specifications

General

Type .... Manual, four or five forward speeds and reverse. Synchromesh on all forward speeds

	Four-speed transmission	Five-speed transmission
Designation:		
Petrol engines	MA4	MA5
Diesel engines:		
Up to April 1989	BE1/4	BE1/5
From April 1989	BE3/4 (until 1990)	BE3/5

Gear ratios

	Ratio	Teeth
MA4 transmission (petrol models):		
1st	3.417 : 1	12/41
2nd	1.810 : 1	21/38
3rd	1.129 : 1	31/35
4th		
up to September 1991	0.814 : 1	43/35
from September 1991	0.854 : 1	41/35
Reverse	3.583 : 1	12/43
Final drive	4.538 : 1	13/59
MA5 transmission (petrol models):		
1st	3.417 : 1	12/41
2nd	1.810 : 1	21/38
3rd	1.276 : 1	29/37
4th	0.975 : 1	40/39
5th	0.767 : 1	43/33
Reverse	3.583 : 1	12/43
Final drive	4.538 : 1	13/59
BE1/4 transmission (diesel models, typical values):		
1st	3.250 : 1	12/39
2nd	1.850 : 1	20/37
3rd	1.148 : 1	27/31
4th	0.800 : 1	35/28
Reverse	3.333 : 1	12/40
Final drive	3.588 : 1	17/61
BE1/5 transmission (diesel models, typical values):		
1st	3.250 : 1	12/39
2nd	1.850 : 1	20/37
3rd	1.280 : 1	25/32
4th	0.969 : 1	32/31
5th	0.757 : 1	37/28
Reverse	3.333 : 1	12/40
Final drive	4.188 : 1	16/67

Gear ratios (continued)

	Ratio	Teeth
BE3/4 transmission (diesel models, typical values):		
1st	3.455 : 1	11/38
2nd	1.850 : 1	20/37
3rd	1.148 : 1	27/31
4th	0.800 : 1	35/28
Reverse	3.333 : 1	12/40
Final drive	3.588 : 1	17/61
BE3/5 transmission (diesel models, typical values):		
1st	3.455 : 1	11/38
2nd:		
up to April 1993	1.850 : 1	20/37
from April 1993	1.870 : 1	23/43
3rd	1.280 : 1	25/32
4th	0.969 : 1	32/31
5th	0.757 : 1	37/28
Reverse	3.333 : 1	12/40
Final drive	4.188 : 1	16/67

Lubrication

Recommended oil	see "Lubricants and fluids"
Capacity*	see Chapter 1A or 1B

*Oil capacity varies according to model and production date. Fill the transmission until oil runs out of the filler/level hole, then check the level as described in Chapter 1A or 1B.

Gearchange linkage

Petrol models:	
Selection link rod*	125 ± 1 mm
Engagement link rod*	205 ± 1 mm
Diesel models:	
BE1 transmissions up to April 1989:	
Selection link rod*	135 ± 1 mm
Engagement link rod*	300 ± 1 mm
Torque link rod*	56.5 ± 1 mm
BE3 transmissions from April 1989:	
Selection link rod*	130 ± 1 mm
Engagement link rod*	300 ± 1 mm
Torque link rod (later models) *	51 mm to 61 mm
Distance from top of gear lever knob to centre-line on front face of radio aperture in neutral	195 mm

*Take the old rod to your Citroën dealer to obtain the correct replacement part.

Torque wrench settings

	Nm	lbf ft
Clutch release bearing guide sleeve bolts:		
Petrol models	6	4
Diesel models	12	9
Engine-to-transmission bolts	40	30
Engine/transmission left-hand mounting:		
Rubber bush centre nut*	35	26
Rubber bush support-to-mounting bracket nuts	18	13
Mounting bracket-to-body bolts	18	13
Stud bracket-to-transmission nuts (petrol models)	18	13
Stud-to-transmission (diesel models)**	50	37
Gear engagement rod bellcrank pivot nut and bolt*	60	44
Gear selection rod balljoint-to-front subframe nut*	20	15
Oil filler/level plug:		
Petrol models	25	18
Diesel models	20	15
Oil drain plug:		
Petrol models	25	18
Diesel models	30	22
Reversing light switch	25	18

*Use new self-locking nut
**Use locking fluid

1 General information

1 The transmission is contained in a cast aluminium alloy casing bolted to the engine's left-hand end, and consists of the gearbox and final drive differential **(see illustrations)**.
2 Drive is transmitted from the crankshaft via the clutch to the input shaft, which rotates in sealed ball-bearings, and has a splined extension to accept the clutch friction plate. From the input shaft, drive is transmitted to the output shaft, which rotates in a roller bearing at its right-hand end, and a sealed ball-bearing at its left-hand end. From the output shaft, the drive is transmitted to the differential crownwheel, which rotates with the differential and planetary gears, thus driving the sun gears and driveshafts. The rotation of the planetary gears on their shaft allows the inner roadwheel to rotate at a slower speed than the outer roadwheel when the van is cornering.
3 The input and output shafts are arranged side by side, parallel to the crankshaft and driveshafts, so that their gear pinion teeth are in constant mesh. In the neutral position, the output shaft gear pinions rotate freely, so that drive cannot be transmitted to the crownwheel.
4 Gear selection is via a floor-mounted lever and linkage mechanism. The linkage causes the appropriate selector fork to move its respective synchro-sleeve along the shaft, to

2.3a Remove the filler/level plug and recover the sealing washer - petrol model

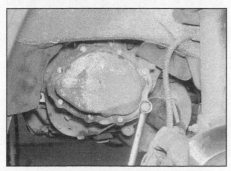

2.3b Remove the filler/level plug - diesel model

1.1a BE3 transmission unit at left-hand end of diesel engine, viewed from the front . . .

lock the gear pinion to the synchro-hub. Since the synchro-hubs are splined to the output shaft, this locks the pinion to the shaft so that drive is transmitted. To ensure that gear-changing can be made quickly and quietly, a synchromesh system is fitted to all forward gears, consisting of baulk rings and spring-loaded fingers, as well as the gear pinions and synchro-hubs. The synchromesh cones are formed on the mating faces of the baulk rings and gear pinions.
5 The transmissions fitted to petrol models have remained substantially the same except for a change in the fourth gear ratio on the four-speed transmission in 1991. The majority of diesel models are fitted with a five-speed BE3 transmission, but a four-speed version was available until 1990. The earlier BE1 transmission was available in four and five speeds until April 1989, with a reverse-stop collar which has to be pulled up on the gearstick to select reverse gear at the upper left position. All other transmissions have reverse gear at the lower right position and no reverse-stop collar.

2 Transmission oil - draining and refilling

Note: *A suitable square-section wrench may be required to undo the transmission filler/level and drain plugs on some petrol models. These wrenches can be obtained from most motor factors, or from your Citroën dealer.*

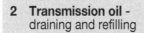

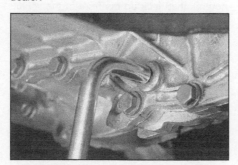

2.4a Using the special square-section wrench to remove the drain plug from the differential housing - petrol model

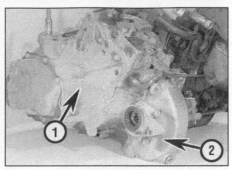

1.1b . . . and from the rear

1 *Gearbox*
2 *Final drive differential*

Note: *New sealing washers are required for the filler/level and drain plugs.*
1 This operation is much quicker and more effective if the van is first taken on a journey of sufficient length to warm the engine/transmission up to normal operating temperature.
2 Park the van on level ground, switch off the engine, and firmly apply the handbrake. For improved access, jack up the front of the van and support it securely on axle stands (see *"Jacking and Vehicle Support"*). Note that the van must be lowered to the ground, and level, to ensure accuracy, when refilling and checking the oil level.
3 Wipe clean the area around the filler/level plug, which is situated on the left-hand end of the transmission. On petrol models it is next to the end cover, and on diesel models it is the largest bolt securing the end cover to the transmission. Unscrew the filler/level plug from the transmission, and recover the sealing washer **(see illustrations)**.
4 Position a suitable container under the drain plug, situated on the differential housing, and unscrew the plug **(see illustrations)**.
5 Allow the oil to drain completely into the container. If the oil is hot, take precautions against scalding. Clean both the filler/level and the drain plugs, being especially careful to wipe any metallic particles off the magnetic inserts.
6 When the oil has finished draining, clean the threads on the drain plug and differential housing. Fit a new sealing washer and refit the

2.4b Transmission drain plug (arrowed) - diesel model (the precise location may vary slightly on some later models)

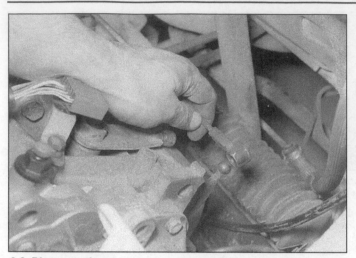

3.8 Disconnecting the gear engagement link rod from its balljoint on the transmission

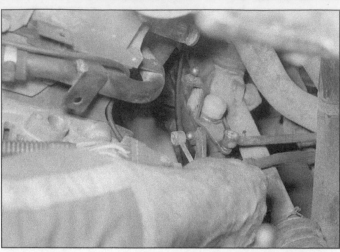

4.3 Disconnecting the inner end of a gearchange link rod from its balljoint

drain plug, tightening it to the specified torque. lt the van was raised for the draining operation, now lower it to the ground.

7 Refilling the transmission is an extremely awkward operation. Above all, allow plenty of time for the oil level to settle properly before checking it. Note that the van must be parked on flat level ground when checking the oil level.

8 Refill the transmission with the specified type and quantity of oil, then check the oil level as described in Chapter 1A or 1B. If the correct amount was poured into the transmission, and a large amount flows out on checking the level, refit the filler/level plug and take the van on a short journey. When the new oil is distributed fully around the transmission components, check the level again on your return, and top-up if necessary. Fit a new sealing washer to the filler/level plug and tighten it to the specified torque.

Note: *On diesel models from January 1991 onwards, there is a web at the lower edge of the filler/level hole, to increase the oil capacity. Make sure the oil delivery nozzle goes over the web and into the casing when filling the transmission.*

3 Gearchange linkage - general information and adjustment

General information

1 The gearstick is connected to the transmission linkage by means of two rods:
  a) *The selection rod which rotates in response to the sideways movement of the gearstick in the neutral position, and connects to the selection lever on the transmission via the selection link rod.*
  b) *The engagement rod which moves forwards and backwards in response to the forward/backward movement of the gearstick, and connects to the*

engagement lever on the transmission via the engagement link rod.

2 All models have at least two link rods - the selection link rod and engagement link rod. Models with BE1 transmissions and some later models with BE3 transmissions also have a torque link rod which connects the selection linkage to a bracket on the transmission casing.

3 If a stiff, sloppy or imprecise gearchange leads you to suspect that a fault exists within the linkage, dismantle it completely and check it for wear or damage as described in Section 4. Reassemble the linkage, applying a smear of multi-purpose grease to all bearing surfaces.

4 If this does not cure the fault, the van should be examined by an expert, as the fault must lie within the transmission itself. There is no adjustment as such in the linkage.

5 While the length of the link rods can be altered as described below, this is for initial setting-up only, and is not intended to provide a form of compensation for wear. If the link rods have been renewed, or if the length of the originals is incorrect, adjust them as follows.

Adjustment

6 Firmly apply the handbrake and select neutral, then jack up the front of the vehicle and support it on axle stands (see "*Jacking and vehicle support*"). Access to the link rods is poor, but they can be reached both from above and below the vehicle.

7 Working in (or under) the engine compartment, measure the length of each link rod and compare it to the length given in the Specifications. Note that the measurements given are the distances between the centre points of the link rod balljoints, and not the total length of the rod.

8 If adjustment is necessary, carefully disconnect the relevant link rod from its balljoint on the transmission **(see illustration)**. Use a screwdriver to lever it off if necessary.

Slacken the locknut and rotate the end of the rod until the specified distance between the balljoint centres is obtained, then press the end of the rod firmly back onto its balljoint and tighten the locknut.

9 The adjustment of the link rods will affect the position of the gearstick within the vehicle. If the length of the selection link rod is correct, the gearstick should point upwards along the centre-line of the vehicle when in neutral, leaning neither to the left nor the right. If the length of the engagement link rod is correct, the top of the gear lever knob should be the specified distance from the centre-line on the front face of the radio aperture, when in neutral.

10 Once the adjustments have been made, check that all gears can be selected, and that the gearchange lever returns to its correct neutral position. Where necessary, refit any components which were removed to improve access.

4 Gearchange linkage - removal and refitting

Removal

1 Firmly apply the handbrake, then jack up the front of the vehicle and support it on axle stands (see "*Jacking and vehicle support*").

2 Disconnect the gearchange link rods from their balljoints on the transmission. There are two or three link rods, as described in Section 3.

3 If required, disconnect the gearchange link rods from their balljoints at their inner ends, and remove them from the vehicle **(see illustration)**.

4 The gearstick connects to the transmission linkage by two rods. If these are to be removed, disconnect them at their front ends by prising them off their balljoints. The engagement rod connects to a balljoint on the bellcrank attached to the subframe. If the bellcrank needs to be removed, carefully pull

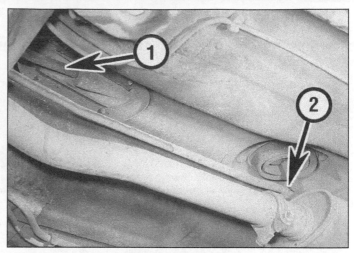

4.4 Gearchange linkage components (engine removed for clarity)

1 Bellcrank pivot nut and bolt
2 Engagement rod balljoint on bellcrank
3 Selection rod balljoint on subframe

4.5 Gearchange rods

1 Selection rod
2 Engagement rod balljoint on gearstick

4.6a Unscrew the gear lever knob . . .

4.6b . . . remove the spacer . . .

4.6c . . . pull the gearstick outer column and boot upwards off the gearstick . . .

off the plastic cap, undo the pivot nut and bolt and recover the washers and bushes. The selection rod connects to a balljoint on the subframe, and on models with BE1 transmissions it may be necessary to pull out a spring clip. If required, undo the nut and remove the balljoint **(see illustration)**.

5 The rear end of the engagement rod connects to a balljoint at the bottom of the gearstick, above the exhaust pipe **(see illustration)**. The balljoint needs to be disconnected to remove either the engagement

rod or the gearstick. Prise off the balljoint and remove the rod from underneath the vehicle.

6 To access the gearstick components, unscrew the gear lever knob and remove the spacer, then detach the rubber boot from the plastic moulding on the centre console, releasing it from its clips. Pull the gearstick outer column and boot upwards off the gearstick **(see illustrations)**. **Note:** On models with BE1 transmissions, the gearstick outer column engages with a reverse stop underneath the boot. There is a spring which

needs to be removed from the upper end of the gearstick, below the knob.

7 For complete access to the gearstick components, the centre console can be removed. Undo the four screws on the console, then lift it upwards and off the gearstick, and place it on the floor. Take care not to tension the radio speaker wires if they are routed through the console **(see illustrations)**.

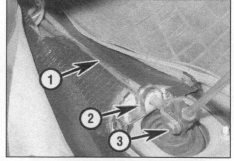

4.7b . . . for complete access to the gearstick components

1 Selection rod
2 Selection rod bearing clamp
3 Selection rod pivot bolt

4.6d . . . to access the gearstick components

4.7a Undo four screws and remove the centre console . . .

5.4 Use a large flat-bladed screwdriver to prise the driveshaft oil seal out of position

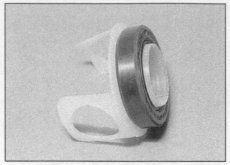

5.5a The new oil seal may be fitted with a plastic protector sleeve

5.5b Oil seal with protector in place

8 To remove the gearstick, make sure the engagement rod is disconnected from the balljoint underneath the vehicle. Then, working within the vehicle, undo the pivot nut and bolt connecting the selection rod to the gearstick, and pull the gearstick upwards out of the lower gaiter.

9 To remove the selection rod, access to the bearing clamp is required, so the centre console must be removed. Undo the front balljoint and the rear pivot nut and bolt (if not done already). Undo the two nuts on the bearing clamp, then withdraw the selection rod from inside the vehicle or from underneath, whichever is easiest.

10 Renew the lower gaiter if there is any sign of damage. Remove the centre console and gearstick as described above, then undo the four nuts on the lower gaiter cover and withdraw the gaiter and cover. Two of the nuts on the gaiter cover also secure the selection rod bearing clamp. **Note:** *On models with BE1 transmissions, there may be a reverse stop plate above the gaiter cover.*

11 Inspect all the linkage components for signs of wear or damage, paying particular attention to the bellcrank pivot bushes and link rod balljoints, and renew worn components as necessary. Examine the lever components for signs of wear or damage, paying particular attention to the rubber gaiters, and renew the components as necessary.

Refitting

12 Refitting is the reverse of removal, noting the following points:

a) *Apply a smear of multi-purpose grease to the gearchange lever pivot bolt, the link rod balljoints, and the bellcrank pivot bushes.*

b) *Ensure that all link rods are securely pressed onto their balljoints.*

c) *Adjust the link rods as described in Section 3.*

d) *If the selector rod balljoint-to-subframe nut, or the engagement rod bellcrank pivot nut and bolt have been removed, fit new self-locking nuts and tighten them to the specified torque.*

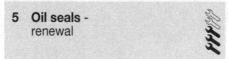

5 Oil seals - renewal

Driveshaft oil seal

1 Chock the rear wheels of the van, firmly apply the handbrake, then jack up the front of the van and support it on axle stands (see *"Jacking and Vehicle Support"*). Remove the appropriate front roadwheel.

2 Drain the transmission oil as described in Section 2.

3 Working as described in Chapter 8, free the inner end of the driveshaft from the transmission, and place it clear of the seal. Follow the instructions for removing the driveshaft, but do not unscrew the hub nut; the driveshaft can be left secured to the hub. Support the driveshaft, to avoid placing any strain on the driveshaft joints or gaiters.

4 Carefully prise the oil seal out of the

transmission, using a large flat-bladed screwdriver **(see illustration)**.

5 Remove all traces of dirt from the area around the oil seal aperture, then apply a smear of grease to the outer lip of the new oil seal. Fit the new seal into its aperture, and drive it squarely into position using a suitable tubular drift, such as a socket, which bears only on the hard outer edge of the seal, until it abuts its locating shoulder. The right-hand driveshaft oil seal may be fitted with a plastic protector sleeve. Remove the protector and fit the seal to the transmission, then apply a little grease to the seal lips and refit the protector **(see illustrations)**.

6 Refit the driveshaft as described in Chapter 8. If the seal is fitted with a protector sleeve, fit the driveshaft with the protector in place, then pull out the protector, which is split so that it will pass over the driveshaft. When refitting a right-hand driveshaft, move the rubber dust seal along the shaft so that it is next to the oil seal **(see illustrations)**.

7 Refill the transmission with the specified type and quantity of oil as described in Section 2.

Input shaft oil seal

8 Remove the transmission from the van as described in Section 8.

9 Undo the three bolts securing the clutch release bearing guide sleeve in position, and slide the guide off the input shaft along with its O-ring or gasket, as applicable **(see illustration)**. Recover any relevant thrust-washers which have stuck to the rear of the guide sleeve, and refit them to the input shaft.

5.6a Fit the driveshaft to the transmission, then pull out the protector

5.6b Make sure the rubber dust seal is next to the oil seal on the right-hand driveshaft

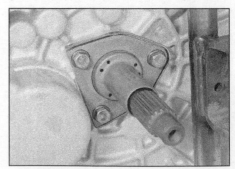

5.9 The clutch release bearing guide sleeve is retained by three bolts

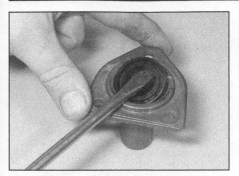

5.10 Remove the input shaft oil seal from the guide sleeve . . .

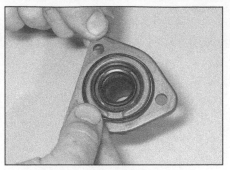

5.12a Fit a new O-ring or gasket (as applicable) . . .

5.12b . . . then carefully refit the guide sleeve over the input shaft

10 Carefully lever the oil seal out of the guide sleeve using a suitable flat-bladed screwdriver **(see illustration)**. On petrol models the seal may not be available as a separate component, in which case you will have to renew the complete guide sleeve. Consult your Citroën dealer for details.

11 Before fitting a new seal, check the input shaft's seal rubbing surface for signs of burrs, scratches or other damage which may have caused the seal to fail in the first place. It may be possible to polish away minor defects of this sort using fine abrasive paper; however, more serious defects will require the renewal of the input shaft. Ensure that the input shaft is clean and greased, to protect the seal lips on refitting.

12 Fit a new O-ring or gasket (as applicable) to the rear of the guide sleeve, then carefully slide the sleeve into position over the input shaft **(see illustrations)**. Refit the retaining bolts, and tighten them to the specified torque.

13 Refit the transmission to the van as described in Section 8.

Selector shaft oil seal

Note: *No specific information is available about removing the selector shaft oil seal from BE1 transmissions.*

MA4 and MA5 transmissions

14 To renew the selector shaft seal, the transmission must be dismantled. This task should therefore be entrusted to a Citroën dealer or transmission specialist.

BE3 transmissions

15 Park the van on level ground, firmly apply the handbrake, then jack up the front of the vehicle and support it on axle stands (see *"Jacking and Vehicle Support"*).

16 Working in the engine compartment, or under the vehicle, carefully lever the link rod balljoint off the transmission selector shaft, using a large flat-bladed screwdriver, and disconnect the link rod.

17 Carefully prise the selector shaft seal out of the housing, and slide it off the end of the shaft **(see illustration)**.

18 Before fitting a new seal, check the selector shaft's seal rubbing surface for signs of burrs, scratches or other damage which

may have caused the seal to fail in the first place. It may be possible to polish away minor defects of this sort using fine abrasive paper; however, more serious defects will require the renewal of the selector shaft.

19 Apply a smear of grease to the new seal's outer edge and sealing lip, then carefully slide the seal along the selector shaft. Press the seal fully into position in the transmission housing.

20 Reconnect the link rod to the selector shaft, ensuring that its balljoint is pressed firmly onto the shaft. Lower the van to the ground.

6 Reversing light switch -
testing, removal and refitting

Testing

1 The reversing light circuit is controlled by a plunger-type switch screwed into the top of the transmission casing **(see illustration)**. If a fault develops in the circuit, first ensure that the circuit fuse has not blown (see Chapter 12).

2 To access the switch, remove the air inlet ducting, and on diesel models remove the air cleaner (see the relevant Part of Chapter 4).

3 To test the switch, disconnect the wiring connector, and use a multi-meter (set to the resistance function) or a battery-and-bulb test circuit to check that there is continuity between the switch terminals only when reverse gear is selected. If this is not the case,

the switch is faulty and must be renewed. If the switch functions correctly, check the associated wiring for possible open or short-circuits.

Removal

4 Disconnect the wiring connector from the switch. Unscrew it from the transmission casing, and remove it along with its sealing washer.

Refitting

5 Fit a new sealing washer to the switch, then screw it back into position in the top of the transmission housing and tighten it to the specified torque. Reconnect the wiring connector and test the operation of the circuit.

6 Refit the air inlet ducting, and on diesel models refit the air cleaner (see the relevant Part of Chapter 4).

7 Speedometer drive -
removal and refitting

Removal

1 Chock the rear wheels of the van, and firmly apply the handbrake. Jack up the front of the van, and support it securely on axle stands (see *"Jacking and vehicle support"*). The speedometer drive is situated on the rear of the transmission housing, next to the inner end of the right-hand driveshaft.

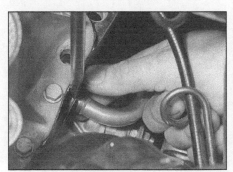

5.17 Use a large flat-bladed screwdriver to prise the selector shaft seal out of position - BE3 transmissions

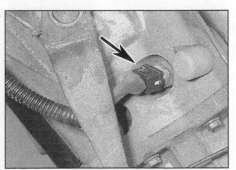

6.1 Reversing light switch (arrowed) - BE3 transmission (air cleaner, ducting, battery and tray removed for clarity)

7.2a Withdraw the rubber retaining pin (arrowed) . . .

7.2b . . . then disconnect the speedometer cable from its drive

7.3a Undo the retaining bolt . . .

7.3b . . . and remove the speedometer drive from the transmission (transmission removed for clarity)

2 Pull out the speedometer cable rubber retaining pin, and disconnect the cable from the speedometer drive **(see illustrations)**.
3 Unscrew and remove the retaining bolt, and withdraw the speedometer drive and driven pinion assembly from the transmission housing, along with its O-ring **(see illustrations)**.
4 Examine the pinion for signs of damage, and renew it if necessary, together with its oil seal. Renew the housing O-ring as a matter of course.
5 If the driven pinion is worn or damaged, also examine the drive pinion in the transmission housing for signs of wear or damage.

7.6 Speedometer drive pinion (arrowed) - MA4 and MA5 transmissions

MA4 and MA5 transmissions

6 To renew the drive pinion on MA4 and MA5 transmissions, the transmission must be dismantled and the differential gear removed. This task should therefore be entrusted to a Citroën dealer or transmission specialist. The drive pinion is on the differential gear **(see illustration)**.

BE1 and BE3 transmissions

7 To remove the drive pinion on BE1 and BE3 transmissions, first disengage the right-hand driveshaft from the transmission, as described in Chapter 8 and place it clear of the oil seal. Follow the instructions for removing the

driveshaft, but do not unscrew the hub nut; the driveshaft can be left secured to the hub. Support the driveshaft, to avoid placing any strain on the driveshaft joints or gaiters.
8 Undo the three retaining bolts and remove the speedometer drive housing from the transmission, along with its O-ring. Remove the drive pinion from the differential gear, and remove any relevant adjustment shims from the gear **(see illustrations)**.

Refitting

BE1 and BE3 transmissions

9 Where the drive pinion has been removed from a BE1 or BE3 transmission, refit the adjustment shims to the differential gear, then locate the speedometer drive on the gear, ensuring it is correctly engaged in the gear slots **(see illustration)**. Fit a new O-ring to the rear of the speedometer drive housing, then refit the housing to the transmission and securely tighten its retaining screws. Inspect the driveshaft oil seal for signs of wear and renew it if necessary. Refit the driveshaft to the transmission as described in Chapter 8.

All transmissions

10 Fit a new O-ring to the speedometer drive. Refit the drive to the transmission, ensuring

7.8a Undo the three retaining bolts . . .

7.8b . . . and remove the housing, O-ring and drive pinion from the transmission (transmission removed for clarity) - BE1 and BE3 transmissions

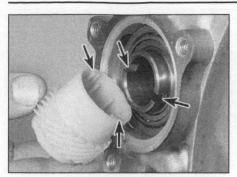

7.9 On refitting, ensure the drive pinion dogs are correctly engaged with the gear slots (arrowed) - BE1 and BE3 transmissions

8.9 Remove the TDC sensor - diesel model

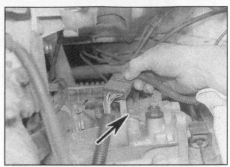

8.10 Disconnect the wiring harness, then undo the two bolts and remove the bracket. One of the bolts (arrowed) secures the earth strap - diesel models

that the drive pinion and driven pinion are correctly engaged.

11 Refit the retaining bolt and tighten it securely.

12 Apply a smear of oil to the speedometer cable O-rings, then reconnect the cable to the drive, securing it in position with the rubber retaining pin. Lower the vehicle to the ground.

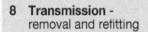

8 Transmission - removal and refitting

Note: *The transmission can be removed on its own, or together with the engine. If the latter method is to be employed, refer to Chapter 2C for details of the removal and separation procedure. This Section describes the removal of the transmission on its own, leaving the engine in the vehicle.*

Note: *A new self-locking nut is required for the left-hand engine mounting.*

Removal

1 Chock the rear wheels, then firmly apply the handbrake. Jack up the front of the vehicle, and support it securely on axle stands (see "*Jacking and vehicle support*"). Remove both front roadwheels.

2 To access the transmission, remove the air cleaner housing and/or intake duct (as applicable), as described in the relevant Part of Chapter 4.

3 Remove the battery as described in

Chapter 5A, then undo the four bolts and remove the battery tray.

4 Drain the transmission oil as described in Section 2, then refit the drain and filler/level plugs, and tighten them to their specified torque.

5 On diesel models with a belt-driven brake vacuum pump mounted on the transmission housing, remove the pump as described in Chapter 9.

6 On diesel models with power steering, it may be necessary to remove the power steering pump. No specific information is available, but the pump is belt-driven from the camshaft and removal is thought to be similar to removing the belt-driven brake vacuum pump. If the hydraulic fluid hoses need to be disconnected, clamp the ends to minimise fluid loss.

7 Disconnect the clutch cable from the release lever and bracket on the transmission (see Chapter 6). Release the cable from any relevant retaining clips, and place it clear of the transmission.

8 Disconnect the wiring connectors from the reversing light switch (see Section 6).

9 Both petrol and diesel models have a TDC sensor at the rear of the transmission, for diagnostic purposes. On petrol models, disconnect the wiring connector from the sensor. On diesel models, the sensor is mounted on a sensor holder, bolted to the engine block. Slacken the bolts and remove the sensor, then undo the bolts and remove the sensor holder **(see illustration)**.

10 Disconnect the earth strap and any other wiring from the top of the transmission. On diesel models, the engine wiring harness connector is mounted on a bracket, secured to the transmission by two bolts. One of the bolts also secures the earth strap. Disconnect the multi-plug, then undo the bolts on the mounting bracket **(see illustration)**.

11 Withdraw the rubber retaining pin, and disconnect the speedometer cable from the drive housing, freeing it from any relevant retaining clips, with reference to Section 7.

12 Using a flat-bladed screwdriver, carefully lever the gearchange mechanism link rods off their respective balljoints on the transmission. Position the rods clear of the transmission. There are two or three link rods, as described in Section 3.

13 Remove the starter motor as described in Chapter 5A.

14 On diesel engines, unbolt the flywheel cover plate from the base of the transmission, and remove it from the vehicle **(see illustrations)**.

15 Working as described in Chapter 8, free the inner end of each driveshaft from the differential housing and position them clear of the transmission **(see illustrations)**. Follow the instructions for removing the driveshaft, but do not unscrew the hub nuts; each driveshaft can be left secured to the hub. Support the driveshafts, to avoid placing any strain on the driveshaft joints or gaiters.

16 Place a piece of thin board over the radiator to protect it from possible damage.

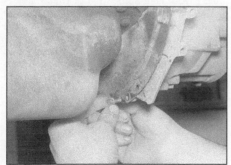

8.14a Undo the bolts . . .

8.14b . . . and remove the flywheel cover plate - diesel model

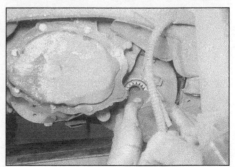

8.15a Disconnect the left-hand driveshaft from the transmission . . .

8.15b . . . and the right-hand driveshaft

8.19a Undo the nut (arrowed) from the left-hand engine/transmission mounting . . .

8.19b . . . then undo the two bolts and remove the mounting bracket

17 Place a jack with interposed block of wood beneath the engine, or attach a hoist or support bar to the engine lifting eyes, to take the weight of the engine.

18 Place a trolley jack and block of wood beneath the transmission, and raise the jack to take the weight of the transmission.

19 Slacken and remove the centre nut and washer from the left-hand engine/trans-mission mounting. Undo the two bolts securing the mounting bracket assembly to the vehicle body, then lower the engine slightly and remove the mounting bracket assembly. Recover the spacer from the stud **(see illustrations).**

20 Remove the left-hand mounting stud from the transmission. On petrol models, undo the three nuts on the mounting stud support bracket. On diesel models, unscrew the stud from the transmission. If it is tight, use a universal stud extractor **(see illustration).**

21 With the jack beneath the transmission taking its weight, slacken and remove the bolts securing the transmission housing to the engine. Note the correct fitted positions of each bolt, and the necessary brackets, as they are removed, to use as a reference on refitting.

8.20 If the mounting stud is tight, use a universal stud extractor to unscrew it

22 Make a final check that all necessary components have been disconnected, and are positioned clear of the transmission so that they will not hinder the removal procedure.

23 With the bolts removed, move the trolley jack and transmission to the left, to free it from its locating dowels.

24 Once the transmission is free, lower the jack and manoeuvre the unit out from under the van. If the locating dowels are loose, remove them from the engine or transmission and keep them in a safe place.

Refitting

25 Refitting is the reverse of removal, bearing in mind the following points:

a) *Apply a little high-melting point grease to the splines of the transmission input shaft. Do not apply too much, otherwise there is a possibility of the grease contaminating the clutch friction plate.*

b) *Ensure that the locating dowels are correctly positioned prior to installation.*

c) *Tighten all nuts and bolts to their specified torque (where given). Use a new self-locking nut on the mounting rubber. On diesel models, apply locking fluid when fitting the mounting stud to the transmission.*

d) *Renew the driveshaft oil seals (see Section 5).*

e) *On completion, refill the transmission with the specified type and quantity of oil, as described in Section 2.*

f) *Check and if necessary adjust the clutch cable (see Chapter 6).*

g) *On models with power steering, it will be necessary to bleed the hydraulic fluid circuit if any hoses have been disconnected. No specific instructions are available, and you are advised to consult your Citroën dealer.*

9 Transmission overhaul - general information

1 Overhauling a manual transmission unit is a difficult and involved job for the DIY home mechanic. In addition to dismantling and reassembling many small parts, clearances must be precisely measured and, if necessary, changed by selecting shims and spacers. Internal transmission components are also often difficult to obtain and, in many instances, extremely expensive. Because of this, if the transmission develops a fault or becomes noisy, the best course of action is to have the unit overhauled by a specialist repairer, or to obtain an exchange reconditioned unit.

2 Nevertheless, it is not impossible for the more experienced mechanic to overhaul the transmission, provided the special tools are available, and the job is done in a deliberate step-by-step manner, so that nothing is overlooked.

3 The tools necessary for an overhaul include internal and external circlip pliers, bearing pullers, a slide hammer, a set of pin punches, a dial test indicator, and possibly a hydraulic press. In addition, a large, sturdy workbench and a vice will be required.

4 During dismantling of the transmission, make careful notes of how each component is fitted, to make reassembly easier and more accurate.

5 Before dismantling the transmission, it will help if you have some idea of which area is malfunctioning. Certain problems can be closely related to specific areas in the transmission, which can reduce the amount of dismantling and component examination required. Refer to *"Fault finding"* in the Reference Section at the end of this manual for more information.

Chapter 8
Driveshafts

Contents

Degrees of difficulty

Easy, suitable for novice with little experience		Fairly easy, suitable for beginner with some experience		Fairly difficult, suitable for competent DIY mechanic		Difficult, suitable for experienced DIY mechanic		Very difficult, suitable for expert DIY or professional	

Specifications

General

Type	Unequal-length solid steel shafts, splined to inner and outer constant velocity joints. Intermediate bearing on right-hand driveshaft.
Lubrication	Use only special grease supplied in sachets in the correct quantity with gaiter kits - joints are otherwise pre-packed with grease and sealed.

Torque wrench settings

	Nm	lbf ft
Driveshaft intermediate bearing securing nuts	10	7
Hub/driveshaft nut*	250	185
Track control arm outer balljoint clamp nut and bolt*	35	26

*Use a new nut.

1 General information

1 Drive is transmitted from the differential to the front wheels by means of two unequal-length driveshafts.

2 Each driveshaft is fitted with an inner and outer constant velocity (CV) joint. Each outer joint is splined to engage with the wheel hub, and is retained by a large nut. Each inner joint is splined to engage with the differential sun gear.

3 The inner joints are of the tripod type consisting of a central spider and three supporting roller-bearings. The outer joints are the six-ball and cage type.

4 On the right-hand driveshaft, the inner constant velocity joint is situated approximately half-way along the shaft, and an intermediate bearing is supported by the rear engine mounting bracket which is bolted to the cylinder block. The inner section of the driveshaft passes through the bearing.

5 There are two different types of driveshaft, with or without a cylindrical inner joint housing cover supplied with the gaiter. The procedures for removing and refitting the driveshafts are the same, but the procedures for gaiter renewal are different.

2.2 Relieve the staking on the hub nut with a suitable cold chisel

2 Driveshaft joints - checking

1 If any of the checks described in Chapter 1A or 1B reveal wear in any driveshaft joint, first check that the hub nut is correctly tightened. Remove the roadwheel trim or centre cap (as appropriate), then remove the hub nut dust cover (if applicable), to expose the hub nut. It may be necessary to remove the roadwheel to access the dust cover, in which case you should chock the rear wheels, apply the handbrake, jack up the front of the vehicle and support it securely on axle stands (see *"Jacking and vehicle support"*).

2 With the hub nut exposed, check that the staking is still firmly in the driveshaft groove. If so, the driveshaft nut will be correctly fitted and should not need to be re-tightened. If in doubt, relieve the staking using a hammer and a suitable cold chisel or punch **(see illustration)**, then tighten the nut to the specified torque. Use a suitable counterhold **(see Tool Tip)** to prevent the hub from turning. With the nut fully tightened, restake it into the driveshaft groove. Refit the roadwheel, if removed, and repeat the check on the other driveshaft nut.

⚠️ *Warning: The hub nut has to be tightened to a very high torque. You can pull the vehicle off the axle stands by attempting to tighten it without a counterhold. Do not use the*

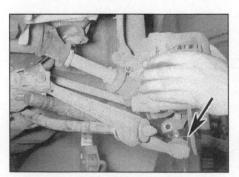

3.5 Use a lever to separate the track control arm outer joint and recover the protector shield (arrowed)

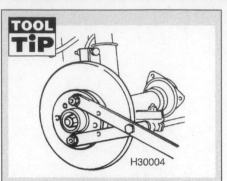

TOOL TiP

H30004

A tool to hold the hub stationary can be fabricated from two lengths of steel strip (one long, one short) held together by a nut and bolt. Use two wheel bolts to hold the tool in position on the hub and use it as a counterhold while tightening the hub nut. Alternatively, insert two adjacent wheel bolts into the hub and use a metal bar between them as a counterhold.

footbrake with the wheel bolts removed because it might damage the brake disc retaining screws.

HAYNES HiNT *If you do not have a counterhold, refit the roadwheel and lower the vehicle to the ground. Apply the handbrake, then tighten the hub nut.*

⚠️ *Warning: Wear suitable eye protection when staking the hub nut.*

3 Road-test the vehicle, and listen for a metallic clicking from the front as the vehicle is driven slowly in a circle on full-lock. If a clicking noise is heard, this indicates wear in the outer constant velocity joint.

4 If vibration, consistent with road speed, is felt through the van when accelerating, there is a possibility of wear in the inner constant velocity joints.

5 To check the joints comprehensively for wear, remove the driveshafts (see Section 3), then remove the gaiters (see Section 4). If any wear or free play is found, it will usually be

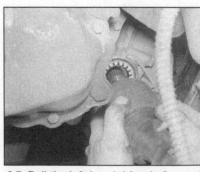

3.7 Pull the left-hand driveshaft out of the differential housing

necessary to renew the complete driveshaft assembly. See your Citroën dealer for further information.

Note: *The driveshaft joints are pre-packed with grease and sealed. Once disturbed, they need to be fitted with new gaiters which are supplied in kits together with the appropriate springs, clips and seals, and the correct amount of special grease.*

3 Driveshafts - removal and refitting

Note: *A new hub nut must be used on refitting, and new nuts on the track control arm to hub carrier clamp nuts.*

⚠️ *Warning: Observe the warnings given in Section 2 when removing and refitting the hub nuts.*

Removal

1 Chock the rear wheels, apply the handbrake, then jack up the front of the vehicle and support it securely on axle stands (see *"Jacking and vehicle support"*). Remove the appropriate roadwheel.

2 Remove the hub nut dust cover (if applicable), then relieve the staking on the hub nut using a hammer and a suitable cold chisel or punch.

3 Slacken the hub nut using a socket and extension bar but do not remove it at this stage. Use a suitable counterhold (see Section 2) to prevent the hub from turning. Do not use the footbrake because it might damage the brake disc retaining screws.

4 Drain the transmission oil as described in Chapter 7.

5 Undo the clamp nut and bolt from the track control arm outer joint, where it meets the hub carrier, noting which way round it is fitted. If necessary, use a screwdriver to open up the clamp slightly. Then use a lever between the anti-roll bar and track control arm to pull the joint downwards and disconnect it. Have an assistant hold the hub steady and pull it off the joint. Recover the protector shield **(see illustration)**.

6 Release the hub from the driveshaft splines by pulling the hub carrier assembly outwards. If necessary, the shaft can be tapped out of the hub using a soft-faced mallet. Support the driveshaft and do not allow it to hang down.

Left-hand driveshaft

7 Withdraw the inner constant velocity joint from the transmission, taking care not to damage the driveshaft oil seal **(see illustration)**. Remove the driveshaft from the vehicle, then plug or tape over the differential aperture to prevent the entry of dirt.

Right-hand driveshaft with intermediate bearing

8 Slacken the nuts that secure the right-hand driveshaft intermediate bearing to the rear engine mounting bracket, until the nuts are

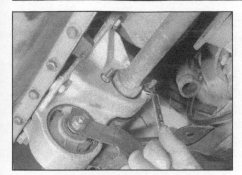

3.8a Slacken the nuts securing the intermediate bearing . . .

3.8b . . . until the nuts are almost at the end of their threads

3.8c Push the bolts in and turn them by 90° to disengage their offset heads (arrowed) from the bearing (shown with driveshaft removed)

almost at the end of their threads. Then push the bolts in and turn them by 90° so that they can't be pulled out again. At this position, the intermediate bearing is unlocked from the mounting bracket **(see illustrations)**.

9 Support the outer end of the driveshaft, then pull on the inner end of the shaft to free the intermediate bearing from the mounting bracket.

10 Pull the end of the driveshaft out of the differential housing and slide the dust seal off the shaft, noting which way round it is fitted **(see illustration)**. Remove the driveshaft from the vehicle, then plug or tape over the differential aperture to prevent the entry of dirt.

Refitting

11 Before installing the driveshaft, examine the driveshaft oil seal in the transmission for signs of damage or deterioration and, if necessary, renew it as described in Chapter 7. It is advisable to renew the seal as a matter of course.

12 Thoroughly clean the driveshaft splines, and the apertures in the transmission and hub assembly. Apply a thin film of grease to the oil seal lips, and to the driveshaft splines and shoulders. Check that all driveshaft gaiter clips are securely fastened.

Left-hand driveshaft

13 Offer up the driveshaft, and engage the joint splines with those of the differential sun gear, taking great care not to damage the oil seal. Push the joint fully into position. Support

the driveshaft and do not allow it to hang down.

14 Ensure that the driveshaft splines and the corresponding splines in the hub are clean, then engage the driveshaft with the hub, and fit a new hub nut. Do not tighten the nut at this stage.

15 Ensure that the protector plate is in place, then reconnect the track control arm to the hub carrier **(see illustration)**. If necessary, use a screwdriver to open up the joint slightly, and use a lever to pull the track control arm downwards so that it can be inserted into the clamp. Fit the clamp nut and bolt the correct way round as noted during removal. Use a new nut and tighten it to the specified torque.

16 Grease the face of the hub bearing where it contacts the hub nut, and grease the threads of the hub nut. Tighten the hub nut to the specified torque, using a counterhold as described in Section 2, and stake the nut in position.

17 Refill the transmission with oil (see Chapter 7).

18 When finished, refit the roadwheels with the centre trim and lower the vehicle to the ground.

Right-hand driveshaft with intermediate bearing

19 Check that the intermediate bearing rotates smoothly, without any sign of roughness, or of undue free play between its inner and outer races. If necessary, renew the bearing as described in Section 5. Examine

the dust seal for signs of damage or deterioration, and renew it if necessary.

20 Apply a smear of grease to the outer race of the intermediate bearing, and to the inner lip of the dust seal.

21 Pass the inner end of the intermediate driveshaft through the engine mounting bracket, then carefully slide the dust seal into position on the shaft, ensuring that its flat surface is facing the differential.

22 Carefully engage the driveshaft splines with the differential sun gear, taking care not to damage the oil seal. Align the intermediate bearing with the mounting bracket, then push the driveshaft fully into position. If a new oil seal with a protector has been fitted, leave the protector in place while inserting the driveshaft, then remove the protector **(see illustration)**.

23 Ensure that the driveshaft splines and the splines in the hub are clean, then engage the driveshaft with the hub, and fit a new hub nut. Do not tighten the nut at this stage.

24 Ensure that the intermediate bearing is correctly seated, then rotate its retaining bolts back through 90° and pull them out so that their offset heads are resting against the bearing outer race. Tighten the retaining nuts to the specified torque.

25 Ensure that the dust seal is tight against the driveshaft oil seal.

26 Carry out the operations described in paragraphs 15 to 18.

3.10 Pull the right-hand driveshaft out of the differential housing and slide the dust seal (arrowed) off the shaft

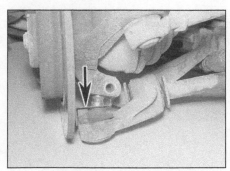

3.15 Ensure that the protector plate (arrowed) is in place, then refit the track control arm to the hub carrier

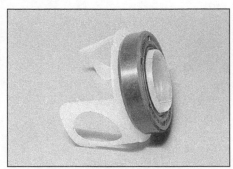

3.22 Right-hand driveshaft oil seal with protector

4.4 Peel back the lip of the inner joint housing cover

4 Driveshaft rubber gaiters - renewal

General information

1 There are two different types of driveshaft, with different procedures for renewing the gaiters:

a) *Driveshaft with inner joint housing cover supplied with gaiter. This type is identified by the cylindrical housing cover that fits entirely over the inner joint. The inner joint gaiter and housing cover have to be removed first, then the outer joint gaiter.*

b) *Driveshaft without inner joint housing cover. This type is identified by an inner joint housing that is partly cylindrical, but has flat faces on the circumference. The outer joint gaiter has to be removed first, then the inner joint gaiter.*

2 For both types of driveshaft, new gaiters are supplied in kits together with the appropriate springs, clips and seals, and the correct amount of special grease. Normally, only the components supplied in the gaiter kits can be renewed. If you remove a gaiter and discover that the joint bearings are worn or damaged, it will usually be necessary to renew the complete driveshaft assembly. See your Citroën dealer for further information.

Driveshaft with inner joint housing cover

Note: *During 1995, the neoprene gaiters on*

the outer constant velocity joints for this type of driveshaft were replaced with thermoplastic gaiters. The change occurred in April 1995 on petrol models, and in May 1995 on diesel models. The driveshafts were also changed to accommodate the new gaiters, and the two types of gaiter are not interchangeable on the same driveshaft. The old neoprene gaiters were only made available until stocks ran out, and if you are unable to obtain the correct replacement part you will have to buy a new driveshaft complete with the new thermoplastic gaiter. The gaiters on the inner constant velocity joint remained unchanged.

Inner joint

3 Remove the driveshaft as described in Section 3.

4 Secure the driveshaft in a vice equipped with soft jaws. Using a suitable pair of pliers, carefully peel back the lip of the inner joint housing cover **(see illustration)**.

5 Once the lip of the cover is fully released, pull the joint out from the cover, and recover the spring and thrust cap from the end of the shaft. Remove the O-ring from the outside of the joint housing, and discard it.

6 Fold the gaiter back, and wipe away the excess grease from the tripod joint. If the rollers are not secured to the joint with circlips, wrap adhesive tape around the joint to hold them in position.

7 Using a dab of paint, or a hammer and punch, mark the relative position of the tripod joint in relation to the driveshaft. Using circlip pliers, extract the circlip securing the joint to the driveshaft **(see illustration)**.

8 The tripod joint can now be removed. If it is tight, draw the joint off the driveshaft end, using a two or three-legged bearing puller. Ensure that the legs of the puller are located behind the joint inner member, and do not contact the joint rollers **(see illustrations)**. Alternatively, support the inner member of the tripod joint, and press the shaft out of the joint using a hydraulic press, ensuring that no load is applied to the joint rollers.

9 With the tripod joint removed, remove the small retaining clip, and slide the gaiter and inner retaining collar off the end of the driveshaft.

4.7 Remove the circlip from the inner tripod joint

10 Thoroughly clean the constant velocity joint components using paraffin, or a suitable solvent, and dry them thoroughly. Take great care not to remove the alignment marks made on dismantling, especially if paint was used. Carry out a visual inspection of the joint.

11 Examine the tripod joint, rollers and housing for any signs of scoring or wear, and for smoothness of movement of the rollers on the tripod stems. If any component is worn, the complete driveshaft assembly must be renewed, as no joint components are available separately. If the joint components are in good condition (and the driveshaft does not have to be renewed because of problems with the outer CV joint), obtain a repair kit from your Citroën dealer, consisting of a new rubber gaiter and housing cover, circlip, thrust cap, spring, O-ring and the correct quantity of the special grease **(see illustration)**.

12 If required, renew the outer joint gaiter, and at the same time inspect the outer joint (see paragraphs 23 to 30).

13 Begin reassembly by sliding the inner CV joint gaiter into position inside the metal outer cover (if not already fitted). Tape over the splines on the end of the driveshaft, and carefully slide the retaining clip and gaiter/cover assembly onto the shaft **(see illustrations)**. Fit the clip loosely onto the gaiter.

14 Remove the tape, align the marks made on dismantling, and engage the tripod joint with the driveshaft splines. Use a hammer and soft metal drift to tap the joint onto the shaft, taking great care not to damage the driveshaft splines or joint rollers **(see illustration)**.

4.8a Use a three-legged puller to release the inner tripod joint

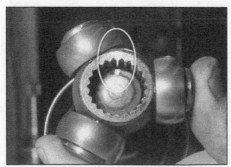

4.8b Withdraw the inner tripod joint. Note the alignment marks

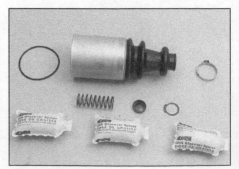

4.11 Driveshaft inner joint gaiter repair kit components

4.13a Slide on the retaining clip . . .

4.13b . . . and the gaiter/cover assembly

4.14 Use a hammer and socket to tap the joint onto the driveshaft

15 Secure the tripod joint in position with the new circlip, ensuring that it is correctly located in the driveshaft groove.

16 Remove the tape (where fitted), and evenly distribute the special grease contained

4.16 Pack the tripod joint and gaiter/cover with grease

in the repair kit around the tripod joint. Pack the gaiter/cover with the remainder, then draw the cover over the tripod joint **(see illustration)**. Leave some of the grease (typically one sachet in a multi-sachet kit) to lubricate the housing as the joint is fitted.

17 Fit the new O-ring, spring and thrust cap to the joint housing **(see illustrations)**.

18 Slide the housing cover partly over the housing, and insert the tripod joint into the housing so that the end of the driveshaft is located on the thrust cap. Apply the remainder of the grease to the joint, then push the tripod joint fully into the housing, compressing the spring, and slide the cover fully over the housing. Secure the cover in position by peening over the end, evenly around the edge of the joint **(see illustrations)**.

19 Briefly lift the inner gaiter lip, using a blunt instrument such as a knitting needle or

welding rod, to equalise the air pressure within the gaiter. Then tighten the gaiter clip.

20 Check that the constant velocity joint moves freely in all directions, then refit the driveshaft to the vehicle as described in Section 3.

Outer joint

21 Remove the driveshaft as described in Section 3.

22 Remove the inner constant velocity joint and gaiter as described in paragraphs 4 to 11. It is recommended that the inner gaiter is also renewed, regardless of its apparent condition.

23 Release the two outer gaiter retaining clips, then slide the gaiter off the inner end of the driveshaft.

24 Thoroughly clean the outer constant velocity joint using paraffin, or a suitable solvent, and dry it thoroughly. Carry out a visual inspection of the joint.

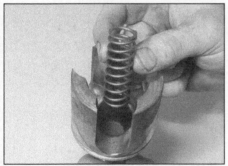

4.17a Fit the new spring . . .

4.17b . . . and thrust cap . . .

4.17c . . . and O-ring

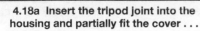

4.18a Insert the tripod joint into the housing and partially fit the cover . . .

4.18b . . . then apply the remainder of the grease . . .

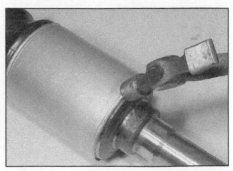

4.18c . . . and peen over the end of the cover

4.26 Driveshaft outer joint gaiter repair kit components

4.27 Slide the outer joint gaiter onto the driveshaft

4.28 Pack the joint with the grease supplied in the repair kit

25 Check the driveshaft spider and outer member yoke for signs of wear, pitting or scuffing on their bearing surfaces. Also check that the outer member pivots smoothly and easily, with no traces of roughness.

26 If, on inspection, the spider or outer member reveal signs of wear or damage, it will be necessary to renew the complete driveshaft as an assembly, since no components are available separately. If the joint components are in satisfactory condition, obtain a repair kit from your Citroën dealer, consisting of a new gaiter, retaining clips, and the correct type and quantity of grease **(see illustration)**.

27 Begin reassembly by taping over the splines on the inner end of the driveshaft, then carefully slide the outer gaiter onto the shaft **(see illustration)**.

28 Pack the joint with the grease supplied in the repair kit **(see illustration)**. Work the grease well into the bearing tracks while twisting the joint, and fill the rubber gaiter with any excess.

29 Ease the gaiter over the joint, and ensure that the gaiter lips are correctly located in the grooves on both the driveshaft and constant velocity joint. Lift the outer sealing lip of the gaiter, to equalise the air pressure within the gaiter.

30 Fit the large metal retaining clip to the gaiter and tighten it by carefully compressing the raised section of the clip. In the absence of the special tool, a pair of side cutters or pincers may be used. Secure the small retaining clip using the same procedure **(see illustration)**. Check that the constant velocity joint moves freely in all directions.

31 Refit the inner constant velocity joint as described in paragraphs 13 to 20.

Driveshaft without inner joint housing cover

Outer joint

32 Remove the driveshaft as described in Section 3.

33 Loosen the clips on the outer joint gaiter. If plastic straps are fitted, cut them free with snips **(see illustration)**.

34 Prise the gaiter large diameter from the outer joint housing **(see illustration)**, then tap the centre hub outwards using a soft metal drift to release it from the retaining circlip. Slide the outer joint off the driveshaft splines.

35 Extract the circlip from the groove in the driveshaft **(see illustration)**.

36 Withdraw the gaiter from the driveshaft, and if necessary, remove the plastic bush from the recess in the driveshaft **(see illustrations)**.

4.30 Secure the gaiter clips with side cutters

4.33 Plastic straps on outer gaiter

4.34 Remove the gaiter from the outer joint housing

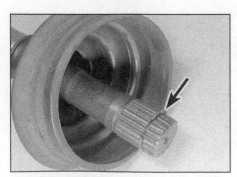

4.35 Driveshaft outer joint retaining circlip (arrowed)

4.36a Remove the outer gaiter from the driveshaft

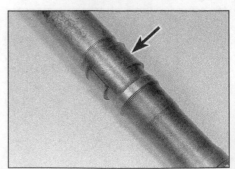

4.36b Plastic bush (arrowed) for the outer gaiter

4.50 Remove the inner gaiter

4.51 Separate the driveshaft and rollers from the inner joint housing

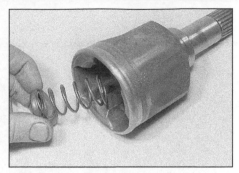

4.52 Remove the pressure pad and spring from the inner joint housing

37 With the outer joint removed from the driveshaft, thoroughly clean the joint with paraffin, or a suitable solvent, and dry it thoroughly. Carry out a visual inspection of the joint.

38 Move the inner splined driving member from side to side, to expose each ball in turn at the top of its track. Examine the balls for cracks, flat spots, or signs of pitting.

39 Inspect the ball tracks on the inner and outer members. If the tracks have widened, the balls will no longer be a tight fit. At the same time, check the ball cage windows for wear or cracking between the windows.

40 If any wear or free play is found, it will be necessary to renew the complete joint assembly, or the complete driveshaft, depending on the availability of parts. See your Citroën dealer for further information. If the joint is in satisfactory condition, obtain a repair kit consisting of a new gaiter, circlip, retaining clips, and the correct quantity of special grease.

41 If required, renew the inner joint gaiter, and at the same time inspect the inner joint (see paragraphs 49 to 61).

42 Begin reassembly by fitting the plastic seating in the driveshaft recess and fitting the new gaiter small diameter on it.

43 Refit the circlip in the driveshaft groove.

44 Inject the required amount of grease into the outer joint, then insert the driveshaft, engage the splines, and press them in until the circlip snaps into the groove.

45 Ease the rubber gaiter onto the outer joint, and fit the two clips. Metal type clips can be tightened using two pliers, by holding the buckle and pulling the clip through. Cut off the excess and bend the clip back under the buckle **(see illustrations 4.61a and 4.61b)**.

46 Check that the constant velocity joint moves freely in all directions, then refit the driveshaft to the vehicle as described in Section 3.

Inner joint

47 Remove the driveshaft as described in Section 3.

48 Remove the outer joint and gaiter, as described in paragraphs 33 to 40.

49 Loosen the clips on the inner gaiter. If plastic straps are fitted, cut them free.

50 Prise the gaiter large diameter from the inner joint housing and slide the gaiter off the outer end of the driveshaft **(see illustration)**. Remove the plastic bush, if fitted.

51 As the gaiter is released, the tri-axe joint will be released from its housing. Mark the driveshaft and inner joint housing in relation to each other, keeping the rollers engaged with their respective spigots, then separate the joint from the housing **(see illustration)**.

52 Remove the pressure pad and spring from inside the inner joint housing **(see illustration)**.

53 Thoroughly clean the joint using paraffin, or a suitable solvent, and dry it thoroughly, but be careful not to remove the marks made

in paragraph 51. Check the tripod joint bearings and housing for signs of wear, pitting or scuffing on their bearing surfaces. Check that the bearing rollers rotate smoothly and easily around the tripod joint, with no traces of roughness.

54 If, on inspection, the tripod joint or housing reveal signs of wear or damage, it will be necessary to renew the complete driveshaft assembly, since the joint is not available separately. If the joint is in satisfactory condition, obtain a kit consisting of a new gaiter, retaining clips, and the correct type and quantity of grease.

55 Retain the rollers using adhesive tape **(see illustration)**.

56 Begin reassembly by inserting the pressure pad and spring into the inner joint housing with the housing mounted upright in a soft-jawed vice.

57 Inject half the required amount of grease into the inner joint housing **(see illustration)**.

58 Locate the new inner gaiter halfway along the driveshaft **(see illustration)**. Refit the plastic bush, if applicable.

59 Remove the adhesive tape and insert the driveshaft into the housing, so that rollers are engaged in the correct spigots, according to the marks made during removal.

60 Inject the remaining amount of grease into the joint.

61 Keep the driveshaft pressed against the internal spring, refit the gaiter and tighten the

4.55 Left-hand driveshaft with rollers retained with adhesive tape

4.57 Apply grease to the inner joint housing

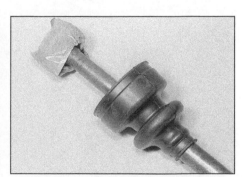

4.58 Fit the new inner gaiter on the driveshaft

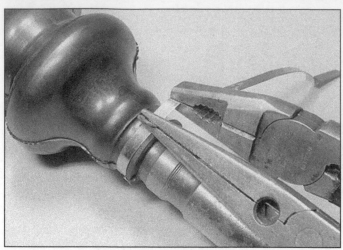

4.61a Tighten the metal clip . . .

4.61b . . . and bend it back under the buckle

clips. Metal type clips can be tightened using two pliers, by holding the buckle and pulling the clip through. Cut off the excess and bend the clip back under the buckle **(see illustrations)**. Check that the constant velocity joint moves freely in all directions.

62 Refit the outer joint and gaiter, as described in paragraphs 42 to 46.

---

5 Right-hand driveshaft intermediate bearing - renewal

Note: *A suitable bearing puller will be required, to draw the bearing and collar off the driveshaft end.*

1 Remove the right-hand driveshaft as described in Section 3.

2 Check that the bearing outer race rotates smoothly and easily, without any signs of roughness or undue free play between the inner and outer races. If necessary, renew the bearing as follows.

3 Using a long-reach universal bearing puller, carefully draw the collar and intermediate bearing off the driveshaft inner end **(see illustration)**. Apply a smear of grease to the inner race of the new bearing, then fit the bearing over the end of the driveshaft. Using a hammer and a suitable long piece of tubing which bears only on the bearing inner race, tap the new bearing into position on the driveshaft, until it abuts the constant velocity outer joint assembly. Once the bearing is correctly positioned, tap the bearing collar onto the shaft until it contacts the bearing inner race.

4 Check that the bearing rotates freely, then refit the driveshaft as described in Section 3.

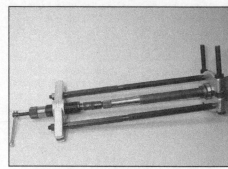

5.3 Use a long-reach bearing puller to remove the intermediate bearing from the right-hand driveshaft

Chapter 9
Braking system

Contents

Degrees of difficulty

Easy, suitable for novice with little experience	**Fairly easy,** suitable for beginner with some experience	**Fairly difficult,** suitable for competent DIY mechanic 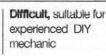	**Difficult,** suitable for experienced DIY mechanic	**Very difficult,** suitable for expert DIY or professional

Specifications

Brake system
Type . . . . Dual hydraulic circuit, split front and rear. Front disc brakes and rear drum brakes. Vacuum servo assisted. Vacuum provided by camshaft-driven pump on diesel models, either directly or from a belt. Cable-operated handbrake acting on rear wheels.

Front brakes
Type . . . . Solid disc, with single-piston sliding caliper
Disc diameter . . . . 247.0 mm
Disc thickness:
   New . . . . 10.0 mm
   Minimum thickness . . . . 8.0 mm
Maximum disc run-out . . . . 0.2 mm
Caliper:
   Make . . . . Girling
   Piston diameter . . . . 48.0 mm
Brake pad friction material minimum thickness . . . . 2.0 mm

Rear brakes
Type . . . . Drum with leading and trailing shoes
Drum diameter:
   New . . . . 228.3 ± 0.3 mm
   Maximum diameter after machining . . . . 229. 8 ± 0.2 mm
Brake shoe friction material minimum thickness . . . . 1.0 mm
Clearance between link rod and brake shoe . . . . 1.0 mm to 1.2 mm
Proportioning valve clearance A (see Section 11) . . . . 4.0 mm to 5.0 mm

Stop-lamp switch
Clearance in front of threaded portion . . . . 3.0 mm to 5.0 mm

Torque wrench settings

	Nm	lbf ft
Stop-lamp switch locknut	10	7
Master cylinder-to-servo unit	8	6
Vacuum servo unit-to-bracket	8	6
Vacuum pump, direct driven from camshaft (diesel models)	25	18
Cross-tube brackets-to-bulkhead (diesel models)	14	10
Hydraulic system bleed screws	3	2
Brake fluid pipe union nuts	15	11
Front brake caliper-to-hub carrier bolts*	90	66
Roadwheel bolts:		
3 bolts (pre-launch petrol models before July 1988)	70	52
4 bolts	80	59

*Use new bolts coated in thread-locking compound.

1 General information and precautions

General information

1 The braking system is the dual-circuit hydraulic type fitted with a vacuum servo. The arrangement of the hydraulic system is such that each circuit operates the front brakes or the rear brakes from a tandem master cylinder. Under normal circumstances, both circuits operate in unison. However, in the event of hydraulic failure in one circuit, full braking force will still be available at either the front or rear wheels **(see illustration)**.
2 All models are fitted with front disc brakes and rear drum brakes.
3 The front disc brakes are actuated by single-piston sliding type calipers, which ensure that equal pressure is applied to each disc pad.
4 The rear drum brakes incorporate leading and trailing shoes which are actuated by twin-piston wheel cylinders. A self-adjust mechanism automatically compensates for rear brake shoe wear. The correct lining-to-drum clearance is maintained by a lever on the leading shoe which is repositioned on a self-adjusting ratchet.
5 A suspension-mounted proportioning valve is fitted in the hydraulic lines to the rear wheels, to help prevent rear wheel lock-up during emergency braking.
6 The handbrake provides an independent mechanical means of rear brake application.
7 On diesel models, there is insufficient vacuum in the inlet manifold to operate the brake vacuum servo unit, so a vacuum pump is fitted. The pump is driven from the camshaft, either directly or from a belt.

Precautions

⚠ *Warning: When servicing any part of the system, work carefully and methodically. Observe scrupulous cleanliness when overhauling any part of the hydraulic system. Always renew components in axle sets where applicable, if in doubt about their condition, and use only genuine Citroën replacement parts, or at least those of known good quality. Note the warnings given in "Safety first!" at the start of this manual, and at relevant points in this Chapter, concerning the dangers of asbestos dust and hydraulic fluid.*

2 Hydraulic system - bleeding

⚠ *Warning: Hydraulic fluid is poisonous; wash off immediately and thoroughly in the case of skin contact, and seek immediate medical advice if any fluid is swallowed or gets into the eyes. Certain types of hydraulic fluid are inflammable, and may ignite when allowed into contact with hot components. When servicing the hydraulic system, it is safest to assume that the fluid is inflammable, and to take precautions against the risk of fire as though it is petrol that is being handled. Hydraulic fluid is also an effective paint stripper, and will attack plastics. If any is spilt, it should be washed off immediately, using copious quantities of fresh water. Finally, it is hygroscopic (it absorbs moisture from the air) - old fluid may be contaminated and unfit for further use. When topping-up or renewing the fluid, always use the recommended type, and ensure that it comes from a freshly-opened sealed container.*

General

1 The correct operation of any hydraulic system is only possible after removing all air from the components and circuit, and this is achieved by bleeding the system.
2 During the bleeding procedure, add only clean, unused hydraulic fluid of the recommended type. Never re-use fluid that has already been bled from the system.

1.1 Braking system layout

1 Master cylinder	*3 Proportioning valve*	*5 Rear drum brake*
2 Vacuum servo	*4 Front disc brake*	

H31041

Ensure that sufficient fluid is available before starting work.

3 If there is any possibility of incorrect fluid being already in the system, the brake components and circuit must be flushed completely with uncontaminated, correct fluid, and new seals should be fitted throughout the system.

4 If hydraulic fluid has been lost from the system, or air has entered because of a leak, ensure that the fault is cured before proceeding further.

5 Park the vehicle on level ground and apply the handbrake. Switch off the engine, then depress the brake pedal several times to dissipate the vacuum from the servo unit.

6 Check that all pipes and hoses are secure, unions tight and bleed screws closed. Remove the dust caps (where applicable), and clean any dirt from around the bleed screws **(see illustration)**.

7 Unscrew the master cylinder reservoir cap, and top-up the master cylinder reservoir to the "MAX" level line **(see illustration)**.

Note: *Remember to maintain the fluid level at least above the "MIN" level line throughout the procedure, otherwise there is a risk of further air entering the system.*

8 There are a number of one-man, do-it-yourself brake bleeding kits currently available from motor accessory shops. It is recommended that one of these kits is used whenever possible, as they greatly simplify the bleeding operation, and also reduce the risk of expelled air and fluid being drawn back into the system. If such a kit is not available, the basic (two-man) method must be used, which is described in detail below.

9 If a kit is to be used, prepare the vehicle as described previously, and follow the kit manufacturer's instructions, as the procedure may vary slightly according to the type being used. Generally, they are as outlined below in the relevant sub-section.

10 Whichever method is used, the same sequence must be followed (paragraphs 11 and 12) to ensure the removal of all air from the system.

Bleeding sequence

11 If the system has been only partially disconnected, and suitable precautions were taken to minimise fluid loss, it should be necessary to bleed only that part of the system (i.e. the front or rear circuit).

12 If the complete system is to be bled, then it should be done working in the following sequence:

a) Right-hand front wheel.
b) Left-hand front wheel.
c) Right-hand rear wheel.
d) Left-hand rear wheel.

Bleeding - basic (two-man) method

13 Collect a clean glass jar, a suitable length of plastic or rubber tubing which is a tight fit over the bleed screw, and a ring spanner to fit the screw. The help of an assistant will also be required.

2.6 Rear brake bleed screw (arrowed)

14 Remove the dust cap from the first screw in the sequence (if not already done). Fit a suitable spanner and tube to the screw, place the other end of the tube in the jar, and pour in sufficient fluid to cover the end of the tube.

15 Ensure that the master cylinder reservoir fluid level is maintained at least above the "MIN" level throughout the procedure.

16 Have the assistant fully depress the brake pedal several times to build up pressure, then maintain it down on the final downstroke.

17 While pedal pressure is maintained, unscrew the bleed screw (approximately one turn) and allow the compressed fluid and air to flow into the jar. The assistant should maintain pedal pressure, following the pedal down to the floor if necessary, and should not release the pedal until instructed to do so. When the flow stops, tighten the bleed screw again. Have the assistant release the pedal slowly, and recheck the reservoir fluid level.

18 Repeat the steps given in paragraphs 16 and 17 until the fluid emerging from the bleed screw is free from air bubbles. If the master cylinder has been drained and refilled, and air is being bled from the first screw in the sequence, allow at least five seconds between cycles for the master cylinder passages to refill.

19 When no more air bubbles appear, tighten the bleed screw securely, remove the tube and spanner, and refit the dust cap (where applicable). Do not overtighten the bleed screw.

20 Repeat the procedure on the remaining screws in the sequence, until all air is removed from the system and the brake pedal feels firm again.

Bleeding - using a one-way valve kit

21 As their name implies, these kits consist of a length of tubing with a one-way valve fitted, to prevent expelled air and fluid being drawn back into the system. Some kits include a translucent container, which can be positioned so that the air bubbles can be more easily seen flowing from the end of the tube.

22 The kit is connected to the bleed screw, which is then opened **(see illustration)**. The user returns to the driver's seat, depresses the brake pedal with a smooth, steady stroke, and slowly releases it. This is repeated until the expelled fluid is clear of air bubbles.

2.7 Topping-up the hydraulic fluid level in the reservoir

23 Note that these kits simplify work so much that it is easy to forget the master cylinder reservoir fluid level. Ensure that this is maintained at least above the "MIN" level at all times.

Bleeding - using a pressure-bleeding kit

24 These kits are usually operated by the reservoir of pressurised air contained in the spare tyre. However, note that it will probably be necessary to reduce the pressure to a lower level than normal. Refer to the instructions supplied with the kit.

25 By connecting a pressurised, fluid-filled container to the master cylinder reservoir, bleeding can be carried out simply by opening each bleed screw in turn (in the specified sequence), and allowing the fluid to flow out until no more air bubbles can be seen in the expelled fluid.

26 This method has the advantage that the large reservoir of fluid provides an additional safeguard against air being drawn into the system during bleeding.

27 Pressure-bleeding is particularly effective when bleeding "difficult" systems, or when bleeding the complete system at the time of routine fluid renewal.

All methods

28 When bleeding is complete, and firm pedal feel is restored, wash off any spilt fluid, tighten the bleed screws to the specified torque, and refit their dust caps.

29 Check the hydraulic fluid level in the master cylinder reservoir, and top-up if necessary (see *"Weekly checks"*).

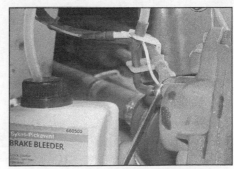

2.22 Bleeding a front brake caliper using a one-way valve kit

30 Discard any hydraulic fluid that has been bled from the system, as it will not be fit for re-use.

31 Check the feel of the brake pedal. If it feels at all spongy, air must still be present in the system, and further bleeding is required. Failure to bleed satisfactorily after a reasonable repetition of the bleeding procedure may be due to worn master cylinder seals.

3 Hydraulic pipes and hoses - renewal

Note: *Before starting work, refer to the warning at the beginning of Section 2 concerning the dangers of hydraulic fluid.*

1 If any pipe or hose is to be renewed, minimise fluid loss by first removing the master cylinder reservoir cap, then tightening it down onto a piece of polythene to obtain an airtight seal. Alternatively, flexible hoses can be sealed, if required, using a proprietary brake hose clamp. Metal brake pipe unions can be plugged (if care is taken not to allow dirt into the system) or capped immediately they are disconnected. Place a wad of rag under any union that is to be disconnected, to catch any spilt fluid.

2 If a flexible hose is to be disconnected, unscrew the brake pipe union nut before removing the spring clip which secures the hose to its mounting bracket **(see illustration)**.

3 To unscrew the union nuts, it is preferable to obtain a brake pipe spanner of the correct size, available from most large motor accessory shops. Failing this, a close-fitting open-ended spanner will be required, though if the nuts are tight or corroded, their flats may be rounded-off if the spanner slips. In such a case, a self-locking wrench is often the only way to unscrew a stubborn union, but it follows that the pipe and the damaged nuts must be renewed on reassembly. Always clean a union

and surrounding area before disconnecting it. If disconnecting a component with more than one union, make a careful note of the connections before disturbing any of them.

4 If a brake pipe is to be renewed, it can be obtained, cut to length with the union nuts and end flares in place, from Citroën dealers. All that is then necessary is to bend it to shape, following the line of the original, before fitting it to the vehicle. Alternatively, most motor accessory shops can make up brake pipes from kits, but this requires very careful measurement of the original, to ensure that the replacement is of the correct length. The safest answer is usually to take the original to the shop as a pattern.

5 On refitting, do not overtighten the union nuts. It is not necessary to exercise brute force to obtain a sound joint.

6 Ensure that the pipes and hoses are correctly routed, with no kinks, and that they are secured in the clips or brackets provided. After fitting, remove the polythene from the reservoir, and bleed the hydraulic system as described in Section 2. Wash off any spilt fluid, and check carefully for fluid leaks.

4 Front brake pads - inspection and renewal

⚠️ *Warning: Renew BOTH sets of front brake pads at the same time - NEVER renew the pads on only one wheel, as uneven braking may result. Note that the dust created by wear of the pads may contain asbestos, which is a health hazard. Never blow it out with compressed air, and don't inhale any of it. An approved filtering mask should be worn when working on the brakes. DO NOT use petrol or petroleum-based solvents to clean brake parts; use brake cleaner or methylated spirit only.*

Removal and inspection

1 Apply the handbrake, then jack up the front of the vehicle and support it on axle stands (see *"Jacking and vehicle support"*). Remove the front roadwheels.

2 Trace the brake pad wear sensor wiring back from the pads and disconnect it from the wiring connector. Note the routing of the wiring and free it from any retaining clips.

3 Before removing the pads, make a note of the correct fitted positions of the pad springs **(see illustration)**.

a) *The retaining pin spring clip passes though a hole at the end of each retaining pin. The brake sensor wiring may pass under the clip or through the centre loop.*

b) *The inner pad anti-rattle spring fits onto a rivet on the pad and clips over one of the retaining pins.*

c) *The outer pad anti-rattle spring fits onto a rivet on the pad, and the ends fit underneath the two retaining pins.*

4 Push the piston into its bore by pulling the caliper outwards.

5 Using a small screwdriver, prise the spring clip out of the holes in each retaining pin and remove the clip. Also prise the inner pad anti-rattle spring off the retaining pin.

6 Using a hammer and punch, tap both of the retaining pins out of the caliper.

7 Pull the caliper outwards, and withdraw the outer pad. Push the caliper inwards to withdraw the inner pad. Withdraw the metal shim, if fitted, from between the inner pad and the caliper piston.

8 Measure the thickness of each brake pad's friction material. If either pad is worn at any point to the specified minimum thickness or less, all four pads must be renewed. Also, the pads should be renewed if any are fouled with oil or grease. There is no satisfactory way of degreasing friction material, once contaminated. If any of the brake pads are worn unevenly, or are fouled with oil or grease, trace and rectify the cause before reassembly.

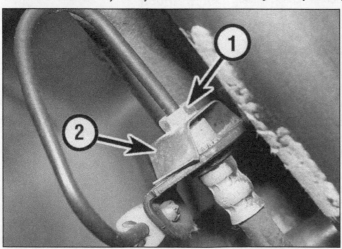

3.2 Brake pipe union nut (1) and hose spring clip (2)

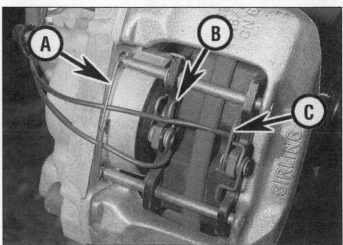

4.3 Front brake caliper with pads and springs

| A | Retaining pin spring clip | C | Outer pad anti-rattle spring |
| B | Inner pad anti-rattle spring | | |

4.14a Tap in the retaining pins, ensuring that the outer pad anti-rattle spring (arrowed) is correctly located under both pins . . .

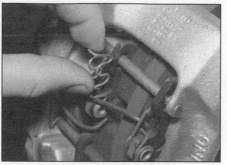

4.14b . . . then clip the inner pad anti-rattle spring onto the retaining pin

4.15 Ensure that the retaining pin spring clip ends are correctly located in the pin holes (arrowed)

New brake pads and spring kits are available from Citroën dealers.

9 If the brake pads are still serviceable, carefully clean them using a clean, fine wire brush or similar, paying particular attention to the sides and back of the metal backing. Clean out the grooves in the friction material, and pick out any large embedded particles of dirt or debris. Carefully clean the pad locations in the caliper body/mounting bracket.

10 Prior to fitting the pads, check that the guide pins are free to slide easily in the caliper body/mounting bracket, and check that the rubber guide pin gaiters are undamaged. Brush the dust and dirt from the caliper and piston, but *do not* inhale it, as it is a health hazard. Inspect the dust seal around the piston for damage, and the piston for evidence of fluid leaks, corrosion or damage. If attention to any of these components is necessary, refer to Section 8.

11 If new brake pads are to be fitted, the caliper piston must be pushed back into the cylinder to make room for them. Either use a G-clamp or similar tool, or use suitable pieces of wood as levers. Provided that the master cylinder reservoir has not been overfilled with hydraulic fluid, there should be no spillage, but keep a careful watch on the fluid level while retracting the piston. If the fluid level rises above the "MAX" level at any time, the surplus should be siphoned off or ejected via a plastic tube connected to the bleed screw (see Section 2).

Warning: Do not syphon the fluid by mouth, as it is poisonous. Use a syringe or an old poultry baster.

Refitting

12 Apply a little anti-squeal brake grease to the pad backing plates, but take great care not to allow any grease onto the pad friction linings.

13 Slide the pads into position in the caliper, ensuring that the inner and outer pads are fitted in their correct positions, and the metal shim (where fitted) is correctly positioned. Also ensure that the friction material of each pad is against the brake disc.

14 Fit the retaining pins into the caliper, through the brake pad holes and the hole in the metal shim (where fitted). Make sure the outer anti-rattle spring goes under both pins. Tap the pins firmly into position, and clip the inner anti-rattle spring over the appropriate pin **(see illustrations)**.

15 Secure the retaining pins into position with the spring clip, making sure that the clip ends are correctly located in the retaining pin holes. Pass the brake pad wear sensor wires under the clip or through the centre loop, as noted before removal **(see illustration)**.

16 Reconnect the brake pad wear sensor wiring connectors and secure the wiring with any fasteners as noted before removal.

17 Depress the brake pedal repeatedly, until the pads are pressed into firm contact with

the brake disc, and normal (non-assisted) pedal pressure is restored.

18 Repeat the above procedure on the remaining front brake caliper.

19 Refit the roadwheels, then lower the vehicle to the ground and tighten the roadwheel bolts to the specified torque.

20 Check the hydraulic fluid level (see "Weekly checks").

5 Rear brake shoes - inspection and renewal

Note: This section applies to the DBA type rear braking system with the levers on top of the brake shoes **(see illustration 5.1)**. Consult your Citroën dealer if any other type of rear braking system is fitted.

Warning: Renew BOTH sets of rear brake shoes at the same time - NEVER renew the shoes on only one wheel, as uneven braking may result. Note that the dust created by wear of the shoes may contain asbestos, which is a health hazard. Never blow it out with compressed air, and don't inhale any of it. An approved filtering mask should be worn when working on the brakes. DO NOT use petrol or petroleum-based solvents to clean brake parts; use brake cleaner or methylated spirit only.

Note: The following new components are required for refitting:
a) New clips for the handbrake levers;
b) New clips and retaining rings for the adjuster levers;
c) New steady springs for the brake shoes.

1 The rear brake assembly consists of a pair of shoes operated by a hydraulic cylinder which prises them apart at the top when the footbrake is pressed. There are two levers, the self-adjustment lever on the leading shoe and the handbrake lever on the trailing shoe. The two levers pivot on a link rod so that when the handbrake operates they are pulled apart at the top. The lower end of the adjuster lever rests on a ratchet and maintains the correct position to compensate for brake wear. The shoes are held in place by return springs and steady springs **(see illustration)**.

5.1 Rear brake assembly (right side)

1 Wheel cylinder
2 Link rod
3 Leading brake shoe
4 Adjuster lever
5 Adjuster ratchet
6 Handbrake cable
7 Handbrake lever
8 Steady spring
9 Trailing brake shoe

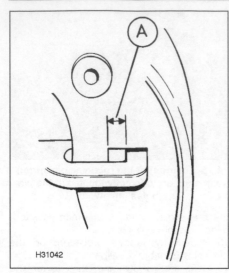

H31042

5.6 Check the clearance (A) between the link rod and brake shoe

Note: The remaining illustrations in this Section show the hub removed for clarity, although it is not necessary to remove it.

Removal and inspection

2 Chock the front wheels, then jack up the rear of the vehicle and support it on axle stands (see *"Jacking and vehicle support"*). Remove both rear wheels.

3 Remove the rear brake drums, as described in Section 7, then work on each side of the vehicle as follows.

4 Brush the dirt and dust from the brake backplate and drum.

 Warning: Do not inhale the dust, as it may be injurious to health.

5 Remove the long spring under the wheel cylinder.

6 Check the clearance between the link rod and the brake shoe **(see illustration)**. If the tolerance is outside the values given in the Specifications, the worn parts must be renewed.

7 Remove the steady springs from both shoes by pushing a large bolt or Allen key into the springs to release them from their anchorage **(see illustrations)**. These springs are prone to corrosion and should be renewed on refitting.

5.7a Insert a large bolt or Allen key into the steady springs . . .

8 Disconnect the handbrake cable from its lever on the trailing shoe **(see illustration)**.

9 Push the adjuster lever backwards to disengage it from the ratchet on the leading shoe and disconnect the link rod, then allow the lever to return to its position on the ratchet. Identify the link rods for refitting, as they are different on each side.

10 Remove both shoes by prising them apart at the bottom against the spring pressure.

11 Place a rubber band or cable-tie over the wheel cylinder, to prevent the pistons from being ejected. If there is any evidence of fluid leakage from the wheel cylinder, renew it or overhaul it as described in Section 9.

12 When removing the brake shoe assemblies, make sure that each shoe can be identified for refitting and do not mix them up. The two trailing shoes are identical but the leading shoes with the ratchet assemblies are different. When renewing the shoes, make sure you get the correct ones for the vehicle, and preferably take the old ones with you to the dealer. Check the adjuster levers and ratchets for wear and renew them if necessary.

Refitting

13 Working on each of the brake assemblies, transfer the adjuster lever and ratchet to the new leading shoe (or fit a new lever and ratchet as required). Fit the adjuster lever using a new clip, then fit the ratchet and spring, securing them with a new retaining ring **(see illustration)**.

5.7b . . . and remove them

14 Transfer the handbrake lever to the new trailing shoe and lock it in place with a new clip. Fit the link rod to the shoe, with the curved edges facing upwards, and fit the spring on the back **(see illustration)**. Make sure you fit the correct link rod, as they are different on the left and right brake assemblies.

15 Clean the backplate and apply brake grease sparingly to all the shoe contact areas on the backplate, wheel cylinder pistons and lower mounting block. Do not allow any grease to foul the brake friction material. Where applicable, remove the rubber band or cable tie from the wheel cylinder.

16 Attach the bottom spring to the leading and trailing shoes so that it lies on the outside faces, then offer the assembly to the backplate, making sure that the bottom spring goes behind the backplate bracket. Push the adjusting lever to the rear, towards the stub axle, and connect the link rod. Work the assembly into position on the backplate and fit the upper ends of the shoes onto the wheel cylinder pistons.

17 Use a pair of long-nosed pliers to refit the upper spring, then push the adjuster lever fully forwards against the brake shoe.

18 Push the handbrake lever forwards and connect the cable to the bottom end.

19 Fit new steady springs using a long bolt or Allen key to hook them into their anchorages.

20 With the brakes re-assembled on both sides of the vehicle, refit the brake drums (see Section 7).

21 Operate the footbrake several times to position the self-adjuster ratchet.

5.8 Disconnect the handbrake cable from the lever

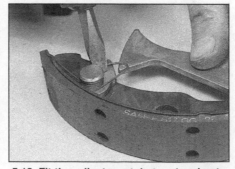

5.13 Fit the adjuster ratchet and spring to the leading shoe

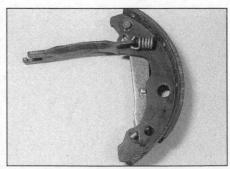

5.14 Fit the handbrake lever, link rod and spring to the trailing shoe

6.3 Using a micrometer to measure disc thickness

6.4 Checking disc run-out using a dial gauge

6.7 Remove the two screws (arrowed) securing the disc to the hub

22 Check the handbrake adjustment as described in Section 14.
23 Take care when driving the vehicle with the new shoes fitted, as they need some time to bed down before they achieve full efficiency.

6 Front brake disc - inspection, removal and refitting

Note: *Before starting work, refer to the warning at the beginning of Section 4 concerning the dangers of asbestos dust.*
Note: *If either disc requires renewal, BOTH should be renewed at the same time, to ensure even and consistent braking. New brake pads should also be fitted.*

Inspection

1 Apply the handbrake, then jack up the front of the vehicle and support it on axle stands (see *"Jacking and vehicle support"*). Remove the appropriate front roadwheel.
2 Slowly rotate the brake disc so that the full area of both sides can be checked. Remove the brake pads (see Section 4) if better access is required to the inboard surface. Light scoring is normal in the area swept by the brake pads, but if heavy scoring or cracks are found, the disc must be renewed.
3 It is normal to find a lip of rust and brake dust around the disc's perimeter; this can be scraped off if required. If, however, a lip has formed due to excessive wear of the brake pad swept area, then the disc's thickness must be measured using a micrometer **(see illustration)**. Take measurements at several places around the disc, at the inside and outside of the pad swept area. If the disc has worn at any point to the specified minimum thickness or less, the disc must be renewed.
4 If the disc is thought to be warped, it can be checked for run-out. Either use a dial gauge mounted on any convenient fixed point, while the disc is slowly rotated, or use feeler blades to measure (at several points all around the disc) the clearance between the disc and a fixed point, such as the caliper mounting bracket **(see illustration)**. If the measurements obtained are at the specified

maximum or beyond, the disc is excessively warped, and must be renewed. However, it is worth checking first that the hub bearing is in good condition (see Chapters 1 and/or 10). Also try the effect of removing the disc and turning it through 180°, to reposition it on the hub. If the run-out is still excessive, the disc must be renewed.
5 Check the disc for cracks (especially around the wheel bolt holes) and for any other wear or damage, and renew it if necessary.

Removal

6 Remove the brake pads as described in Section 4.
7 Use chalk or paint to mark the relationship of the brake disc to the hub, then remove the two screws securing the disc to the hub **(see illustration)**.

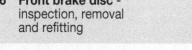

> **HAYNES HINT** *If the screws are tight, use an impact screwdriver to remove them. If the disc does not come easily off the hub, lightly tap the rear face of the disc with a hide or plastic mallet.*

8 Withdraw the disc and manoeuvre it out, tilting it as necessary to clear the hub and caliper.
9 If the disc cannot be manoeuvred out, place it back on the hub and lightly fit the screws. Then undo the two bolts securing the brake caliper assembly to the swivel hub (see Section 8) but do not disconnect the hydraulic fluid hose. Instead, tie the caliper to the suspension strut to avoid placing any strain on the hose. Then undo the screws and remove the brake disc.

Refitting

10 Ensure that the mating faces of the disc and the hub are clean and flat. If necessary, wipe the mating surfaces clean.
11 Refit the disc in the correct position, according to the chalk mark, then refit and securely tighten the disc retaining screws.
12 If a new disc has been fitted, use a suitable solvent to wipe any preservative coating from the disc.
13 Refit the caliper if it has been removed, using new bolts coated in thread-locking fluid.

14 Refit the brake pads as described in Section 4.

7 Rear brake drum - removal, inspection and refitting

Note: *Before starting work, refer to the warning at the beginning of Section 5 concerning the dangers of asbestos dust.*

Removal

1 Chock the front wheels, then jack up the rear of the vehicle and support it on axle stands (see *"Jacking and vehicle support"*). Remove the appropriate rear wheel.
2 Undo the retaining screw **(see illustration)** and withdraw the brake drum from the wheel hub. If the drum does not come off because the brake shoes are binding, first check that the handbrake is fully released, then proceed as follows.
3 Referring to the handbrake adjustment procedure in Section 14, fully slacken the handbrake cable adjuster nut, to obtain maximum free play in the cable.
4 Prise the plug from behind the backplate and insert a screwdriver through the hole so that it contacts the handbrake operating lever on the trailing brake shoe. Push the lever so that the stop-peg slips in front of the brake shoe web, allowing the brake shoes to retract fully. The brake drum can now be withdrawn. Refit the plug in the backplate hole **(see illustrations)**.

7.2 Undo the retaining screw (arrowed) and remove the brake drum

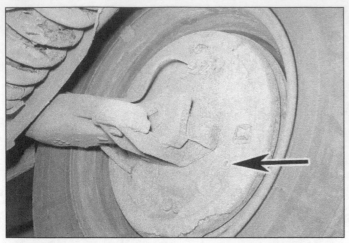

7.4a Prise the plug (arrowed) from behind the backplate and insert a screwdriver . . .

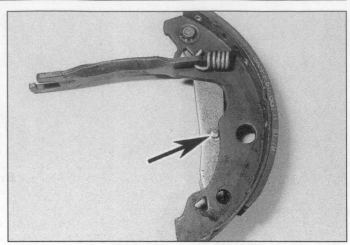

7.4b . . . and push the lever so that the stop-peg (arrowed) slips in front of the brake shoe web

Inspection

Note: *If either drum requires renewal, BOTH should be renewed at the same time, to ensure even and consistent braking. New brake shoes should also be fitted.*

5 Working carefully, remove all traces of brake dust from the drum, but *avoid inhaling the dust, as it is injurious to health.*

6 Clean the outside of the drum, and check it for obvious signs of wear or damage, such as cracks around the roadwheel bolt holes. Renew the drum if necessary.

7 Carefully examine the inside of the drum. Light scoring of the friction surface is normal, but if heavy scoring is found, the drum must be renewed. It is usual to find a lip on the outer edge of the drum which consists of a mixture of rust and brake dust. This should be scraped away, to leave a smooth surface which can be polished with fine (120 to 150 grade) emery paper. If, however, the lip is due to the friction surface being recessed by excessive wear, then the drum must be renewed.

8 If the drum is thought to be excessively worn, or oval, its internal diameter must be measured at several points using an internal micrometer. Take measurements in pairs, the second at right-angles to the first, and compare the two, to check for signs of ovality. Provided that it does not enlarge the drum to beyond the specified maximum diameter, it may be possible to have the drum refinished by skimming or grinding. If this is not possible, the drums on BOTH sides must be renewed. Note that if the drum is to be skimmed, BOTH drums must be refinished, to maintain a consistent internal diameter on both sides.

Refitting

9 If a new brake drum is to be installed, use a suitable solvent to remove any preservative coating that may have been applied to its internal friction surfaces.

10 If the handbrake lever was released while removing the drum, check that the lever is correctly repositioned with the stop-peg against the edge of the brake shoe web.

11 Fit the drum over the brake shoe assembly, then fit the retaining screw. It may be necessary to release the adjuster ratchet from the lower end of the lever on the leading shoe, to allow the drum to pass over the shoes.

12 Depress the footbrake several times to operate the self-adjusting mechanism.

13 Repeat the above procedure on the remaining rear brake assembly (where necessary). Check and, if necessary, adjust the handbrake cable as described in Section 14.

14 On completion, refit the roadwheel(s), then lower the vehicle to the ground and tighten the wheel bolts to the specified torque.

8 Front brake caliper - removal, overhaul and refitting

Note: *Before starting work, refer to the warning at the beginning of Section 2 concerning the dangers of hydraulic fluid, and to the warning at the beginning of Section 4 concerning the dangers of asbestos dust.*

Note: *New brake caliper securing bolts, coated in locking compound, are required for refitting.*

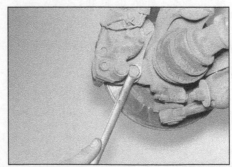

8.5 Undo the two bolts securing the front brake caliper assembly to the swivel hub

Removal

1 Apply the handbrake, then jack up the front of the vehicle and support it on axle stands (see *"Jacking and vehicle support"*). Remove the appropriate roadwheel.

2 To minimise fluid loss, first remove the master cylinder reservoir cap, and then tighten it down onto a piece of polythene, to obtain an airtight seal. Alternatively, use a brake hose clamp, a G-clamp or a similar suitable tool to clamp the flexible hose running to the caliper.

3 Remove the brake pads as described in Section 4.

4 Clean the area around the union, then loosen the brake hose union nut.

5 Undo the two bolts securing the caliper assembly to the swivel hub **(see illustration)**.

6 Lift the caliper assembly away from the brake disc, and unscrew it from the end of the brake hose.

Overhaul

Note: *All seals, dust covers and caps that are removed should be renewed.*

7 With the caliper on the bench, wipe away all traces of dust and dirt, but *avoid inhaling the dust, as it is injurious to health.*

8 Where necessary, use a small flat-bladed screwdriver to carefully prise the dust seal retaining clip out of the caliper bore.

9 Withdraw the partially-ejected piston from the caliper body, and remove the dust seal. The piston can be withdrawn by hand, or if necessary pushed out by applying compressed air to the brake hose union hole. Only low pressure should be required, such as is generated by a foot pump.

Caution: Be careful as the piston may be ejected with some force.

10 Using a small screwdriver, extract the piston hydraulic seal, taking great care not to damage the caliper bore.

11 Slide the mounting bracket out of the caliper body, and remove the guide pins and rubber gaiters.

12 Thoroughly clean all components, using only methylated spirit or clean hydraulic fluid as a cleaning medium. Never use mineral-based solvents such as petrol or paraffin, as they will attack the hydraulic system's rubber components. Dry the components immediately, using compressed air or a clean, lint-free cloth. Use compressed air to blow clear the fluid passages.

13 Check all components, and renew any that are worn or damaged. If there is any doubt about the condition of any component, renew it. Check particularly the cylinder bore and piston, and if they are scratched, worn or corroded in any way, renew the complete caliper body assembly. Similarly check the condition of the guide pins and their bores in the caliper body and mounting bracket. Both pins should be undamaged and (when cleaned) they should be a reasonably tight sliding fit in their bores. If the guide pins or their bores are worn, renew the complete caliper and mounting bracket assembly.

14 If the assembly is fit for further use, obtain the appropriate repair kit. The components are available from Citroën dealers in various combinations.

15 Renew all rubber seals, dust covers and caps disturbed on dismantling as a matter of course; these should never be re-used.

16 On reassembly, ensure that all components are absolutely clean and dry.

17 Dip the piston and the new piston (fluid) seal in clean hydraulic fluid. Smear clean fluid on the cylinder bore surface.

18 Fit the new piston (fluid) seal, using only your fingers (no tools) to manipulate it into the cylinder bore groove. Fit the new dust seal to the piston, and refit the piston to the cylinder bore using a twisting motion, and ensure that the piston enters squarely into the bore. Press the piston fully into the bore, then press the dust seal into the caliper body.

19 Where fitted, install the dust seal retaining clip, ensuring that it is correctly seated in the caliper groove.

20 Apply the grease supplied in the repair kit, or a good quality high-temperature brake grease or anti-seize compound to the guide pins and their bores. Fit the gaiters to the mounting bracket pins, then refit the bracket to the caliper, ensuring that the gaiters are correctly located in the grooves on the bracket and caliper body.

Refitting

21 Screw the caliper fully onto the flexible hose union, then position the caliper over the brake disc.

22 Fit new caliper mounting bolts, pre-coated with locking compound. If the bolts are not pre-coated, apply a suitable locking compound. Tighten the bolts to the specified torque.

23 Securely tighten the brake hose union nut, then refit the brake pads as described in Section 4.

24 Remove the brake hose clamp, or remove the polythene from the fluid reservoir (as applicable), and bleed the hydraulic system as described in Section 2. Note that, providing the precautions described were taken to minimise brake fluid loss, it should only be necessary to bleed the relevant front brake.

25 Refit the roadwheel, then lower the vehicle to the ground and tighten the roadwheel bolts to the specified torque.

---

9 Rear wheel cylinder - removal, overhaul and refitting

Note: *Before starting work, refer to the warning at the beginning of Section 2 concerning the dangers of hydraulic fluid.*

Removal

1 Remove the brake drum as described in Section 7.

2 Using pliers, carefully unhook the upper brake shoe return spring, and remove it from both brake shoes (note the orientation of the spring, to ensure correct refitting). Pull the upper ends of the shoes away from the wheel cylinder to disengage them from the pistons.

 HAYNES HiNT *Pull the handbrake lever on the trailing shoe fully forwards, so that the upper ends of the shoes are clear of the wheel cylinder. Wedge the lever in this position using a block of wood.*

3 To minimise fluid loss, first remove the master cylinder reservoir cap, then tighten it down onto a piece of polythene, to obtain an airtight seal. Alternatively, use a brake hose clamp, a G-clamp or a similar tool to clamp the flexible hose (connected between the metal pipe sections on the rear axle and trailing arm) at the nearest convenient point to the wheel cylinder.

4 Wipe away all traces of dirt around the brake pipe union at the rear of the wheel cylinder, and unscrew the union nut **(see illustration)**. Carefully ease the pipe out of the wheel cylinder, and plug or tape over its end, to prevent the entry of dirt. Wipe off any spilt fluid immediately.

5 Unscrew the two wheel cylinder retaining bolts from the rear of the backplate, and remove the cylinder, taking great care not to allow surplus hydraulic fluid to contaminate the brake shoe linings.

Overhaul

6 Clean the exterior of the cylinder to remove all traces of dirt and brake dust.

7 Pull the dust seals from the ends of the cylinder.

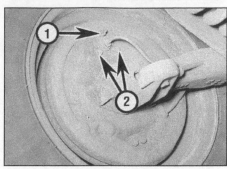

9.4 Wheel cylinder brake pipe union nut (1) and wheel cylinder retaining bolts (2)

8 Extract the pistons and seals, and return spring, noting the locations of all components to ensure correct refitting.

9 Examine the surfaces of the cylinder bore and pistons for signs of scoring and corrosion; if evident, renew the complete wheel cylinder. If the components are in good condition, discard the seals and obtain a repair kit which will contain all the necessary renewable components. If the correct type of repair kit is not available, renew the complete wheel cylinder.

10 Clean the pistons and the cylinder with methylated spirit or clean brake fluid, and reassemble in reverse order, making sure that the components are fitted in the correct sequence and orientated correctly, as noted before removal. Ensure that the lips of the seals face into the cylinder.

11 On completion, wipe the outer surfaces of the dust seals to remove any excess brake fluid.

Refitting

12 Clean the backplate, then place the wheel cylinder in position, and tighten the securing bolts.

13 Reconnect and tighten the brake pipe to the rear of the wheel cylinder, taking care not to allow dirt into the system.

14 Where applicable, release the handbrake lever on the trailing brake shoe, and reposition the upper ends of the shoes to engage them with the wheel cylinder pistons.

15 Refit the upper brake shoe return spring, ensuring that it is orientated as noted before removal.

16 Refit the brake drum as described in Section 7.

17 On completion, remove the brake hose clamp, or remove the polythene from the fluid reservoir (as applicable), and bleed the hydraulic system as described in Section 2. Note that, providing the precautions described were taken to minimise brake fluid loss, it should only be necessary to bleed the relevant rear brake.

10 Master cylinder - removal, overhaul and refitting

Note: *Before starting work, refer to the warning at the beginning of Section 2 concerning the dangers of hydraulic fluid.*
Note: *A new seal is required to refit the master cylinder to the brake servo unit.*

Removal

1 Disconnect the wiring connectors from the brake fluid level sender unit **(see illustration)**. There are two connectors, one for the switch feed and the other for earth.
2 Remove the master cylinder fluid reservoir cap, and syphon the hydraulic fluid from the reservoir. Alternatively, open any convenient bleed screw in the system, and gently pump the brake pedal to expel the fluid through a plastic tube connected to the screw (see Section 2).

⚠ **Warning: Do not syphon the fluid by mouth, as it is poisonous. Use a syringe or an old poultry baster.**

3 At this stage, the reservoir may be prised from the master cylinder and the seals extracted, if required.
4 Wipe clean the area around the brake pipe unions on the master cylinder, and place absorbent rags beneath the pipe unions to catch any surplus fluid. Make a note of the correct fitted positions of the unions, then unscrew the union nuts and carefully withdraw the pipes. Plug or tape over the pipe ends and master cylinder orifices, to minimise the loss of brake fluid, and to prevent the entry of dirt into the system. Wash off any spilt fluid immediately with cold water.
5 Slacken and remove the two nuts securing the master cylinder to the vacuum servo unit, then withdraw the unit from the engine compartment.
6 Recover the seal from the rear of the master cylinder, and discard it.

Overhaul

7 Pull the fluid reservoir from the top of the master cylinder. Prise the reservoir seals from the reservoir or the master cylinder, as applicable.
8 Using a wooden dowel (or suitable instrument), press the piston assembly into the master cylinder body, then extract the circlip from the end of the master cylinder bore.
9 Noting the order of removal and the direction of fitting of each component, withdraw the washer and the piston assemblies with their springs and seals, tapping the body on to a clean wooden surface to dislodge them. If necessary, clamp the master cylinder body in a vice (fitted with soft jaw covers) and use compressed air (applied through the secondary circuit fluid port) to assist the removal of the secondary piston assembly.
10 Thoroughly clean all components, using only methylated spirit or clean hydraulic fluid as a cleaning medium. Never use mineral-based solvents such as petrol or paraffin, as they will attack the hydraulic system rubber components. Dry the components immediately, using compressed air or a clean, lint-free cloth.
11 Check all components, and renew any that are worn or damaged. Check particularly the cylinder bores and pistons, and renew the complete assembly if these are scratched, worn or corroded. If there is any doubt about the condition of the assembly or of any of its components, renew it. Check that the cylinder body fluid passages are clear.
12 If the assembly is fit for further use, obtain a repair kit from your Citroën dealer. The kit consists of both piston assemblies and springs, complete with all seals and washers. Never re-use the old components.
13 Before reassembly, dip the pistons and the new seals in clean hydraulic fluid. Smear clean fluid onto the cylinder bore.
14 Insert the piston assemblies into the cylinder bore, using a twisting motion to avoid

trapping the seal lips. Ensure that all components are refitted in the correct order and the right way round, then fit the washer to the end of the primary piston. Where applicable, follow the assembly instructions supplied with the repair kit.
15 Press the piston assemblies fully into the bore using a clean wooden dowel, and secure them in position with the new circlip. Ensure that the circlip is correctly located in the groove in the cylinder bore.
16 Examine the fluid reservoir seals, and if necessary renew them. Fit the reservoir seals to the master cylinder body, then refit the reservoir.

Refitting

17 Remove all traces of dirt from the master cylinder and servo unit mating surfaces, and fit a new seal to the master cylinder.
18 Fit the master cylinder to the servo unit, ensuring that the pushrod enters the master cylinder bore centrally. Refit the master cylinder mounting nuts, and tighten them to the specified torque.
19 Wipe clean the brake pipe unions, then refit them to the correct master cylinder ports, as noted before removal, and tighten the union nuts securely.
20 Refill the master cylinder reservoir with new fluid, and bleed the complete hydraulic system as described in Section 2.

11 Rear brake proportioning valve - testing, adjustment and renewal

Note: *Before starting work, refer to the warning at the beginning of Section 2 concerning the dangers of hydraulic fluid.*

Testing and adjustment

1 The valve is located on the rear axle **(see illustration)**. Its purpose is to maintain equal braking effect on the front and rear wheels, and to prevent the rear brakes from locking up. It is connected by a spring to a lever on the anti-roll bar, to increase the braking limit for heavy loads on the rear axle.
2 To check and adjust the valve, apply the footbrake hard to close the valve, then check that the clearance (A) is within the limits given in the Specifications **(see illustration)**. Adjust the position of the clamp if necessary. **Do not** adjust the position of the factory-set screw.
3 If, after adjusting the valve correctly, you feel that the rear wheels are not being braked sufficiently, or you find that they are locking up under heavy braking, it is likely that the proportioning valve is defective. Take the vehicle to a Citroën dealer and have the unit pressure-tested using special equipment, and renew it if necessary. A table of values for pressure testing is given in the engine compartment next to the left-hand suspension strut **(see illustration)**. No attempt must be made to dismantle the unit.

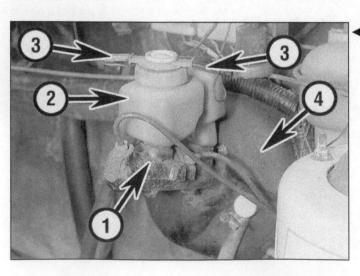

10.1 Brake master cylinder assembly and vacuum servo

1 Brake master cylinder
2 Fluid reservoir
3 Level sender wiring connectors
4 Vacuum servo unit

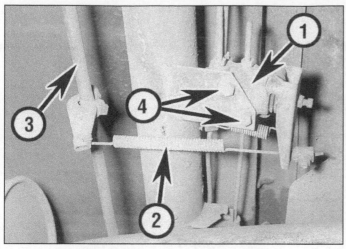

11.1 Rear brake proportioning valve

1 Valve	3 Anti-roll bar
2 Valve spring	4 Mounting bolts

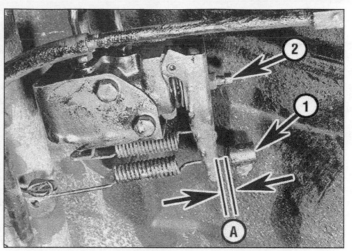

11.2 Rear brake proportioning valve adjustment

With the brake pedal depressed, adjust the clearance (A) by moving the clamp (1). **Do not** *alter the position of the factory-set screw (2)*

Removal

4 Chock the front wheels, then jack up the rear of the vehicle and support it on axle stands (see *"Jacking and vehicle support"*).

5 To minimise fluid loss, first remove the master cylinder reservoir cap, and then tighten it down onto a piece of polythene, to obtain an airtight seal.

6 Undo the union nuts and carefully pull out the pipes so that they are clear of the unit.

7 Unhook the spring from its lever on the anti-roll bar.

8 Undo the two bolts and remove the unit from its mounting bracket on the rear axle.

Refitting

9 Refitting is the reverse of removal. Tighten the union nuts to the specified torque, and tighten the mounting bolts securely.

---

12 Brake pedal - general, removal and refitting

General

1 The brake pedal operates a pushrod on the brake servo unit which is mounted on the engine compartment bulkhead.

2 On carburettor petrol models and on all left-hand drive models, the brake servo unit is directly in front of the brake pedal.

3 On diesel models and monopoint petrol injection models, the brake servo unit is always on the left side of the bulkhead, regardless of whether the vehicle is left-hand or right-hand drive. On right-hand drive models, a cross-tube connects the brake pedal to the servo unit pushrod (see Section 13).

Left-hand drive diesel models

Removal

4 Remove the parcel shelf from below the steering column (see Chapter 11).

5 Release the return spring from the brake pedal and bracket.

6 Remove the clevis pin and disconnect the servo unit pushrod from the brake pedal.

7 Remove the circlip from the end of the pivot pin. Remove the pin and withdraw the pedal.

Refitting

8 Refitting is the reverse of removal. Lubricate the pivot pin, and on completion adjust the stop-lamp switch as described in Section 18.

Right-hand drive diesel models

Removal

9 On right-hand drive diesel models, the brake and accelerator pedals have to be removed with their mounting bracket as a complete assembly **(see illustration)**. There is no stop-lamp switch on the pedal assembly. Instead it is at the servo end of the cross-tube.

10 Remove the parcel shelf from below the steering column (see Chapter 11).

11 Disconnect the accelerator cable from the pedal.

12 Disconnect the clutch pedal spring from the brake pedal mounting bracket.

13 Undo the nuts and detach the pedal bracket from the bulkhead.

14 Extract the spring clip and disconnect the link from the brake pedal assembly and cross-tube, then withdraw the brake pedal assembly.

Refitting

15 Refitting is the reverse of removal. Tighten the mounting bracket nuts to the specified torque, and adjust the clutch and accelerator cables.

Petrol models

16 The brake pedal design for petrol models is different from diesel models, for both left- and right-hand drive. No specific information is available, but the procedure for removal and refitting should be self-evident.

11.3 Table of values for pressure-testing the valve

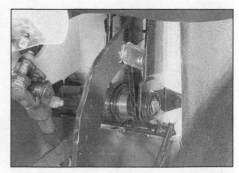

12.9 Brake pedal assembly and cross-tube

13 Brake linkage cross-tube (RH drive models) - removal and refitting

1 On diesel models and monopoint petrol injection models, the brake servo unit is on the left side of the bulkhead, regardless of whether the vehicle is left- or right-hand drive. On right-hand drive models, a cross-tube operates the servo unit pushrod (see illustration).

Removal

2 Disconnect the battery negative lead.
3 Remove the left and right parcel shelves (see Chapter 11).
4 Remove the steering column (see Chapter 10).
5 Remove the brake pedal assembly (see Section 12).
6 Disconnect the cable from the clutch pedal.
7 Remove the clevis pin and disconnect the servo unit pushrod from the cross-tube.
8 Disconnect the wiring from the stop-lamp switch.
9 Undo the nuts and detach the left-hand mounting bracket from the bulkhead.
10 Extract the spring clip and withdraw the bracket from the cross-tube, then withdraw the cross-tube from the vehicle.

Refitting

11 Refitting is the reverse of removal. Tighten the left- and right-hand mounting bracket nuts to their specified torque, and adjust the clutch and accelerator cables.

14 Handbrake - adjustment

1 Chock the front wheels, then jack up the rear of the vehicle and support it on axle stands (see "Jacking and vehicle support").
2 Depress the brake pedal firmly several times to set the self-adjusting rear brake shoe mechanism.

3 Pull up the boot from the handbrake to gain access to the equaliser and adjusting nut. The equaliser is a short metal bar at the rear of the lever, which lies transversely across the vehicle and pulls on both handbrake cables simultaneously.
4 Apply the handbrake lever and check that the equaliser and cables move freely and smoothly, then set the lever on the third notch on the ratchet mechanism. With the lever in this position, slacken the locknut and turn the adjusting nut until only a slight drag can be felt when the rear wheels are turned.
5 Make sure there is equal tension on the left- and right-hand cables. If the tensions are unequal, the equaliser bar will be inclined at an angle. If necessary, adjust the cables individually until the equaliser becomes transverse. This can only be done if the cables are fitted with in-line adjusters (normally underneath the vehicle). If the cables cannot be equalised they will need to be renewed (see Section 16).
6 Fully release the handbrake lever and check that the wheels rotate freely. Then apply the handbrake lever one notch at a time until both rear wheels are locked. The adjustment is correct if the wheels become locked at 5 notches. Turn the central adjustment nut on the equaliser bar to achieve the correct adjustment, then tighten the locknut.
7 Refit the boot to the handbrake and lower the vehicle to the ground.

15 Handbrake lever - removal and refitting

Removal

1 Chock the front wheels, then jack up the rear of the vehicle and support it on axle stands (see "Jacking and vehicle support"). Release the handbrake lever.
2 Pull up the boot from the handbrake to gain access to the equaliser and adjusting nut.
3 Slacken the locknut and turn the adjuster

so that the cables become slack, then detach the cables from the equaliser.
4 Undo the bolts securing the handbrake assembly to the floor and remove it.

Refitting

5 Refitting is the reverse of removal, but adjust the cables as described in Section 14 and check for correct operation.

16 Handbrake cables - removal and refitting

Note: Before starting work, refer to the warning at the beginning of Section 5 concerning the dangers of asbestos dust.

Removal

1 Chock the front wheels, then jack up the rear of the vehicle and support it on axle stands (see "Jacking and vehicle support"). Release the handbrake lever.
2 Pull up the boot from the handbrake to access to the equaliser and central adjusting nut.
3 Slacken the locknut and turn the adjuster so that the cables become slack, then detach the cables from the equaliser.
4 Remove the rear brake drums (see Section 7), then disconnect the handbrake cables from the brake shoe operating levers.
5 Undo the clips securing the cables to the vehicle underbody, then remove the cables by pulling them out from their front end locations and from the rear brake backplates.

Refitting

6 Refitting is the reverse of removal, but adjust the cables as described in Section 14, and check for correct operation. If new cables are fitted, make sure they have in-line adjusters so that the tension can be equalised.

17 Handbrake "on" warning light switch - removal and refitting

Removal

1 Disconnect the battery negative lead.
2 Pull up the boot from the handbrake to gain to access to the switch.
3 Remove the securing screw, then withdraw the switch from the floor and disconnect the wiring.

> **HAYNES HINT** Tape the wiring to the floor, to prevent it falling back down the hole. Alternatively, tie a piece of string to the wiring to retrieve it.

Refitting

4 Refitting is the reverse of removal, ensuring (where applicable) that the rubber gaiter is correctly seated on the switch.

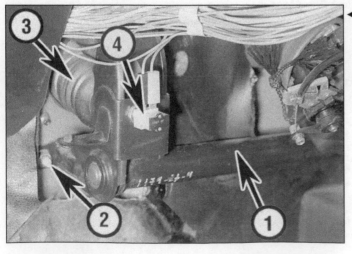

13.1 Brake linkage cross-tube

1 Cross-tube
2 Mounting bracket nut
3 Servo unit push-rod with gaiter
4 Stop-lamp switch

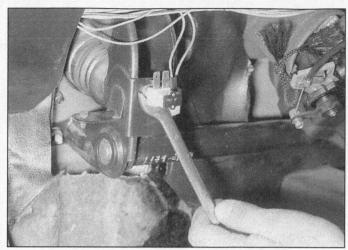

18.3 Slacken the locknut on the stop-lamp switch (right-hand drive diesel model with cross-tube)

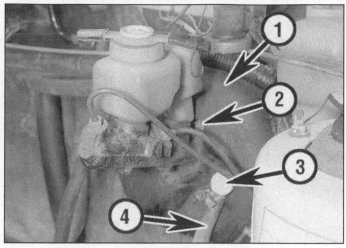

19.6 Servo unit and associated components

1 Servo unit	3 Check valve
2 Nut	4 Vacuum hose

18 Stop-lamp switch -
adjustment, removal and refitting

Note: *These instructions apply to right-hand drive diesel models with a brake linkage cross-tube. A similar type of switch is mounted above the brake pedal on left-hand drive diesel models. No specific instructions are available about the type of switch fitted to petrol models.*

Adjustment

1 Remove the left parcel shelf (see Chapter 11) to access the switch.
2 With the brake pedal released, the clearance in front of the threaded portion of the switch should be according to the Specifications.
3 If adjustment is necessary, slacken the locknut **(see illustration)**, then rotate the switch as required until the clearance is correct. If necessary, to avoid twisting the wires, disconnect the battery negative lead, then disconnect the wiring from the switch. Tighten the locknut on completion, and reconnect the wiring and battery.
4 Check that the stop-lights operate when the pedal is depressed, and that they extinguish when the pedal is released.

Removal

5 Disconnect the battery negative lead.
6 Remove the left parcel shelf (see Chapter 11).
7 Disconnect the wiring connectors from the switch.
8 Slacken the locknut, then unscrew the switch and remove it from the bracket.

Refitting

9 Refitting is the reverse of removal, making sure that the adjustment is correct and the lamp operates when the brake pedal is depressed. Tighten the locknut to the specified torque.

19 Vacuum servo unit -
testing, removal and refitting

Testing

1 To test the operation of the servo unit, depress the footbrake several times to exhaust the vacuum, then start the engine whilst keeping the pedal firmly depressed. As the engine starts, there should be a noticeable "give" in the brake pedal as the vacuum builds up. Allow the engine to run for about one minute, then switch it off. If the brake pedal is now depressed it should feel normal, but further applications should result in the pedal feeling firmer, with the pedal stroke decreasing with each application.
2 If the servo does not operate as described, first inspect the servo unit check valve as described in Section 20. On diesel engine models, also check the operation of the vacuum pump as described in Section 23.
3 If the servo unit still fails to operate satisfactorily, the fault lies within the unit itself. Repairs to the unit are not possible - if faulty, the servo unit must be renewed.

Removal

Note: *A new gasket is required for refitting.*
4 To improve access if necessary, remove any wiring and hoses from around the servo unit and brake master cylinder.
5 Disconnect the wiring connectors from the brake fluid level sender unit on the reservoir cap. There are two connectors, one for the switch feed and the other for earth.
6 Loosen the securing clip and disconnect the vacuum hose from the servo **(see illustration)**.
7 Unscrew and remove the nuts securing the master cylinder to the vacuum servo.
8 Working under the servo, unclip the master cylinder brake fluid pipes from the bulkhead.
9 Carefully pull the master cylinder forwards from the servo, taking care not to strain the fluid pipes. Recover the gasket.
10 Working inside the van, remove the appropriate parcel shelf, then remove the clevis pin attaching the brake servo pushrod to the brake pedal or cross-tube (as applicable).
11 Undo the mounting nuts and withdraw the servo unit from the bulkhead **(see illustration)**. Recover the gasket and spacer.
12 Do not attempt to adjust the servo unit pushrods as they are factory set. If in doubt, take the servo unit to your Citroën dealer.

Refitting

13 Refitting is the reverse of removal, using a new gasket. Tighten all nuts and bolts to the specified torque.

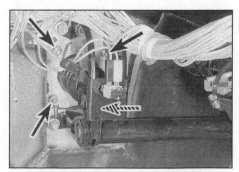

19.11 Servo unit mounting nuts (arrowed) - right-hand drive diesel model with cross-tube

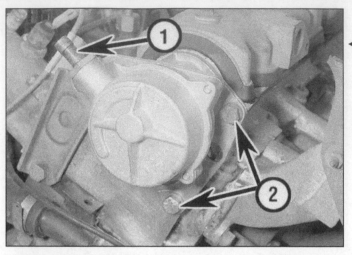

22.6 Brake vacuum pump on transmission end of cylinder head - diesel engine model

1 *Vacuum outlet*
2 *Mounting bolts*

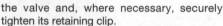

20 Vacuum servo unit check valve - removal, testing and refitting

Removal

1 Slacken the securing clip (where fitted), and disconnect the vacuum hose from the servo unit check valve.
2 Withdraw the valve from its rubber sealing grommet, using a pulling and twisting motion. Remove the grommet from the servo.

Testing

3 Examine the check valve for signs of damage, and renew if necessary. The valve may be tested by blowing through it in both directions. Air should flow through the valve in one direction only - when blown through from the servo unit end of the valve. Renew the valve if this is not the case.
4 Examine the rubber sealing grommet and flexible vacuum hose for signs of damage or deterioration, and renew them as necessary.

Refitting

5 Fit the sealing grommet into position in the servo unit.
6 Carefully ease the check valve into position, taking great care not to displace or damage the grommet. Reconnect the vacuum hose to

the valve and, where necessary, securely tighten its retaining clip.
7 On completion, start the engine and check the servo as described in Section 19.

21 Vacuum pump (diesel engine models) - general

1 Diesel engine models do not have sufficient vacuum in the inlet manifold to operate the brake servo unit, therefore they require a vacuum pump. A diaphragm pump was used up to December 1989, mounted on the transmission and driven by a belt from the camshaft.
2 From December 1989 to January 1991, a Pierburg vane-type pump was used, driven directly from the transmission end of the camshaft. This modification resulted in a modified cylinder head, with mounting bolt holes for the pump, and an oil channel that passes oil from the engine lubrication system to the pump. The camshaft is shorter, because it no longer requires a belt drive pulley, and the end of the camshaft has a slot for engagement with the pump drive dog.
3 On models between January 1991 and early 1993, the original (longer) camshaft is fitted, with the original belt-driven vacuum

pump driven from a pulley on the end of the camshaft.
4 From early 1993, an improved (second generation) direct-driven vane-type brake vacuum pump was introduced, and the engine was modified as described in paragraph 2.
5 In December 1993, power assisted steering was introduced as an option, and became standard on the heavier 765 kg payload van. A new Pierburg diaphragm vacuum pump was introduced on all models with power steering, driven by a belt from the camshaft. The camshaft was lengthened to accommodate a double pulley which drives both the brake vacuum pump and the power steering pump. The new diaphragm pump is not interchangeable with the old one and cannot be fitted to earlier models with manual steering. From December 1993 onwards, models with manual steering continued to use the vane-type pump described in paragraph 4.

22 Vacuum pump (diesel engine models) - removal and refitting

Belt-driven diaphragm pump
Removal

1 Remove the air cleaner and ducting.
2 Disconnect the inlet and outlet hoses.
3 Loosen the pivot and adjustment link nuts and bolts, swivel the vacuum pump upwards and slip the drivebelt from the pulleys.
4 Unscrew the bolts and remove the vacuum pump from the mounting bracket and adjustment link.

Refitting

5 Refitting is the reverse of removal, but swivel the pump downwards until the drivebelt tension is as given in the Specifications (see Chapter 1B), then tighten the pivot and adjustment link nuts and bolts. With the vehicle on level ground, unscrew the filler/level plug and check that the oil level is up to the bottom of the hole. If not, top-up with the correct grade of oil then refit and tighten the plug.

Direct-driven vane-type pump
Removal

Note: *New O-rings should be used on refitting.*

6 The pump is located at the transmission end of the cylinder head **(see illustration)**.
7 Loosen the securing clip and disconnect the vacuum hose from the pump.
8 Undo the three mounting bolts and remove the brake vacuum pump from the end of the cylinder head.
9 Extract the two O-rings from the grooves in the pump **(see illustrations)**.
10 Using a small screwdriver, extract the filter from the oil lubrication channel in the vacuum pump.

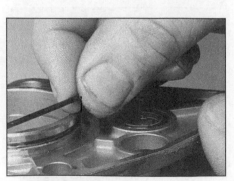

22.9a Remove the large O-ring ...

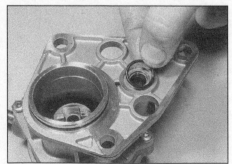

22.9b ... and the small O-ring from the grooves in the pump

Refitting

11 Before refitting the pump, clean the O-ring grooves, and also clean the mating surfaces of the pump and cylinder head. Clean the filter, or if necessary renew it.

12 Locate the filter in the oil lubrication channel.

13 Fit new O-rings in the grooves on the pump, and lightly oil them.

14 Locate the pump on the end of the cylinder head, making sure that the dog engages correctly with the end of the camshaft. To avoid the O-rings being displaced, align the slot in the end of the camshaft with the dog on the vacuum pump before refitting the pump. If necessary, put the engine in gear, then hold the pump steady in line with the bolt holes and rock the vehicle to rotate the camshaft.

15 Insert the mounting bolts, and tighten them to the specified torque.

16 Connect the vacuum hose and tighten the clip.

17 Start the engine, and check that the brake pedal operates correctly, with assistance from the vacuum pump. Check around the pump for signs of oil leakage.

23 Vacuum pump (diesel engine models) - testing and overhaul

Testing

1 The operation of the brake system vacuum pump can be checked using a suitable vacuum gauge.

2 Disconnect the vacuum hose from the pump, and connect the gauge to the pump union using a suitable length of hose.

3 Start the engine and allow it to idle, then measure the vacuum created by the pump. As a guide, after one minute, a minimum of approximately 500 mm Hg should be recorded. If the vacuum registered is significantly less than this, it is likely that the pump is faulty. However, seek the advice of a Citroën dealer before condemning the pump.

Overhaul

4 Overhaul of the camshaft-driven vacuum pump, mounted on the cylinder head, is not possible because no components are available separately. If faulty, the complete pump assembly must be renewed. Consult your Citroën dealer for availability of parts for belt-driven vacuum pumps.

Chapter 10
Suspension and steering

Contents

Degrees of difficulty

Easy, suitable for novice with little experience	Fairly easy, suitable for beginner with some experience	Fairly difficult, suitable for competent DIY mechanic	Difficult, suitable for experienced DIY mechanic	Very difficult, suitable for expert DIY or professional 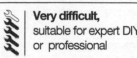

Specifications

Front suspension
Type .... Independent MacPherson struts, with shock absorbers and coil springs. Front subframe carries track control arms, steering gear and anti-roll bar.

Rear suspension
Type .... Independent trailing arms mounted on pivot bearings. Suspension struts with shock absorbers and coil springs housed in rear subframe. Anti-roll bar connects to trailing arms.

Steering
Type .... Rack-and-pinion, with power-assisted steering available on some models.

Wheel alignment and steering angles
All measurements to be taken with vehicle at kerb weight (vehicle unladen with 5 litres of fuel in tank)

Front wheel camber angle
Petrol models:
475 and 600 kg payload .... 0°37' ± 30'
765 kg payload .... 0°34' ± 30'
Diesel models:
  Manual steering:
    600 kg payload .... 0°30' ± 30'
    765 kg payload .... 0°41' ± 30'
  Power-assisted steering .... 0°30' ± 30'

Front wheel castor angle
Petrol models:
475 kg payload .... 0°56' ± 30'
600 kg payload .... 0°57' ± 30'
765 kg payload .... 0°43' ± 30'
Diesel models:
  Manual steering:
    600 kg payload .... 0°55' ± 30'
    765 kg payload .... 1°47' ± 30'
  Power-assisted steering .... 0°55' ± 30'

Wheel alignment and steering angles (continued)

Front wheel king pin inclination
Petrol models:
475 and 600 kg payload	8°39' ± 40'
765 kg payload	8°45' ± 40'

Diesel models:
Manual steering:	
600 kg payload	8°50' ± 40'
765 kg payload	8°34' ± 40'
Power-assisted steering	8°50' ± 40'

Front wheel toe setting
Petrol models:
Carburettor	2.0 mm toe-in
Fuel injection:	
600 kg payload	3.7 mm toe-in
765 kg payload	3.4 mm toe-in

Diesel models:
Manual steering	2.0 mm toe-in
Power-assisted steering	3.0 mm toe-in

Rear wheel camber angle (all models)	0°9' ± 20'
Rear wheel toe setting (all models)	1.6 mm to 5.0 mm toe-in

Suspension height
All measurements to be taken with vehicle at kerb weight (vehicle unladen with 5 litres of fuel in tank)

Front suspension height (see Section 14)
Petrol models:
475 kg payload	193 mm ± 10 mm
600 kg payload	204 mm ± 10 mm
765 kg payload	211 mm ± 10 mm

Diesel models:
600 kg payload	203 mm ± 10 mm
765 kg payload	214 mm ± 10 mm

Rear suspension height (see Section 14)
Petrol models:
475 kg payload	500 mm ± 10 mm
600 kg payload	510 mm ± 10 mm
765 kg payload	529 mm ± 10 mm

Diesel models:
600 kg payload	515 mm ± 10 mm
765 kg payload	527 mm ± 10 mm
Height difference between both sides (petrol and diesel)	4 mm maximum

Coil springs

Front spring
Free height:
475 and 600 kg payload	441.6 mm
765 kg payload	396.2 mm

Colour code:
475 and 600 kg payload	Blue/grey or yellow/grey
765 kg payload	Green or yellow

Rear spring
Free height	263.5 mm

Colour code:
475 and 600 kg payload	Green (on first coil)
765 kg payload	Yellow (on first coil)

Shock absorbers
Front shock absorber reference:
475 and 600 kg payload	9B
765 kg payload	2D or 3D

Rear shock absorber reference:
475 and 600 kg payload	1O
765 kg payload	1D

Anti-roll bars

Front anti-roll bar diameter:
475 and 600 kg payload .... 23 mm
765 kg payload .... 24 mm
Front anti-roll guide bar adjustment length (see Section 6) .... 330 mm
Rear anti-roll bar diameter:
Early models (specific shape, measured near exhaust pipe):
475 and 600 kg payload .... 15 mm
765 kg payload .... 18 mm
Later models:
475 and 600 kg payload .... 18 mm
765 kg payload .... 19 mm

Roadwheels

Type .... Pressed-steel
Size:
Grey .... 4.50 B 13 FHC 4.30
Black (available on diesel models) .... 4.50 B 13 FH 4.30

Tyres

Standard tyres (Michelin):
Petrol, 475 kg payload .... 135 R 13 MX (72 R) reinforced
Petrol, 600 kg payload .... 145 or 155 R 13 MX (78 R) reinforced
Petrol/diesel, 765 kg payload .... 155 R 13 XCA
Diesel, 600 kg payload .... 155 R 13 MX (78 S)
Authorised fitting:
Petrol, 475 and 600 kg payload .... 145 R 13 (M + S)
Petrol/diesel, 765 kg payload .... 155 R 13 XCA (M + S)
Diesel, 600 kg payload .... 155 R 13 (M + S)

Tyre pressures .... See end of "Weekly checks" on page 0•16

Torque wrench settings

	Nm	lbf ft
Front suspension		
Anti-roll bar to track control arm	65	48
Anti-roll bar mounting clamp bolts to subframe	35	26
Anti-roll guide bar to anti-roll bar	45	33
Anti-roll guide bar to subframe	75	55
Suspension strut upper mounting-to-body nuts*	20	15
Suspension strut shock absorber rod nut*	70	52
Suspension strut to hub carrier clamp nut and bolt	75	55
Track control arm outer balljoint clamp nut and bolt*	35	26
Track control arm inner pivot nut and bolt	35	26
Rear suspension		
Anti-roll bar bracket to trailing arm	90	67
Hub nut*	275	204
Subframe front mounting rubber bush, central nut and bolt	65	48
Shock absorber rod nut	17	13
Trailing arm pivot shaft and nut	130	96
Steering		
Steering column intermediate clamp nut and bolt**	20	15
Steering column mounting nuts	15	11
Steering shaft to steering gear pinion nut and bolt**	15	11
Steering gear-to-subframe bolts	47	35
Steering wheel nut*	50	37
Track rod end balljoint nut*	35	26
Track rod end locknut	45	33
Track rod inner collar:		
Manual steering	50	37
Power steering	55	41
Roadwheels		
Roadwheel bolts:		
3 bolts (pre-launch petrol models before July 1988)	70	52
4 bolts	80	59

*Use a new nut.
**Use a new nut and bolt.

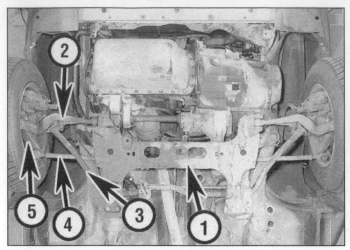

1.1a Front underbody

1 Subframe	4 Steering track rod
2 Track control arm	5 Steering arm on hub
3 Anti-roll bar	carrier

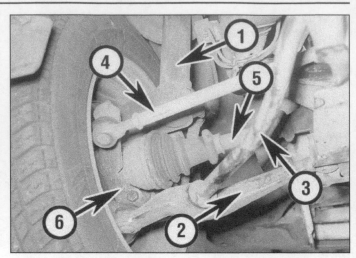

1.1b Left front wheel suspension viewed from rear

1 Suspension strut	4 Steering track rod
2 Track control arm	5 Driveshaft
3 Anti-roll bar	6 Hub carrier

1 General information

1 The front suspension is the independent MacPherson strut type, incorporating coil springs and integral telescopic shock absorbers. The lower ends of the MacPherson struts are connected by clamp bolts to the hub carriers, which carry the wheel bearings, brake calipers and the discs. The hub carriers are connected by balljoints to the outer ends of the track control arms, which have rubber mounting bushes at their inner ends. The front anti-roll bar is fitted by rubber bearings to the subframe, and is connected by rubber mountings to the track control arms at each end **(see illustrations)**.

2 The rear suspension has two independent trailing arms with pivot bearings, connected to a subframe which consists of a tubular crossmember and two suspension strut housings. Each suspension strut consists of a telescopic shock absorber and coil spring.

The rear anti-roll bar connects to the upper ends of the trailing arms **(see illustration)**.

3 The rear subframe assembly is mounted onto the vehicle underbody by four rubber mountings.

4 The steering column has intermediate and lower universal joints. The lower joint is clamped to the steering gear pinion by means of a clamp nut and bolt.

5 The steering gear is mounted on the subframe in front of the engine compartment bulkhead. It is connected by two track rods, with balljoints at their outer ends and flexible bearings at their inner ends, to the steering arms projecting rearwards from the hub carriers. The track rod outer ends are threaded, to facilitate steering angle adjustment.

6 Power-assisted steering was available on diesel models from December 1993-on and in the UK it was fitted as standard to the 765 kg payload Champ vans. The hydraulic system is powered by a pump, which is belt-driven from a pulley on the camshaft. The hydraulic fluid is cooled by passing it through a single-bore cooling tube located on the underbody.

2 Front hub carrier - removal and refitting

Note: *A balljoint separator tool will be required for this operation, together with the following components:*
 a) A new hub nut;
 b) A new track-rod end to steering arm nut;
 c) A new hub carrier to track control arm balljoint clamp nut;
 d) New brake caliper to hub carrier bolts.

Removal

1 Chock the rear wheels, apply the handbrake, then jack up the front of the vehicle and support it securely on axle stands (see *"Jacking and vehicle support"*). Remove the appropriate roadwheel.

2 Remove the hub nut dust cover (if applicable), then relieve the staking on the hub nut using a hammer and a suitable cold chisel or punch **(see illustration)**.

⚠ *Warning: Wear suitable eye protection.*

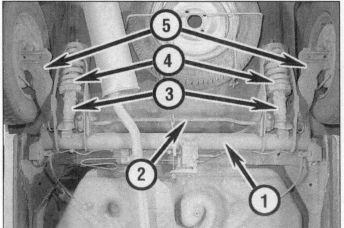

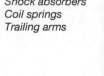

1.2 Rear underbody

1 Crossmember
2 Anti-roll bar
3 Shock absorbers
4 Coil springs
5 Trailing arms

2.2 Relieve the staking on the hub nut using a chisel

3 Slacken the hub nut using a socket and extension bar but do not remove it at this stage. Use a suitable counterhold (**see Tool Tip**) to prevent the hub from turning. Do not use the footbrake because it might damage the brake disc retaining screws.

⚠️ *Warning: The hub nut is very tight. You can pull the vehicle off the axle stands by attempting to slacken it without a counterhold.*

HAYNES HINT *If you do not have a counterhold, refit the roadwheel and lower the vehicle to the ground. Apply the handbrake, then slacken the hub nut.*

4 Undo the nut securing the track-rod end to the steering arm on the hub carrier. Using a balljoint separator tool, separate the track-rod end from the steering arm. Remove the nut and washer **(see illustration)**.

5 Undo the clamp nut and bolt from the track control arm outer joint, where it meets the hub carrier, noting which way round it is fitted. If necessary, use a screwdriver to open up the clamp slightly. Then use a lever between the anti-roll bar and track control arm to pull the joint downwards and disconnect it. Have an assistant hold the hub steady and pull it off the joint. Note the location of the protector shield **(see illustration)**.

6 Release the hub from the driveshaft splines by pulling the hub carrier assembly outwards. If necessary, the shaft can be tapped out of the hub using a soft-faced mallet. Support the driveshaft by suspending it from the body using wire or string, and do not allow it to hang down. When disengaging the hub from the left-hand driveshaft (which has no intermediate bearing) take care not to pull the driveshaft out of the transmission.

7 Remove the front brake caliper and disc (see Chapter 9). First remove the brake pads, then undo the two bolts securing the brake caliper assembly to the hub. Slide the caliper off the disc, and suspend it from the bodywork using wire or string, or place it on a stand to avoid placing any strain on the hydraulic brake hose. Use chalk or paint to mark the relationship of the brake disc to the hub, then undo the two screws and remove the brake disc.

8 Undo the clamp nut and bolt **(see illustration)**, then drive a wedge into the slot to open up the clamp. Slide the hub carrier down from the bottom of the suspension strut and remove it from the vehicle.

Refitting

9 Refit the hub carrier to the suspension strut and tighten the clamp nut and bolt to the specified torque.

10 Ensure that the driveshaft splines and the corresponding splines in the hub are clean. Grease the threads of a new hub nut, and

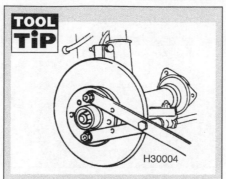

TOOL TiP
H30004

A tool to hold the hub stationary can be fabricated from two lengths of steel strip (one long, one short) held together by a nut and bolt. Use two wheel bolts to hold the tool in position on the hub and use it as a counterhold while slackening the hub nut. Alternatively, insert two adjacent wheel bolts into the hub and use a metal bar between them as a counterhold.

grease the face of the hub bearing where it contacts the nut. Engage the driveshaft with the hub and fit the nut, but do not tighten it at this stage.

11 Ensure that the protector shield is in place, then reconnect the track control arm to the hub carrier. If necessary, use a screwdriver to open up the joint slightly, and use a lever to pull the track control arm downwards so that it can be inserted into the clamp. Fit the clamp nut and bolt the correct way round, using a new nut, and tighten it to the specified torque.

12 Reconnect the track-rod end balljoint to the steering arm, and refit the washer and a new nut. Tighten the nut to the specified torque.

13 Refit the brake disc in the correct position, according to the chalk mark, and securely tighten the screws. Refit the brake caliper using new bolts coated in thread-locking fluid. Refit the brake pads (see Chapter 9).

14 Tighten the hub nut to the specified torque and stake the nut in position. Use a counterhold, according to the Tool Tip mentioned earlier in this Section.

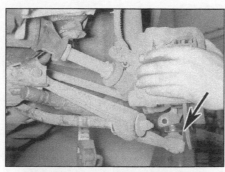

2.5 Use a lever to separate the track control arm outer joint and recover the protector shield (arrowed)

2.4 Undo the track-rod end balljoint using a separator tool

15 Refit the hub nut dust cover (if applicable), then refit the roadwheels.

3 Front hub bearings - renewal

Note: *The bearing is a sealed, pre-adjusted and pre-lubricated, double-row roller type, and is intended to last the vehicle's entire service life without maintenance or attention.*

Note: *Citroën special tools are available to carry out this operation, in which case the task can be carried out without removing the hub. The procedure described in this Section assumes that the special tools are not available - in this case, the task can be accomplished using improvised tools, but only once the hub has been removed.*

Note: *A press will be required to dismantle and rebuild the assembly. If such a tool is not available, a large bench vice and spacers (such as large sockets) will serve as an adequate substitute. The bearing's inner races are an interference fit on the hub. If the inner race remains on the hub when it is pressed out of the hub carrier, a bearing puller will be required to remove it.*

Note: *A new hub bearing circlip will be required.*

1 Remove the hub carrier as described in Section 2.

2 If not already done, remove the securing screws and withdraw the brake disc from the hub (refer to Chapter 9 if necessary).

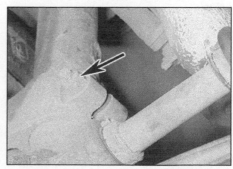

2.8 Clamp nut and bolt (arrowed) at bottom of suspension strut

3.3 Driving the hub flange from the front hub bearing

3.4 Front hub bearing retaining circlip (arrowed)

3 Support the hub carrier securely on blocks, or in a vice. Using a tubular spacer which bears only on the inner end of the hub flange, press the hub flange out of the bearing **(see illustration)**. If the bearing outboard inner race remains on the hub, remove it using a bearing puller (see note above).

4 Extract the bearing retaining circlip from the inner end of the hub carrier **(see illustration)**.

5 Where necessary, refit the bearing inner race back in position over the ball cage, and securely support the inner face of the hub carrier. Using a tubular spacer which bears only on the bearing inner race, press the complete bearing assembly out of the hub carrier.

6 Thoroughly clean the hub and hub carrier, removing all traces of dirt and grease, and polish away any burrs or raised edges which might hinder reassembly. Check the components for cracks or any other signs of wear or damage, and renew them if necessary. Renew the bearing retaining circlip, regardless of its apparent condition.

7 Commence reassembly by applying a light film of oil to the bearing outer race and the contact faces of the hub, to aid installation of the bearing.

8 Securely support the hub carrier, and locate the bearing in the hub. Press the bearing fully into position, ensuring that it enters the hub squarely, using a tubular spacer which bears only on the bearing outer race.

9 Once the bearing is correctly seated, fit a new bearing retaining circlip, ensuring that it is correctly located in the groove in the hub carrier.

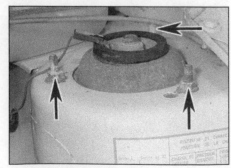

4.7 Suspension strut upper mounting nuts (arrowed). Do not undo the large centre nut

10 Securely support the outer face of the hub, and locate the hub carrier bearing inner race over the end of the hub. Press the bearing onto the hub, using a tubular spacer which bears only on the inner race of the bearing, until the bearing seats against the hub shoulder. Check that the hub rotates freely, and wipe off any excess oil or grease.

11 Refit the hub carrier as described in Section 2, including the brake disc.

4 Front suspension strut - removal, overhaul and refitting

⚠️ *Warning: Before attempting to remove or dismantle the front suspension strut, a suitable tool to hold the coil spring in compression must be obtained. Adjustable coil spring compressors are readily-available, and are recommended for this operation. Any attempt to remove or dismantle the strut without such a tool is likely to result in damage or personal injury.*

Note: *Front shock absorbers and springs are colour-coded for the vehicle payload (see Specifications). Make sure you get the correct components when renewing.*

Note: *New suspension strut upper mounting nuts will be required.*

Removal

1 Chock the rear wheels, apply the handbrake, then jack up the front of the vehicle and support on axle stands (see *"Jacking and vehicle support"*).

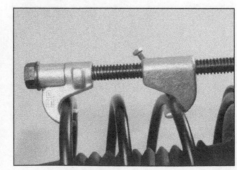

4.9 Spring compressor

2 Remove the appropriate roadwheel.

3 Release any hoses or wiring from their clips or brackets on the suspension strut, and move them to one side. Note the locations of any clips or brackets.

4 Undo the clamp bolt securing the hub carrier to the bottom of the suspension strut.

5 To prevent any damage to the driveshaft joints, fit a length of wire from the top of the hub carrier to the subframe.

6 Drive a wedge into the slot on the hub carrier and slide the carrier from the bottom of the strut.

7 Open the bonnet and disconnect the earth wire from the top of the suspension strut. Have an assistant support the strut from under the wheel arch, then undo the three upper mounting nuts and recover the washers **(see illustration)**. Withdraw the strut from under the wheel arch.

⚠️ *Warning: Do not unscrew the centre nut from the top of the shock absorber without first removing the suspension strut and compressing the spring.*

Overhaul

Note: *A new shock absorber rod nut must be used on reassembly.*

8 With the strut removed from the vehicle, clean away all external dirt, then mount the strut upright in a vice.

9 Fit and tighten the spring compressor(s), and compress the coil spring until all tension is relieved from the upper spring seat **(see illustration)**.

10 Hold the shock absorber rod using a 7 mm Allen key or T40 Torx bit, and unscrew the strut top nut. Withdraw the washer and upper spring seat, followed by the coil spring. Recover any other associated components and note their location and orientation for correct reassembly.

11 If necessary, remove the gaiter and bump rubber and any other associated components from the shock absorber rod. Note the location and orientation of each component for correct reassembly.

12 With the strut assembly now dismantled, examine all the components for wear, damage or deformation, and check the upper bearing for smoothness of operation. Renew any of the components as necessary.

13 Examine the strut for signs of fluid leakage. Check the strut piston for signs of pitting along its entire length, and check the strut body for signs of damage. While holding it in an upright position, test the operation of the strut by moving the piston through a full stroke, and then through short strokes of 50 to 100 mm. In both cases, the resistance felt should be smooth and continuous. If the resistance is jerky, or uneven, or if there is any visible sign of wear or damage to the strut, renewal is necessary.

 HAYNES HiNT *Temporarily refit the upper mounting to the shock absorber rod to test the suspension strut.*

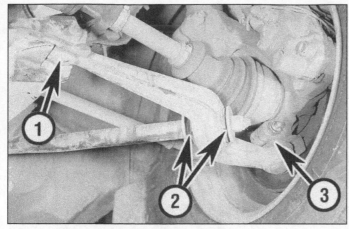

5.1 Track control arm

1 *Inner joint with rubber bush*
2 *Anti-roll bar rubber bushes*
3 *Outer balljoint*

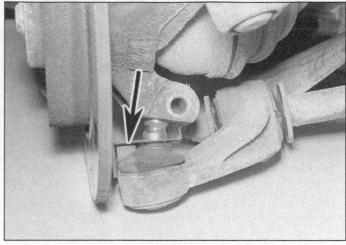

5.4 **Note the location of the protector shield (arrowed) on the outer joint**

14 If any doubt exists about the condition of the coil spring, carefully remove the spring compressors, and check the spring for distortion and signs of cracking. Renew the spring if it is damaged or distorted, or if there is any doubt as to its condition. Refit the spring compressors to the new spring

15 Inspect all other components for signs of damage or deterioration, and renew any that are suspect.

10 To reassemble the strut, refit the components in the reverse order, and note that the bump rubber must be refitted with the largest diameter uppermost. Compress the spring sufficiently to allow the top mounting components to be refitted, and if necessary pull on the end of the piston to extend the shock absorber. Fit a new shock absorber rod nut and tighten it to the specified torque. Release and remove the spring compressor(s).

Refitting

17 Refitting is the reverse of removal, noting the following points:

18 On some later models a lug at the bottom of the suspension strut fits into the gap in the lower clamp. Make sure the strut is fitted to the upper bodywork so that the lug is in the correct position.

19 Replace the three upper mounting nuts with new Nylstop nuts and tighten them to the specified torque.

20 Refit any hoses and wiring that were removed from the suspension strut.

21 Make sure the lower end of the suspension strut fully enters the hub carrier. If there is any doubt, refit the roadwheels and lower the vehicle to the ground, then loosen the clamp bolt so that the weight of the van forces the strut fully into position. Then tighten the clamp bolt to the specified torque. Make sure the lug (if applicable) fits into the slot in the clamp.

5 Front suspension track control arm - removal, overhaul and refitting

Note: *A new track control arm outer balljoint clamp nut must be used on refitting.*

Removal

1 The track control arm inner joint connects to the subframe using a pivot bolt and rubber bush. The outer balljoint connects to hub carrier. The end of the anti-roll bar connects to the track control arm using two rubber bushes **(see illustration)**.

2 Chock the rear wheels, firmly apply the handbrake, then jack up the front of the vehicle and support on axle stands (see *"Jacking and vehicle support"*). Remove the appropriate front roadwheel.

3 Unscrew the nut from the pivot bolt on the inner joint. Have an assistant hold the suspension strut pressed inwards, then remove the bolt and release the strut. Note that the bolt head faces to the rear.

4 Undo the clamp nut and bolt securing the outer balljoint to the hub carrier, noting which way round it is fitted. Use a screwdriver to open up the clamp slightly, then release the balljoint. Note the location of the protector shield **(see illustration)**.

5 Unscrew the nut from the end of the anti-roll bar, remove the washer and withdraw the track control arm.

Overhaul

6 The rubber bushes may be renewed if necessary. Lever or drive out the anti-roll bar bushes. Ideally, the pivot bush should be pressed out from the inner joint using a bench press or flypress. However, it is possible to remove and insert the bush using a long bolt, nut and washers and a metal tube.

Refitting

7 Refitting is the reverse of removal. Fit a new

nut to the clamp bolt securing the outer balljoint to the hub carrier. Tighten the bolts to the specified torque with the weight of the vehicle on the front suspension. On completion, check and if necessary adjust the steering angles and front wheel alignment (see Section 22).

6 Front suspension anti-roll bar - removal and refitting

Note: *The front anti-roll bar diameter depends on the vehicle payload (see Specifications). When renewing, make sure you get the correct identical component.*

Removal

1 Jack up the front of the vehicle and support on axle stands (see *"Jacking and vehicle support"*). Apply the handbrake and remove both roadwheels.

2 Remove one track control arm, referring to Section 5.

3 Unscrew the nut securing the remaining end of the anti-roll bar to the other track control arm and recover the washer.

4 Unbolt the guide bar from the subframe **(see illustration)**.

6.4 **Guide bar showing mounting bolt to subframe (1) and adjustment clamp (2)**

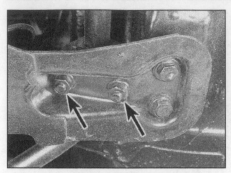

6.5 Anti-roll bar mounting clamp bolts (arrowed)

7.2 Brake drum retaining screw (arrowed)

7.3 Hub nut dust cap (arrowed)

5 Undo the mounting clamp bolts **(see illustration)** and withdraw the anti-roll bar over the subframe. If necessary disconnect the gearchange rods to provide additional working room.

Refitting

6 Examine the rubber bearings for damage and deterioration, and renew them if necessary.
7 Refitting is the reverse of removal. Tighten all nuts and bolts to the specified torque, but delay fully tightening the mounting clamp bolts until the full weight of the vehicle is on the suspension. The guide bar bolts should also remain loosened until after the mounting clamp bolts have been tightened. The guide bar length should then be adjusted so that the transverse distance between the guide bar-to-subframe mounting bolt and the centre of the adjustment clamp is according to the Specifications.

---

7 Rear hub and bearings - removal and refitting

Note: *The outer race of the rear wheel bearing is integral with the hub. The inner race is bonded to a nylon/plastic dust seal and fits onto the stub axle.*

Note: *A new hub nut and dust cap will be required.*

Removal

1 Chock the front wheels, then jack up the rear of the vehicle and support it on axle stands (see *"Jacking and vehicle support"*). Remove the appropriate rear wheel.
2 Undo the brake drum retaining screw **(see illustration)** and withdraw the brake drum, with reference to Chapter 9.
3 Prise off the dust cap using a screwdriver **(see illustration)**. The cap will probably become distorted on removal and will require renewal.
4 Relieve the staking from the hub nut using a hammer and a suitable cold chisel or punch.

⚠ *Warning: Wear suitable eye protection.*

5 Undo the hub nut using a socket and extension bar. To prevent the stub axle from turning, remove the stop from the suspension trailing arm and insert a 19mm Allen key into the aperture **(see illustration)**. Recover the washer.

⚠ *Warning: The hub nut is very tight. You can pull the vehicle off the axle stands by attempting to slacken it.*

6 Remove the hub using a suitable puller or slide hammer. If possible, use the Citroën puller 2400-T and pad 7101-TJ **(see illustration)**.

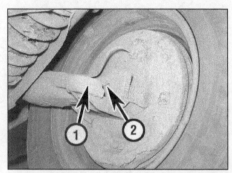

7.5 Lock the stub axle by removing the stop (1) from the suspension trailing arm and insert a 19mm Allen key into the aperture (2).

7 Withdraw the bearing inner race from the stub axle using a puller and withdraw the hub seal thrust cup.

Refitting

8 Clean and inspect the hub and stub axle components. Take care not to get grease onto the brake linings. Renew the hub and bearings if they are worn or damaged. Also check that the stub axle threads are in good condition.
9 Fit the thrust cup into position on the stub axle. Tap it carefully into position using a tubular drift or Citroën special tool 7106T if available **(see illustration)**.
10 Drive the hub inner race into position using the old hub nut, then a suitable tubular drift. Clean and lubricate the race with a suitable hub bearing grease.

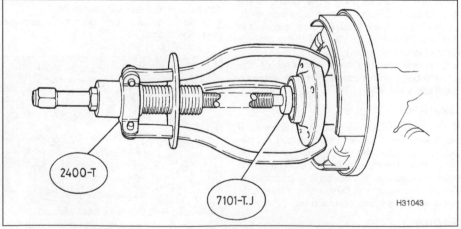

7.6 Remove the rear hub using the special puller and pad

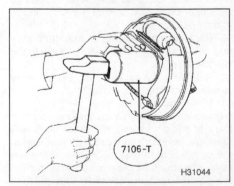

7.9 Fit the rear hub thrust cup using the special tool if available

11 Locate the hub over the stub axle. Partially fit the hub using a socket or tube drift so that some of the hub nut thread is exposed. Fit the old hub nut and tighten it to fully fit the hub onto the stub axle, then remove the nut.

12 Clean the stub axle and bearing, lubricate it with a suitable bearing grease, then fit the hub washer and a new nut. Tighten the nut to the specified torque. Use a hammer and cold chisel to stake the nut into position. Refit the suspension arm stop, if removed.

13 Drive a new dust cap into position over the hub nut, then refit the brake drum. Check that the drum rotates without excessive play.

14 Apply the handbrake, refit the roadwheel and trim, then lower the vehicle to the ground. Tighten the roadwheel bolts to the specified torque.

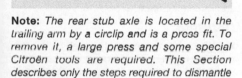

8 Rear stub axle - renewal

Note: *The rear stub axle is located in the trailing arm by a circlip and is a press fit. To remove it, a large press and some special Citroën tools are required. This Section describes only the steps required to dismantle the trailing arm and take it to a Citroën dealer.*

1 Remove the rear suspension trailing arm (see Section 10). Make sure you slacken the hub nut first.

2 Remove the brake drum and shoes (see Chapter 9).

3 Remove the hub and bearings (see Section 7).

4 Undo the three securing bolts and remove the brake backplate.

5 Remove the rubber stop from the end of the trailing arm.

6 Take the trailing arm to a Citroën dealer to have the stub axle renewed.

7 Refitting is the reverse of removal.

9 Rear suspension strut - removal, overhaul and refitting

Note: *The Citroën spring compressor tool 4068-T and clamping jaws 4082-T are required for this job. Do not use any other kind of spring compressor.*

Note: *Rear shock absorbers and springs are colour-coded for the vehicle payload (see Specifications). Make sure you get the correct components when renewing.*

Note: *A new nut is required for the shock absorber rod.*

Removal

1 The rear suspension strut consists of a shock absorber and spring, and is operated by movement of the trailing arm.

2 Chock the front wheels, then jack up the rear of the vehicle so that the roadwheels are clear of the ground, and support the vehicle on axle stands (see *"Jacking and vehicle support"*). Remove the wheel, if required, to get easier access to the suspension strut. Otherwise, leave the wheel fitted so that its weight helps to decompress the suspension.

3 Remove the protector (if fitted) from the end of the shock absorber rod, then unscrew the retaining nut **(see illustration)**. Prevent the rod from turning by inserting an Allen key into its end aperture. If the rod is allowed to turn, the shock absorber piston may be damaged. Withdraw the cup washer, noting the orientation, and the flexible spacer. Remove the centring bush from the rod if accessible externally.

4 Screw the nut back onto the shock absorber rod a few turns, then fit the compressor tool 4068-T with clamping jaws 4082-T. Make sure the clamping jaws are fully positioned on the thrust cups. Compress the spring and release the tapered balljoint from its housing. Tap the balljoint with a soft drift if

9.3 Rear shock absorber retaining nut, cup washer and spacer (arrowed)

necessary. Remove the release stop **(see illustrations)**.

5 Decompress the spring and remove the compressor tool.

6 Remove the nut from the end of the shock absorber rod, then remove the suspension strut.

7 Recover the two half-shell spacers which are wedged between the suspension spring lower retaining cup and a shoulder on the shock absorber. These are used for rear suspension height adjustment (see Section 14) and are likely to fall off when the suspension strut is removed. If the struts are being removed from both sides, take care not to mix the half-shell spacers, as they are matched to their springs.

Overhaul

8 Remove the spring from the suspension strut, then remove the rod protector and all spacers, cups, bushes and washers from the shock absorber rod, noting their precise location and orientation. Remove the suspension spring lower retaining cup. Renew any defective components.

9 Commence reassembly by locating the half-shell spacers in position over the shock absorber and retaining them with elastic bands. The spacers should be positioned so

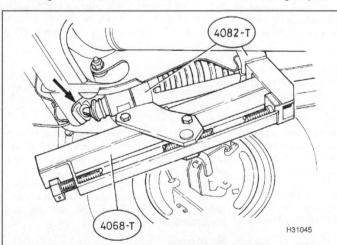

9.4a Compress the rear suspension spring with the Citroën special tool and release the tapered balljoint from its housing (arrowed)

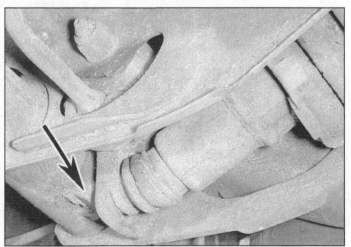

9.4b Remove the spring release stop (arrowed)

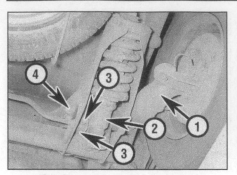

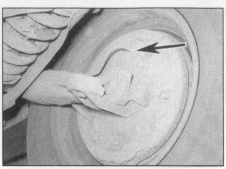

10.1 Rear suspension assembly

1 Trailing arm
2 Shock absorber
3 Anti-roll bar clamp bolts
4 Pivot nut and bolt to rear subframe

that a mark on the shock absorber is visible in the gap between them.

10 Fit the suspension spring lower retaining cup.

11 Fit the components onto the shock absorber rod, in their correct position and orientation as noted during removal.

Refitting

12 Fit the spring, ensuring that it is correctly positioned on the lower cup, then fit the suspension strut into the housing with the mark on the strut (between the two half-shell spacers) facing downwards.

13 Fit all the external components onto the upper end of the shock absorber rod, including the centring bush (if not already fitted), flexible spacer, washer (in its correct orientation) and a new Nylstop nut. Screw the nut hand tight at this stage.

14 Fit the compressor tool 4068-T and clamping jaws 4082-T and compress the spring to the point where the tapered balljoint fits into its recess.

15 Refit the release stop. Carefully release the spring compression and make sure that the tapered balljoint is correctly engaged in its recess, then remove the compression tool.

16 Use an Allen key to prevent the shock absorber rod from turning (as during removal) and tighten the nut to the specified torque.

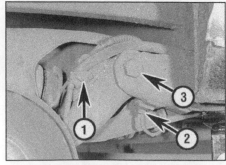

10.8 Trailing arm pivot joint and fittings

1 Brake hose and
   handbrake cable
   clip
2 Brake hose
   union
3 Pivot shaft

10.5 Rear brake hydraulic hose connection (arrowed)

Refit the rod protector (if applicable).
17 Lower the vehicle to the ground.

10 Rear suspension trailing arm - removal and refitting

Note: *The Citroën spring compressor tool 4068-T and clamping jaws 4082-T are required for this job. Do not use any other kind of spring compressor.*

Removal

1 The rear suspension trailing arm is connected by a pivot shaft to the suspension strut housing, which is part of the rear subframe. The trailing arm engages with the tapered balljoint at the end of the shock absorber. It also connects to the anti-roll bar **(see illustration)**.

2 Chock the front wheels, then jack up the rear of the vehicle and support it on axle stands (see *"Jacking and vehicle support"*). Remove the appropriate roadwheel.

> **HAYNES HINT** *If you wish to remove the hub at any stage during this procedure, slacken the hub nut first while the assembly is still fitted to the vehicle, using any of the methods described in Section 7 and observing the appropriate precautions.*

3 Remove the brake drum (if it has not been removed already) and disconnect the handbrake cable from the operating lever, then pull the cable out from the brake backplate (see Chapter 9). Refit the brake drum onto the hub to protect the brake shoes and prevent them from being contaminated with oil and grease. If the hub has been removed, it will be necessary to remove the brake shoes and keep them in a safe place.

4 To minimise fluid loss, first remove the master cylinder reservoir cap, and then tighten it down onto a piece of polythene, to obtain an airtight seal.

5 Undo the brake hydraulic hose connection from the backplate **(see illustration)**. Plug the end of the hose, and the hole in the backplate, to prevent fluid loss and entry of dirt.

6 Fit the compressor tool 4068-T and the clamping jaws 4082-T to the suspension strut, same as in Section 9, but do not remove the nut or any other components from the shock absorber rod. Compress the spring and release the tapered balljoint from its housing. Recover the two half-shell spacers from under the suspension spring lower retaining cup, if they fall off, and do not mix them up on both sides of the vehicle as they are matched to their springs. Remove the release stop from the suspension crossmember.

7 Undo the anti-roll bar bearing bolts and recover the clamp and spacer.

8 Detach the handbrake cable and flexible brake hydraulic hose from the retaining clip on the trailing arm, and move them aside so that they are away from the pivot joint. If necessary, undo the hydraulic hose union under the pivot joint and remove the flexible hose completely, and plug the ends of both the flexible and fixed hoses. Then undo the trailing arm pivot nut. Support the weight of the arm and withdraw the pivot shaft. Lift the trailing arm assembly and withdraw it from the subframe **(see illustration)**.

Refitting

9 Refitting is the reverse of removal. Tighten the retaining nuts and fastenings to the specified torque. Do not remove the spring compression tool until the suspension arm is fully fitted. Top up the brake hydraulic system (see *"Weekly checks"*), bleed the brakes and adjust the handbrake as described in Chapter 9.

11 Rear suspension trailing arm pivot bearings - general

Although the pivot shaft is easily removed, while removing the trailing arm, the renewal of the pivot bearing requires the use of special Citroën tools and workshop procedures. It is recommended that you should remove the trailing arm as described in Section 10 and take it to a Citroën dealer to have the bearing renewed.

12 Rear suspension anti-roll bar - removal and refitting

Note: *The Citroën spring compressor tool 4068-T and clamping jaws 4082-T are required for this job. Do not use any other kind of spring compressor.*
Note: *The rear anti-roll bar diameter depends on the vehicle payload (see Specifications). When renewing, make sure you get the correct identical component.*

Removal

1 The rear anti-roll bar is secured at each end to the left and right trailing arms, using clamps and spacers **(see illustration)**.

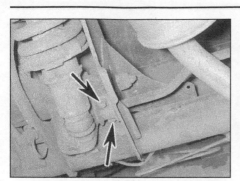

12.1 Anti-roll bar clamp bolts (arrowed)

2 Disconnect the spring from the brake proportioning lever on the anti-roll bar **(see illustration)**.

3 Remove the left and right rear suspension struts, using the compression tools (see Section 9).

4 Undo the clamp bolts at each end of the anti-roll bar and recover the clamps and spacers, then withdraw the anti-roll bar.

Refitting

5 Refitting is the reverse of removal. Tighten the clamp bolts to the specified torque, then refit the suspension struts as described in Section 9. When finished, check that the brake proportioning valve is working correctly, and if necessary adjust it as described in Chapter 9.

13 Rear subframe -
removal and refitting

Note: *Multiple supports are required for this job. Two axle stands are required to support the bodywork, and three jacks to support the subframe and hubs.*

Removal

1 The rear subframe is fitted to the vehicle underbody using four rubber mountings, two at the front, at the ends of the crossmember, and two at the rear above the springs. The two front mountings consist of a mounting bracket with a circular rubber bush and central bolt **(see illustration)**. The rear mountings are rubber pads bolted directly to the underbody.

2 Chock the front wheels, then jack up the rear of the vehicle and support it on axle stands (see *"Jacking and vehicle support"*). Remove the roadwheels on both sides.

3 Remove the brake drums and disconnect the handbrake cables. Pull the cables out of the brake backplates and disconnect them from their clips on the suspension, then refit the brake drums to prevent the shoes from being contaminated with oil and grease (see Chapter 9).

4 Remove the rear brake proportioning valve and keep it in a safe place (see Chapter 9).

12.2 Disconnect the spring from the brake proportioning lever (arrowed)

Plug the ends of the hoses to prevent contamination and loss of fluid. Undo any hydraulic hose clips, upstream and downstream of the proportioning valve, that would prevent the bodywork and suspension from being separated. The upstream hoses should be connected to the bodywork and the downstream hoses should be connected to the suspension.

5 Remove the rear section of the exhaust pipe (see relevant Part of Chapter 4).

6 Place a trolley jack under the crossmember to support the weight of the subframe. Place additional jacks under the left and right trailing arms to support the weight of the hubs.

7 Working within the load compartment, prise off the plastic pads from the floor to access the mounting nuts. There are eight pads altogether, arranged in four rows. Ignore the two rows at the front and rear, and prise off the pads from the two central rows. One of the pads is under the fuel filler trim panel, which will need to be removed by pulling it off the studs. One of the studs is common to the wiring loom trim panel.

8 Use a socket to undo the mounting nuts. There are two nuts on each of the front mountings and one nut on each of the rear mountings.

9 Carefully lower the jacks supporting the subframe and hubs, and withdraw the assembly from the rear of the vehicle.

Refitting

10 Refitting is the reverse of removal, noting the following points.

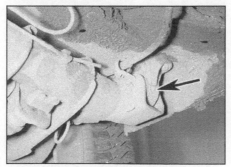

13.1 Rear subframe front mounting (arrowed)

11 Slacken the central nut and bolt on the two front mounting bushes, to allow some movement of the mountings. This will make it easier to raise the assembly and get all six studs into their holes in the bodywork and fit the nuts. Then re-tighten the central nuts and bolts to their specified torque.

12 Bleed the rear brakes, check and adjust the brake proportioning valve, and adjust the handbrake as described in Chapter 9.

14 Suspension height -
checking and adjustment

General

1 The height of the front and rear suspension can be checked by measuring the distance from the ground to the appropriate points on the underbody. The measurements should be taken at the correct tyre pressures with the handbrake released, and with the vehicle at kerb weight defined as follows:
 a) *Vehicle unladen*
 b) *5 litres of fuel in tank*

2 The measurements should be according to the Specifications. If the front suspension is not correct, the vehicle needs to be taken to a Citroën dealer for a detailed examination of the suspension components. If the rear suspension is not correct, it can be adjusted.

Checking front suspension

3 The front suspension can be checked by measuring the distance from the ground to a point on the subframe near the inner joint of the track control arm **(see illustration)**.

Checking rear suspension

4 As described in Section 13, the rear subframe is fitted to the vehicle underbody using four rubber mountings, two at the front, at the ends of the crossmember, and two at the rear above the springs. The height of the rear suspension is measured from the ground to the upper face of the rear rubber mounting where it bolts onto the underbody. The height on both sides, and the difference between them, should be according to the Specifications.

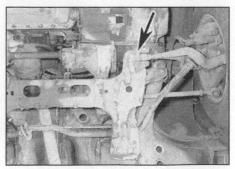

14.3 Measure the height from the ground to the point on the subframe (arrowed)

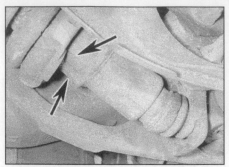

14.5 Half-shell spacers (arrowed) on rear shock absorber

15.1 Remove the steering wheel centre pad

decompress the spring and remove the compressor. Remove the tape from the spacers.

9 Repeat the procedure on the other side of the vehicle if required.

10 Refit the roadwheel and lower the vehicle to the ground, then measure the height on both sides. If this is not according to the Specifications, repeat the procedure with new spacers.

Adjustment

Note: *The Citroën spring compressor tool 4068-T and clamping jaws 4082-T are required for this job. Do not use any other kind of spring compressor.*

5 The height is adjusted by inserting half-shell spacers of the appropriate length between the suspension spring lower retaining cup and a shoulder on the shock absorber **(see illustration)**. There are 24 shells available, from 26 to 47 mm in steps of 1mm. Each increment of 1 mm increases the height of the rear suspension by 3.5 mm.

6 Chock the front wheels, then jack up the rear of the vehicle so that the roadwheels are clear of the ground, and support the vehicle on axle stands (see *"Jacking and vehicle support"*). Remove the wheel, if required, to get easier access to the suspension strut.

Otherwise, leave the wheel fitted so that its weight helps to decompress the suspension.

7 Fit the compressor tool 4068-T and the clamping jaws 4082-T to the suspension strut, same as in Section 9, but do not remove the nut or any other components from the end of the shock absorber rod. Compress the spring and remove the two half-shell spacers, then fit new ones so that the mark on the lower face of the shock absorber is visible in the gap between them. If necessary, secure them with tape until the spring is decompressed. If the balljoint at the front of the suspension strut comes out of its housing at any stage during this procedure, push it back in again. Alternatively, if extra room for the spacers is required, allow the balljoint to drop down and push the suspension strut forwards.

8 With the new spacers fitted, make sure the balljoint is securely fitted to its housing, then

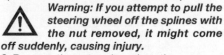

15 Steering wheel - removal and refitting

Note: *A new steering wheel retaining nut is required.*

Removal

1 Carefully prise off the steering wheel centre pad **(see illustration)**.

2 Set the front wheels in the straight-ahead position, and release the steering lock by inserting the ignition key.

3 Using paint or a suitable felt pen, make a vertical mark on the steering column shaft, for correct refitting.

4 Loosen the steering wheel retaining nut, but leave it on the threads for the time being **(see illustration)**.

5 Lift the steering wheel off the column splines so that it hits the back of the nut. If it is tight, twist it from side to side while pulling upwards to release it from the shaft splines.

⚠ *Warning: If you attempt to pull the steering wheel off the splines with the nut removed, it might come off suddenly, causing injury.*

6 Remove the nut and washer, then remove the steering wheel **(see illustrations)**.

Refitting

7 Refitting is the reverse of removal. Make sure the alignment mark is vertical **(see illustration)** and the steering wheel is at its centre position. Use a new retaining nut and tighten it to the specified torque.

15.4 Loosen the steering wheel retaining nut

15.6a Remove the nut . . .

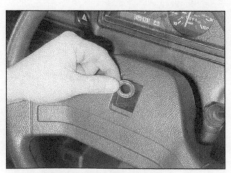

15.6b . . . and the washer . . .

15.6c . . . and remove the steering wheel

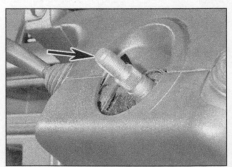

15.7 Use the mark (arrowed) for correct alignment when refitting the steering wheel

16.6 Undo the clamp nut and bolt (arrowed) from the steering column universal joint

16.7 Undo the nuts from the steering column upper mounting (arrowed)

17.3 Remove the rubber cover from the ignition switch

16 Steering column - removal, inspection and refitting

Note: *A new steering column intermediate clamp nut and bolt must be used when refitting.*

Removal

1 Disconnect the battery negative lead and remove the steering wheel (see Section 15).
2 Remove the steering column lower shroud and the driver's side parcel shelf (see Chapter 11). The shroud can be left hanging on the headlamp adjuster cable.
3 Disconnect the multi-plug(s) from the ignition switch wiring and make sure they are identified for refitting.
4 Remove the steering column stalk switches (see Chapter 12). Alternatively, if you wish to remove the steering column with the stalk switches fitted, disconnect the multi-plugs from the switches.
5 Remove the bonnet release cable (see Chapter 11).
6 Undo the intermediate clamp nut and bolt at the lower end of the steering column, where it connects to the universal joint on the lower shaft **(see illustration)**.
7 Undo the two self-locking upper mounting nuts **(see illustration)**, and the lower mounting nuts. Lower the steering column and withdraw it from the universal joint. It may

be necessary to prise open the clamp slightly, using a large screwdriver.

Inspection

8 Check the steering column for signs of wear or free play, and if necessary, renew the column as a complete assembly.

Refitting

9 Refitting is the reverse of removal, noting the following points.
10 Tighten the upper and lower mounting nuts to the specified torque.
11 Make sure the steering wheel is correctly centred with the wheels in the straight-ahead position, then fit a new intermediate clamp nut and bolt at the lower end of the steering column and tighten it to the specified torque.
12 When finished, make sure all lighting and switches are working correctly.

17 Ignition switch/steering column lock - removal and refitting

Removal

1 Disconnect the battery negative lead.
2 Remove the four screws and pull down the steering column lower shroud (see Chapter 11).
3 Remove the rubber cover from the ignition switch **(see illustration)**.
4 Insert the key and turn the switch to the

arrow between positions A and M. Use a screwdriver to release the two lugs under the switch, then withdraw the switch **(see illustration)**.

Refitting

5 Refitting is the reverse of removal.

18 Steering gear assembly (manual steering) - removal, overhaul and refitting

Note: *A new steering column lower clamp nut and bolt, and new track rod end balljoint nuts must be used when refitting.*

Removal

1 Chock the rear wheels, firmly apply the handbrake, then jack up the front of the vehicle and support on axle stands (see *"Jacking and vehicle support"*). Remove both front roadwheels.
2 Undo the nut securing the track-rod end to the steering arm on the hub carrier. Separate the track-rod end from the steering arm using a balljoint separator tool. Remove the nut and washer.
3 Undo the clamp nut and bolt securing the bottom of the steering shaft to the steering gear pinion splines **(see illustration)**.
4 Undo the mounting bolts **(see illustration)**, and withdraw the steering gear sideways from the subframe.

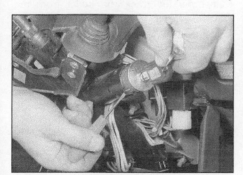

17.4 Release the two lugs and withdraw the switch with the key at the arrow between A and M

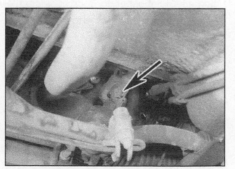

18.3 Clamp nut and bolt (arrowed) securing the steering shaft to the steering gear pinion splines

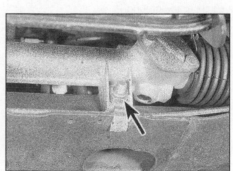

18.4 Steering gear mounting bolt (arrowed)

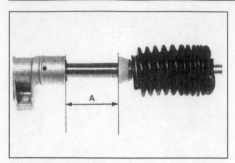

18.6 Steering gear centralising dimension

A = 72.5 mm

Overhaul

5 Examine the steering gear assembly for signs of wear or damage. Check that the rack moves freely throughout the full length of its travel, with no signs of roughness or excessive free play between the steering gear pinion and rack. If the assembly is worn, change it for a new or reconditioned unit. The only components which can be renewed easily by the home mechanic are the steering gear rubber gaiters, the track rod end balljoints and the track rods. These procedures are covered in Sections 19, 20 and 21 respectively.

Refitting

6 Begin refitting by centralising the rack. To do this, disconnect the rubber gaiter and set the rack to the dimension shown **(see illustration)**.
7 With the steering wheel in the straight-ahead position, refit the steering gear and connect the steering shaft to the pinion splines.
8 Refit and tighten the mounting bolts to the specified torque.
9 Insert a new clamp nut and bolt to the steering shaft and tighten it to the specified torque.
10 Reconnect the rubber gaiter to the steering gear.
11 Reconnect the track rod end balljoints to the steering arms and refit the washers and new nuts. Tighten the nuts to the specified torque.
12 Refit the front roadwheels and lower the vehicle to the ground.

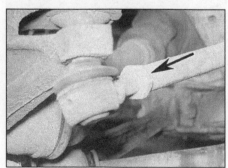

20.2 Track rod end locknut (arrowed)

13 Have the front wheel alignment checked at the earliest opportunity (see Section 22), and check that the steering wheel is centralised. If necessary, the steering wheel position can be altered by removing the wheel and moving it the required number of splines on the column shaft, then refitting (see Section 15).

19 Steering gear rubber gaiters (manual steering) - renewal

1 Remove the track rod end balljoint, making sure that the locknut is slackened by only a quarter of a turn for correct alignment on refitting (see Section 20).
2 Remove the inner and outer clips, then slide the gaiter off the track rod.
3 Fit a new gaiter, using new clips.
4 Refit the track rod end, using the position of the locknut for alignment purposes. Provided the locknut has not been disturbed, the front wheel alignment should not have altered. Nevertheless, you should have it checked by your dealer.

20 Track rod end balljoint - removal and refitting

Note: *A new track rod end balljoint nut is required.*

Removal

1 Chock the rear wheels, firmly apply the handbrake, then jack up the front of the vehicle and support on axle stands (see *"Jacking and vehicle support"*). Remove the relevant front roadwheel.
2 Release the track rod end locknut and unscrew it by a quarter of a turn **(see illustration)**. Do not move the locknut from this position, as it will serve as a handy reference mark on refitting. Flats are provided at the inboard end of the track rod so you can hold the rod steady while the locknut is released.
3 Unscrew the nut securing the track rod end to the steering arm on the suspension strut. Using a balljoint separator tool, separate the

20.3 Typical balljoint separator tool

track rod end from the steering arm **(see illustration)**. To prevent any possible damage to the threaded shank of the balljoint, leave the nut in position until the shank is released.
4 Counting the **exact** number of turns necessary to do so, unscrew the track rod end balljoint from the track rod.

Refitting

5 If a new track rod end is being fitted, grease the threads and locate the locknut in the same position as noted on the old track rod end (i.e. with the same number of threads from the locknut to the end of the track rod end).
6 Screw the track rod end into the track rod the same number of turns as noted during removal.
7 Fit the track rod end balljoint to the steering arm on the suspension strut, then fit a new nut and tighten it to the specified torque.
8 Hold the track rod, then tighten the locknut to the specified torque.
9 Refit the roadwheel, and lower the van to the ground.
10 Have the front wheel alignment checked at the earliest opportunity (see Section 22).

21 Track rod (manual steering) - removal and refitting

Removal

1 Chock the rear wheels, firmly apply the handbrake, then jack up the front of the vehicle and support on axle stands (see *"Jacking and vehicle support"*). Remove the relevant front roadwheel.
2 Undo the nut securing the track rod end balljoint to the steering arm on the suspension strut. Using a balljoint separator tool, separate the track rod end from the steering arm. If the track rod end balljoint is to be removed from the track rod, undo the locknut first and use it for position on refitting, following the procedures given in Section 20.
3 Undo the inner and outer clips and pull back the rubber gaiter to expose the collar at the inner end of the track rod.
4 Unscrew the track rod collar from the steering gear and recover any washers and other components, noting the correct position for refitting. On the passenger side, there is a conical buffer that fits onto the steering gear shaft.

Refitting

5 Refitting is the reverse of removal, noting the following points.
6 Tighten the track rod collar to the specified torque, using a torque wrench fitted with the special Citroën tool 80707-T or other suitable tool.
7 Tighten the track rod end locknut to the specified torque, if the track rod end has been removed.

8 Tighten the track rod end balljoint nut to the specified torque.

9 On completion, have the front wheel alignment checked at the earliest opportunity (see Section 22).

22 Wheel alignment and steering angles - general information

Definitions

1 A vehicle's steering and suspension geometry is defined in four basic settings. All angles are expressed in degrees, and toe settings can also be expressed as a linear measurement. The steering axis is defined as an imaginary line drawn through the axis of the suspension strut, extended where necessary to contact the ground **(see illustration)**.

2 Camber is the angle between each roadwheel and a vertical line drawn through its centre and tyre contact patch, when viewed from the front or rear of the van. "Positive" camber is when the roadwheels are tilted outwards from the vertical at the top; "negative" camber is when they are tilted inwards.

3 Camber is not adjustable, and is given for reference only. It can be checked using a camber checking gauge, and if the figure obtained is significantly different from that specified, the vehicle must be taken for careful checking by a professional, as the fault can only be caused by wear or damage to the body or suspension components.

4 Castor is the angle between the steering axis and a vertical line drawn through each roadwheel centre and tyre contact patch, when viewed from the side of the van. "Positive" castor is when the steering axis is tilted so that it contacts the ground ahead of the vertical. "Negative" castor is when it contacts the ground behind the vertical.

5 Castor is not adjustable, and is given for reference only. It can be checked using a castor checking gauge, and if the figure obtained is significantly different from that specified, the vehicle must be taken for careful checking by a professional, as the fault can only be caused by wear or damage to the body or suspension components.

6 Steering axis inclination/SAI - also known as **kingpin inclination/KPI** - is the angle between the steering axis and a vertical line drawn through each roadwheel centre and tyre contact patch, when viewed from the front of the van.

7 SAI/KPI is not adjustable, and is given for reference only.

8 Toe is the difference between lines drawn through the roadwheel centres and the van's centre-line. "Toe-in" is when the roadwheels point inwards, towards each other at the front, while "toe-out" is when they splay outwards from each other at the front.

9 The front wheel toe setting is adjusted by screwing the track rods onto or away from the track rod ends, to alter the effective length of the track rod assemblies.

10 Rear wheel toe setting is not adjustable, and is given for reference only. While it can be checked, if the figure obtained is significantly different from that specified, the vehicle must be taken for careful checking by a professional, as the fault can only be caused by wear or damage to the body or suspension components.

Checking - general

11 Due to the special measuring equipment necessary to check the wheel alignment, and the skill required to use it properly, the checking and adjustment of these settings is best left to a Citroën dealer or similar expert. Note that most tyre-fitting shops now possess sophisticated checking equipment.

12 For accurate checking, the ride height must be correct (see Section 14) and the vehicle must be at kerb weight defined as follows:

   a) Vehicle unladen

   b) 5 litres of fuel in tank

13 Check that the tyre sizes and types are as specified, then check the tyre pressures and tread wear, the roadwheel run-out, the condition of the hub bearings, and the condition of the front suspension components (see "*Weekly checks*" and Chapter 1A or 1B). Correct any faults found.

14 Park the vehicle on level ground, check that the front roadwheels are in the straight-ahead position, then rock the front and rear ends to settle the suspension. Release the handbrake, and roll the vehicle backwards approximately 1 metre, then forwards again the same distance, to relieve any stresses in the steering and suspension components.

Toe setting - checking and adjusting

Front wheel toe setting

15 The front wheel toe setting is checked using a proprietary toe measurement gauge, available from motor accessory shops.

16 Prepare the vehicle as described in paragraphs 12 to 14 above.

17 Make sure that the lengths of both track rods are equal when the steering is in the straight-ahead position. This can be checked without disturbing the steering rack gaiters by measuring the number of exposed threads on the track rod end balljoint fittings. If necessary, release the gaiters and locknuts (see paragraph 19) and turn the rods.

18 Measure the distance between the inner rims of the wheels, at hub height, at the rear of the wheels. Push the vehicle forwards to rotate the wheels through 180° (half a turn), then take the same measurement at the front of the wheels. This last measurement should be less than the first by the appropriate toe-in (see Specifications).

19 On each side of the vehicle, release the clip from the outboard end of the steering rack gaiter, where it grips the inboard end of the track rod. Hold the track rod steady by gripping the flats at the inboard end, and slacken the locknut at the outboard end.

20 Turn the track rods equally, only a quarter of a turn at a time, and reposition the gaiters so that they are not twisted, then recheck the alignment. When each track rod is viewed from the centre-line of the vehicle, turning the track rods anti-clockwise will increase the toe-in.

21 Always turn both rods in the same direction when viewed from the centre-line of the vehicle, otherwise they will become unequal in length. This will cause the steering wheel to become de-centralised and will cause problems on turns with tyre-scrubbing.

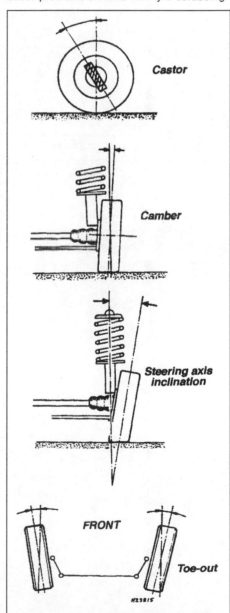

22.1 Wheel alignment and steering angle measurements

22 When the setting is correct, hold the track rods and tighten the locknuts to their specified torque, then tighten the gaiter retaining clips.

Rear wheel toe setting

23 The procedure for checking the rear toe setting is same as described for the front, but the setting is not adjustable. If the figure obtained is significantly different from that specified, the vehicle must be taken for careful checking by a professional, as the fault can only be caused by wear or damage to the body or suspension components.

23 Power steering - general information

1 Power-assisted steering was available on diesel models from December 1993 onwards and was fitted as standard to the 765 kg payload Champ vans. Very few power steering models were sold in the UK and there is no specific information about removing and refitting the components.

2 The hydraulic system is powered by a pump with integral flow regulator, mounted on the transmission and belt-driven from a pulley on the camshaft. A second belt on the same pulley also drives the brake vacuum pump, described in Chapter 9. The belt-driven vacuum pump is fitted as an alternative to the usual camshaft-mounted pump.

3 Fluid from the pump passes to a control valve which is integral with the steering gear drive pinion. The movement of the pinion causes oil to pass to a hydraulic ram, mounted in parallel to the steering gear, and connecting to the central shaft to give assistance to the motion.

4 The hydraulic fluid is cooled by passing it through a single-bore cooling tube located on the underbody.

Chapter 11
Bodywork and fittings

Contents

Degrees of difficulty

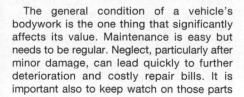

| Easy, suitable for novice with little experience | | Fairly easy, suitable for beginner with some experience | | Fairly difficult, suitable for competent DIY mechanic | | Difficult, suitable for experienced DIY mechanic | | Very difficult, suitable for expert DIY or professional |

---

1 General description

The bodyshell is made of pressed steel sections, with most components welded together, but some use is made of structural adhesives. The front wings are bolted on.

The bonnet, doors and some other vulnerable panels are made of zinc-coated metal, and are further protected by being coated with an anti-chip primer, before being sprayed.

Extensive use is made of plastic materials, mainly in the interior, but also in the exterior components. The front and rear bumpers are injection-moulded from a synthetic material which is very strong and yet light. Plastic components such as wheel arch liners are fitted to the underside of the vehicle, to improve the body's resistance to corrosion.

2 Maintenance - bodywork and underframe

The general condition of a vehicle's bodywork is the one thing that significantly affects its value. Maintenance is easy but needs to be regular. Neglect, particularly after minor damage, can lead quickly to further deterioration and costly repair bills. It is important also to keep watch on those parts of the vehicle not immediately visible, for instance the underside, inside all the wheel arches and the lower part of the engine compartment.

The basic maintenance routine for the bodywork is washing, preferably with a lot of water, from a hose. This will remove all the loose solids that may have stuck to the vehicle. It is important to flush these off in such a way as to prevent grit from scratching the finish. The wheel arches and underframe need washing in the same way, to remove any accumulated mud that will retain moisture and tend to encourage rust. Oddly enough, the best time to clean the underframe and wheel arches is in wet weather when the mud is thoroughly wet and soft. In very wet weather the underframe is usually cleaned of large accumulations automatically and this is a good time for inspection.

Periodically, except on vehicles with a wax-based underbody protective coating, it is a good idea to have the whole of the underframe of the vehicle steam cleaned, engine compartment included, so that a thorough inspection can be carried out to see what minor repairs and renovations are necessary. Steam cleaning is available at many garages and is necessary for removal of the accumulation of oily grime that sometimes is allowed to become thick in certain areas. If steam-cleaning facilities are not available, there are some excellent grease solvents available which can be brush-applied, and the dirt can then be simply hosed off. Note that these methods should not be used on vehicles with wax-based underbody protective coating or the coating will be removed. Such vehicles should be inspected annually, preferably just before winter, when the underbody should be washed down and any damage to the wax coating repaired. Ideally, a completely fresh coat should be applied. It would also be worth considering the use of such wax-based protection for injection into door panels, sills, box sections, etc., as an additional safeguard against rust damage where such protection is not provided by the vehicle manufacturer.

After washing paintwork, wipe off with a chamois leather to give an unspotted clear finish. A coat of clear protective wax polish will give added protection against chemical pollutants in the air. If the paintwork sheen has dulled or oxidised, use a cleaner/polisher combination to restore the brilliance of the shine. This requires a little effort, but such dulling is usually caused because regular washing has been neglected. Care needs to be taken with metallic paintwork, as special non-abrasive cleaner/polisher is required to avoid damage to the finish.

Always check that the door and ventilator opening drain holes and pipes are completely clear so that water can be drained out. Brightwork should be treated in the same way as paintwork. Windscreens and windows can be kept clear of the smeary film that often appears, by using a glass cleaner. Never use any form of wax or other body or chromium polish on glass.

3 Maintenance - upholstery and carpets

Mats and carpets should be brushed or vacuum-cleaned regularly to keep them free of grit. If they are badly stained, remove them from the vehicle for scrubbing or sponging and make quite sure they are dry before refitting. Seats and interior trim panels can be kept clean by wiping with a damp cloth. If they do become stained (which can be more apparent on light coloured upholstery), use a little liquid detergent and a soft nail brush to scour the grime out of the grain of the material. Do not forget to keep the headlining clean in the same way as the upholstery. When using liquid cleaners inside the vehicle, do not over-wet the surfaces being cleaned. Excessive damp could get into the seams and padded interior causing stains, offensive odours or even rot.

Caution: If the inside of the vehicle gets wet accidentally, it is worthwhile taking some trouble to dry it out properly, particularly where carpets are involved. Do not leave oil or electric heaters inside the vehicle for this purpose.

4 Minor body damage - repair

Repairs of minor scratches in bodywork

If the scratch is very superficial, and does not penetrate to the metal of the bodywork, repair is very simple. Lightly rub the area of the scratch with a paintwork renovator, or a very fine cutting paste, to remove loose paint from the scratch and to clear the surrounding bodywork of wax polish. Rinse the area with clean water.

Apply touch-up paint to the scratch using a fine paint brush. Continue to apply fine layers of paint until the surface of the paint in the scratch is level with the surrounding paintwork. Allow the new paint at least two weeks to harden, then blend it into the surrounding paintwork by rubbing the scratch area with a paintwork renovator or a very fine cutting paste. Finally, apply wax polish.

Where the scratch has penetrated right through to the metal of the bodywork, causing the metal to rust, a different repair technique is required. Remove any loose rust from the bottom of the scratch with a penknife, then apply rust inhibiting paint, to prevent the formation of rust in the future. Using a rubber or nylon applicator, fill the scratch with bodystopper paste. If required, this paste can be mixed with cellulose thinners to provide a very thin paste that is ideal for filling narrow scratches. Before the stopper-paste in the

scratch hardens, wrap a piece of smooth cotton rag around the top of a finger. Dip the finger in cellulose thinners and then quickly sweep it across the surface of the stopper-paste in the scratch. This will ensure that the surface of the stopper-paste is slightly hollowed. The scratch can now be painted over as described earlier in this Section.

Repair of dents in bodywork

When deep denting of the vehicle's bodywork has taken place, the first task is to pull the dent out, until the affected bodywork almost attains its original shape. There is little point in trying to restore the original shape completely, as the metal in the damaged area will have stretched on impact and cannot be reshaped fully to its original contour. It is better to bring the level of the dent up to a point that is about 3 mm below the level of the surrounding bodywork. In cases where the dent is very shallow anyway, it is not worth trying to pull it out at all. If the underside of the dent is accessible, it can be hammered out gently from behind, using a mallet with a wooden or plastic head. While doing this, hold a block of wood firmly against the outside of the panel to absorb the impact from the hammer blows and thus prevent a large area of the bodywork from being "belled-out".

Should the dent be in a section of the bodywork that has a double skin or some other factor making it inaccessible from behind, a different technique is called for. Drill several small holes through the metal inside the area particularly in the deeper section. Then screw long self-tapping screws into the holes just sufficiently for them to gain a good grip in the metal. Now the dent can be pulled out by pulling on the protruding heads of the screws with a pair of pliers.

The next stage of the repair is the removal of the paint from the damaged area, and from an inch or so of the surrounding "sound" bodywork. This is accomplished most easily by using a wire brush or abrasive pad on a power drill, although it can be done just as effectively by hand using sheets of abrasive paper. To complete the preparation for filling, score the surface of the bare metal with a screwdriver or the tang of a file, or alternatively, drill small holes in the affected area. This will provide a good "key" for the filler paste.

To complete the repair see the Section on filling and re-spraying.

Repair of rust holes or gashes in bodywork

Remove all paint from the affected area and from an inch or so of the surrounding "sound" bodywork, using an abrasive pad or a wire brush on a power drill. If these are not available a few sheets of abrasive paper will do the job just as effectively. With the paint

removed, you will be able to gauge the severity of the corrosion and therefore decide whether to renew the whole panel (if this is possible) or to repair the affected area. New body panels are not as expensive as most people think and it is often quicker and more satisfactory to fit a new panel than to attempt to repair large areas of corrosion.

Remove all fittings from the affected area except those which will act as a guide to the original shape of the damaged bodywork (e.g. headlamp shells, etc.). Then, using tin snips or a hacksaw blade, remove all loose metal and any other metal badly affected by corrosion. Hammer the edges of the hole inwards to create a slight depression for the filler paste.

Wire-brush the affected area to remove the powdery rust from the surface of the remaining metal. Paint the affected area with rust inhibiting paint. If the back of the rusted area is accessible treat this also.

Before filling can take place, it will be necessary to block the hole in some way. This can be achieved by using aluminium or plastic mesh, or aluminium tape.

Aluminium or plastic mesh, or glass fibre matting, is probably the best material to use for a large hole. Cut a piece to the approximate size and shape of the hole to be filled, then position it in the hole so that its edges are below the level of the surrounding bodywork. It can be retained in position by several blobs of filler paste around its periphery.

Aluminium tape should be used for small or very narrow holes. Pull a piece off the roll and trim it to the approximate size and shape required. Then pull off the backing paper (if used) and stick the tape over the hole. It can be overlapped if the thickness of one piece is insufficient. Burnish down the edges of the tape with the handle of a screwdriver or similar, to ensure that the tape is securely attached to the metal underneath.

Bodywork repairs - filling and respraying

Before using this Section, see the Sections on dent, deep scratch, rust holes and gash repairs.

Many types of bodyfiller are available, but generally those proprietary kits that contain a tin of filler paste and a tube of resin hardener are best for this type of repair. A wide, flexible plastic or nylon applicator will be found invaluable for imparting a smooth and well-contoured finish to the surface of the filler.

Mix up a little filler on a clean piece of card or board - measure the hardener carefully (follow the maker's instructions on the pack) otherwise the filler will set too rapidly or too slowly. Using the applicator, apply the filler paste to the prepared area and draw the applicator across the surface of the filler to achieve the correct contour and level the filler surface. When a contour that approximates the correct one is achieved, stop working the paste - if you carry on too long the paste will become sticky and begin to "pick up" on the applicator. Continue to add thin layers of filler paste at twenty-minute intervals until the level of the filler is just proud of the surrounding bodywork.

Once the filler has hardened, excess can be removed using a metal plane or file. From then on, progressively finer grades of abrasive paper should be used, starting with a 40 grade production paper and finishing with 400 grade wet-and-dry paper. Always wrap the abrasive paper around a flat rubber, cork, or wooden block, otherwise the surface of the filler will not be completely flat. During the smoothing of the filler surface, the wet-and-dry paper should be periodically rinsed in water. This will ensure that a very smooth finish is imparted to the filler at the final stage.

At this stage the "dent" should be surrounded by a ring of bare metal, which in turn should be encircled by the finely "feathered" edge of the good paintwork. Rinse the repair area with clean water, until all the dust produced by the rubbing-down operation has gone.

Spray the whole repair area with a light coat of primer. This will show up any imperfections in the surface of the filler. Repair these imperfections with fresh filler paste or bodystopper, and again smooth the surface with abrasive paper. Repeat this spray and repair procedure until you are satisfied that the surface of the filler, and the feathered edge of the paintwork are perfect. Clean the repair area with clean water and allow to dry fully.

> **HAYNES HINT**
>
> *If bodystopper is used, it can be mixed with cellulose thinners to form a thin paste that is ideal for filling small holes.*

The repair area is now ready for final spraying. Paint spraying must be carried out in a warm, dry, windless and dust free atmosphere. This condition can be created artificially if you have access to a large indoor working area, but if you are forced to work in the open, you will have to pick your day very carefully. If you are working indoors, dousing the floor in the work area with water will help to settle the dust that would otherwise be in the atmosphere. If the repair area is confined to one body panel, mask off the surrounding panels; this will help to minimise the effects of a slight mis-match in paint colours. Bodywork fittings (e.g. chrome strips, door handles, etc.), will also need to be masked off. Use genuine masking tape and several thicknesses of newspaper for the masking operations.

Before beginning to spray, agitate the aerosol can thoroughly, then spray a test area (an old tin, or similar) until the technique is mastered. Cover the repair area with a thick coat of primer; the thickness should be built up using several thin layers of paint rather than one thick one. Using 400 grade wet-and-dry paper, rub down the surface of the primer until it is smooth. While doing this, the work area should be thoroughly doused with water, and the wet-and-dry paper periodically rinsed in water. Allow to dry before spraying on more paint.

Spray on the top coat, again building up the thickness by using several thin layers of paint. Start spraying in the centre of the repair area and then work outwards, with a side-to-side motion, until the whole repair area and about 2 inches of the surrounding original paintwork is covered. Remove all masking material 10 to 15 minutes after spraying on the final coat of paint.

Allow the new paint at least two weeks to harden, then using a paintwork renovator or a very fine cutting paste, blend the edges of the paint into the existing paintwork. Finally, apply wax polish.

Plastic components

With the use of more and more plastic body components (e.g. bumpers, and in some cases major body panels), repair of more serious damage to such items has become a matter of either entrusting repair work to a specialist in this field, or renewing complete components. Repair of such damage by the DIY owner is not feasible owing to the cost of the equipment and materials required for effecting such repairs. The basic technique involves making a groove along the line of the crack in the plastic using a rotary burr in a power drill. The damaged part is then welded back together by using a hot air gun to heat up and fuse a plastic filler rod into the groove. Any excess plastic is then removed and the area rubbed down to a smooth finish. It is important that a filler rod of the correct plastic is used, as body components can be made of a variety of different types (e.g. polycarbonate, ABS, polypropylene).

Damage of a less serious nature (abrasions, minor cracks, etc.), can be repaired by the DIY owner using a two-part epoxy filler repair material. Once mixed in equal proportions, this is used in similar fashion to the bodywork filler used on metal panels. The filler is usually cured in twenty to thirty minutes, ready for sanding and painting.

If the owner is renewing a complete component himself, or if he has repaired it with epoxy filler, he will have a problem of finding a paint for finishing which is compatible with the type of plastic used. At one time the use of a universal paint was not possible owing to the complex range of plastics come across in body component applications. Standard paints, generally, will not bond to plastic or rubber satisfactorily. However, it is now possible to obtain a plastic body parts finishing kit that consists of a pre-primer treatment, a primer and coloured top coat. Full instructions are normally supplied with a kit, but basically, the method of use is to first apply the pre-primer to the component concerned and allow it to dry for up to 30 minutes. Then the primer is applied and left to dry for about an hour before finally applying the special coloured top coat. The result is a correctly coloured component where the paint will flex with the plastic or rubber, a property that standard paint does not normally possess.

5 Major body damage - repair

Where serious damage has occurred, or large areas need renewal due to neglect, it means that complete new panels will need welding-in, and this is best left to professionals. If the damage is due to impact, it will also be necessary to check completely the alignment of the bodyshell, and this can only be carried out effectively by a Citroën dealer using special jigs. If the body is left misaligned, the van will not handle properly, and uneven stresses will be imposed on the steering, suspension and possibly transmission, causing abnormal wear, or complete failure, particularly to such items as the tyres.

6 Bumpers - removal and refitting

Front bumper trim

Removal

1 Chock the rear wheels, apply the handbrake, then jack up the front of the vehicle and support securely on axle stands (see *"Jacking and vehicle support"*). Remove both front roadwheels.

2 Remove the liners from the front of the wheel arches on both sides, by pulling them off their studs **(see illustration)**.

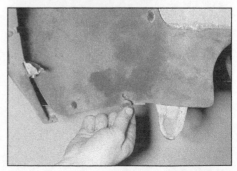

6.2 Remove the liners from the front of the wheel arches

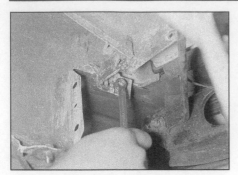

6.3a Undo one bolt . . .

6.3b . . . and one nut and bolt from each side of the front bumper trim

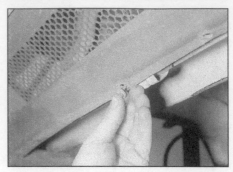

6.4 Undo one bolt from the bottom centre of the bumper trim

3 Undo one bolt, and one nut and bolt, from either side of the bumper trim, under the wheel arch (see illustrations).
4 Undo one bolt from the bottom centre of the trim (see illustration).
5 Remove the front grille and cross panel as described in Section 7.
6 Undo two screws from the top of the trim, to detach it from the bracket.
7 Pull the trim forward and remove it from the vehicle.

Refitting

8 Refitting is the reverse of removal. Make sure the slotted fitting on each side of the bumper trim goes over the lug on the bodywork, where it fits into the outer wheel arch trim (see illustration). Also make sure the weatherstrips under the headlamps are fitted correctly.

Front bumper trim and bracket

Removal

9 Proceed as described in paragraphs 1 to 5.
10 Undo one bolt on the bumper bracket extension arm, under each wheel arch, to detach it from the inner wing (see illustration).
11 Disconnect the wiring connectors from the horn, situated behind the bumper bracket (see illustration).
12 Detach the dim-dip resistor, if fitted, from its clip behind the bumper bracket (see illustration). Note: *This component is fitted to some models to supply a low current to the dipped headlights if the vehicle is driven on sidelights alone.*
13 Remove both front indicator lamps as described in Chapter 12.

14 Undo the bolt from the bumper bracket, behind each of the indicator lamp apertures (see illustration).
15 Pull the trim and bracket assembly forward and remove it from the vehicle (see illustration).

Refitting

16 Refitting is the reverse of removal, following the instructions in paragraph 8.

Rear bumper

Removal

17 Slacken the two bolts on each of the bumper plastic end sections, one under the rear wheel arch and the other behind the tubular crossmember (see illustrations).
18 Pull the plastic end sections out of the crossmember (see illustration).

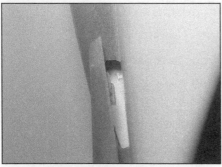

6.8 Make sure the fitting on the bumper trim goes over the lug on the bodywork

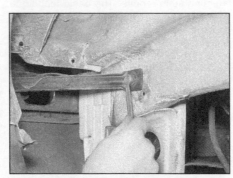

6.10 Undo the bolt on the bumper bracket extension arm

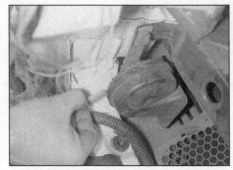

6.11 Disconnect the wiring connectors from the horn (radiator and fans removed)

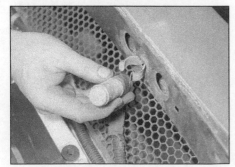

6.12 Detach the dim-dip resistor from its clip behind the bumper bracket

6.14 Undo the bolt from the bumper bracket, behind the indicator lamps

6.15 Front bumper and bracket assembly

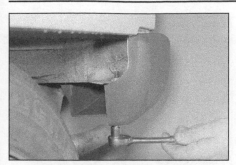

6.17a Slacken the bolts on the rear bumper plastic end sections, one under the wheel arch . . .

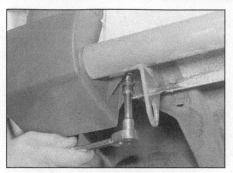

6.17b . . . and the other behind the crossmember

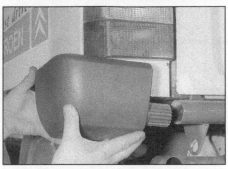

6.18 Pull the plastic end sections out of the crossmember

19 Undo the two bolts on each side of the crossmember and remove it **(see illustration)**.

Refitting

20 Refitting is the reverse of removal.

7 Radiator grille and cross panel - removal and refitting

Removal

Radiator grille

1 To remove the grille without the cross panel, undo the two bolts attaching it to the oroo panel and lift it clear of the slots in the bumper.

Grille and cross panel

2 To remove both the grille and cross panel **(see illustration)**, proceed as follows.
3 Slacken the grubscrew on the looped end of the bonnet release cable and detach the cable from the bonnet lock.
4 Undo three bolts on each side of the cross panel, then lift the assembly clear of the slots in the bumper.

Refitting

5 Refitting is the reverse of removal. If the bonnet release cable has been detached from the lock, adjust the looped end of the cable to take up almost all the slack, then tighten the grub screw and check that the lock operates correctly.

8 Bonnet - removal, refitting and adjustment

Removal

1 Open the bonnet and support it in the fully open position.
2 Place rags or cardboard under the hinges, between the bonnet and the bodywork, to prevent the paintwork from getting scratched.
3 Disconnect the fluid hose from the windscreen washer pump **(see illustration)**. The type of pump varies according to model (see Chapter 12).
4 Using a pencil or felt tip pen, mark the outline of the hinges on the bonnet, as a guide to refitting.
5 With the help of an assistant, support the weight of the bonnet, then remove the two bolts and washers from each hinge **(see illustration)** and lift the bonnet from the vehicle.

Refitting

6 With the aid of an assistant, offer up the bonnet and loosely fit the retaining bolts and washers. Align the hinges with the marks made on removal, then tighten the retaining bolts securely. Remove the rags or other protective materials from under the hinges.
7 Reconnect the fluid hose to the windscreen washer pump.
8 Adjust the alignment of the bonnet as follows.

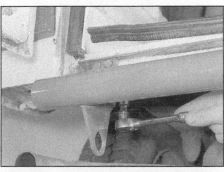

6.19 Undo the bolts on the crossmember

Adjustment

9 Close the bonnet and check for alignment with the adjacent panels. If necessary, slacken the hinge bolts and re-align the bonnet, then tighten the hinge bolts.
10 Check that the front of the bonnet is level with the front of each wing. If not, it is possible to adjust the height of the bonnet by turning the rubber buffers at the two front corners of the bodywork.
11 Once the bonnet is correctly aligned, check that it fastens and releases in a satisfactory manner. If adjustment is necessary, slacken the bonnet lock retaining bolts, adjust the position of the lock on its slotted holes, then tighten the bolt again.

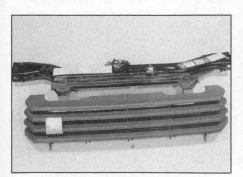

7.2 Radiator grille and cross panel

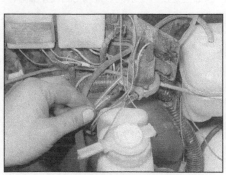

8.3 Disconnect the fluid hose from the windscreen washer pump

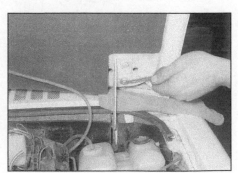

8.5 Remove the bonnet hinge bolts

10.2 Bonnet lock on radiator cross panel

10.6 Lock striker and clip on underside of bonnet

11.3 Remove the hinge pin with a punch

9 Bonnet release cable - removal and refitting

Removal

1 Open and support the bonnet, then slacken the grubscrew on the looped end of the bonnet release cable and detach the cable from the lock.
2 Note the routing of the cable and release it from any clips in the engine compartment.
3 Working inside the vehicle, detach the plastic grommet from the cable support under the steering column and withdraw the cable, feeding it through the grommet in the bulkhead.

Refitting

4 Refitting is the reverse of removal, but ensure that the bulkhead grommet is securely located, and make sure that the cable is routed as noted before removal. Adjust the looped end of the cable to take up almost all the slack, then tighten the grub screw and check that the lock operates correctly.

10 Bonnet lock - removal and refitting

Removal

1 Open and support the bonnet, then slacken the grubscrew on the looped end of the

bonnet release cable and detach the cable from the lock.
2 Undo the two bolts and remove the lock from the radiator cross panel **(see illustration)**.

Refitting

3 Refitting is the reverse of removal, noting the following points.
4 Adjust the looped end of the cable to take up almost all the slack, then tighten the grub screw.
5 Adjust the position of the lock by slackening the two bolts and moving the lock on its slotted holes, then tightening the bolts again.
6 If necessary, adjust the position of the lock striker on the underside of the bonnet **(see illustration)**.

11 Doors and hinges - removal, refitting and adjustment

Front door

Note: *The front door hinges are welded to both the door and the bodywork and cannot be removed.*

Removal

1 Open the door and drive out the roll pin from the check arm. Special hinge pin tool sets are available, but you can use a punch or long improvised tool.
2 Open the door fully and support it on blocks

or on a jack, with a rag under the door to protect the paintwork. Also have an assistant to support the door.
3 Drive out the roll pins from the two hinges **(see illustration)** and remove the door.

Refitting

4 Refitting is the reverse of removal, using new roll pins.

Adjustment

5 The door alignment can be adjusted by bending the top hinge slightly using a wrench or a special Citroën tool No. 8.1305. Be careful not to bend the hinge too much. If in doubt, take the vehicle to a Citroën dealer for adjustment.

Rear door

Note: *All the standard load compartment vans covered in this manual have double rear doors. Some earlier models have a large single rear door, but the method of removal is the same (except that the door is heavier and needs to be supported). This manual does not cover campers or other specialised units that may be fitted to platform cabs.*

Removal

6 If you are removing the door that has the number plate, undo the two electrical connectors from the number plate light **(see illustration)** and pull the cable out of the door.
7 Unclip the door stop **(see illustration)**.
8 Have an assistant hold the door, then undo the single bolt from each hinge and remove the door from the vehicle **(see illustrations)**.

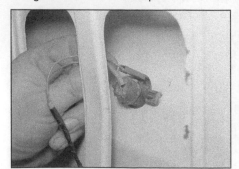

11.6 Disconnect the electrical connectors from the number plate light

11.7 Unclip the door stop

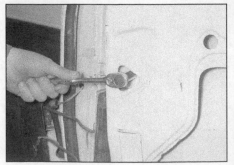

11.8a Undo the hinge bolts . . .

11.8b ... and remove the door from the vehicle

11.12a Remove the plastic cap from the top of the hinge ...

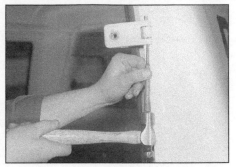

11.12b ... then tap the roll pin upwards with a punch

Refitting

9 Refitting is the reverse of removal. Adjust the door as follows.

Adjustment

10 The rear door hinges have slotted holes for adjustment. Slacken the hinge bolts and move the door within the slotted holes, then tighten the bolts again.

Rear door hinges

Note: *Only the outer part of the hinge can be removed. The inner part is welded to the bodywork.*

Removal

11 Remove the door as described in paragraphs 6 to 8.
12 Remove the plastic cap from the top of the hinge, then tap the roll pin upwards with a punch **(see illustrations)**.

Refitting

13 Refitting is the reverse of removal, using a new roll pin.

---

12 Door inner trim panels - removal and refitting

Front door trim panels

Removal

1 On models with manual windows, insert a thin piece of wire, hooked at one end, and pull out the clip from under the window winder handle. Remove the handle and recover the plastic grommet from the trim panel **(see illustrations)**.
2 Prise off the grommets, then undo the two

screws from the armrest and remove it **(see illustration)**.
3 Undo the six screws and remove the oddments tray from the bottom of the door **(see illustration)**.
4 Carefully pull away the panel so that the studs are detached from the door **(see illustration)**.
5 On models with manual windows, remove the spring from the window winder spindle **(see illustration)**.
6 If you require access to the internal door components, carefully peel off the membranes **(see illustration)**. There are two membranes, a front membrane covering the window winder components and a rear membrane covering the door lock.

Refitting

7 Refitting is the reverse of removal, remembering the following points.

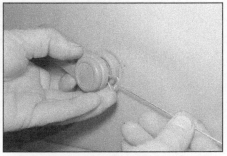

12.1a Pull out the clip from the window winder handle with a hooked wire and remove the handle ...

12.1b ... then recover the plastic grommet from the trim panel

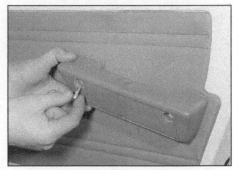

12.2 Undo the screws and remove the armrest

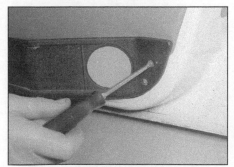

12.3 Undo the screws and remove the oddments tray

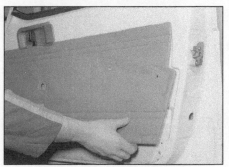

12.4 Carefully pull the trim panel, detaching the studs from the door

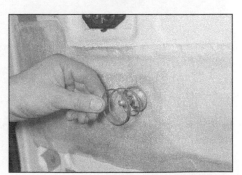

12.5 Remove the spring from the window winder spindle

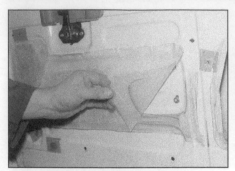

12.6 Peel off the membranes from the door apertures

8 If you have removed the membranes, make sure that they are refitted intact and securely glued to the door. If a sheet is damaged or detached, rainwater may leak into the vehicle or damage the door trim.

9 Renew any studs that were broken during removal of the trim panel, and mount them in the panel together with the other studs. Refit the panel by pressing all the studs into the holes and make sure they engage correctly.

10 The manual window winder should normally be fitted in the vertically upward position with the window closed.

Rear door trim panels

Removal

11 Rear door trim panels, where fitted, are removed by carefully pulling them off their studs.

13.3 Detach the handle from the link rod and remove it from the door

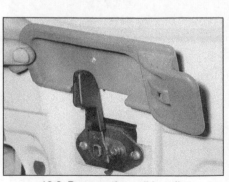

13.8 Remove the pull handle

13.2 Undo the screws from the door handle

Refitting

12 Refitting is the reverse of removal.

13 Door handles and lock components (front) - removal and refitting

Front door interior handle

Removal

1 Remove the door inner trim panel and the front membrane as described in Section 12.

2 Undo the two screws from the door handle (see illustration).

3 Detach the handle from the link rod, inside the door panel, and remove it from the door (see illustration).

13.6 Remove the screw from the pull handle behind the door handle (trim panel removed)

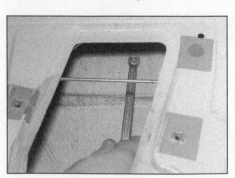

13.11a Undo the nuts securing the exterior door handle . . .

Refitting

4 Refitting is the reverse of removal, but check that the mechanism works correctly before refitting the door inner trim panel.

Front door inner pull handle

Removal

5 The pull handle can be removed from behind the door handle, without removing either the door handle or the trim panel as follows.

6 Undo one screw from behind the door handle (see illustration).

7 Prise off the plastic panel from the door pull (see illustration) and undo the two screws underneath.

8 Lift the pull handle upwards and over the door handle (see illustration).

Refitting

9 Refitting is the reverse of removal.

Front door exterior handle

Removal

10 Remove the door inner trim panel and the rear membrane as described in Section 12.

11 Working within the door aperture, undo the two nuts securing the exterior door handle, then remove the handle (see illustrations).

Refitting

12 Refitting is the reverse of removal.

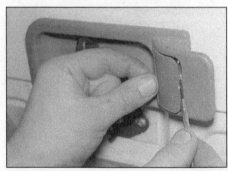

13.7 Prise off the plastic panel from the door pull

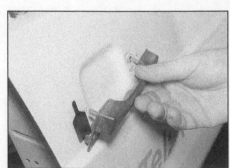

13.11b . . . then remove the handle

13.14 Door lock mechanism and link rod (viewed with mirror)

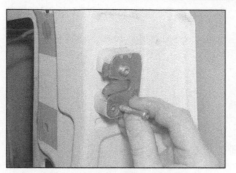

13.15 Undo the two bolts and withdraw the lock catch

13.16a Pull the clip (arrowed) from the lock barrel . . .

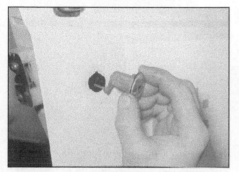

13.16b . . . and withdraw the barrel from the door

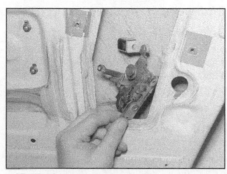

13.17 Manoeuvre the lock assembly out from the door aperture

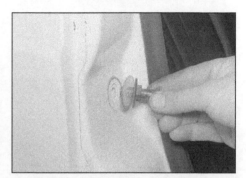

13.19 Remove the lock stub from the body pillar

Front door lock (without central locking)

Note: On models with central locking, there is a locking motor below the lock mechanism, connected by a vertical link rod. Otherwise, no specific information is available.

Removal

13 Remove the door inner trim panel and the rear membrane as described in Section 12.

14 Working within the door aperture, detach the horizontal link rod from the door lock mechanism **(see illustration)**.

15 Undo the two bolts with an Allen key and withdraw the lock catch from the edge of the door **(see illustration)**.

16 Pull the clip from the lock barrel, then withdraw the barrel from the door **(see illustrations)**.

17 Manoeuvre the lock assembly out from the door aperture **(see illustration)**.

Refitting

18 Refitting is the reverse of removal. When refitting the lock barrel, make sure the pin engages with the lever in the lock mechanism.

Front door lock stub

Removal

19 Unscrew the lock stub from the body pillar and recover the washer **(see illustration)**.

Refitting

20 Refitting is the reverse of removal.

14 Door handles and lock components (rear) - removal and refitting

Note: This section covers only the standard load compartment van with double rear doors.

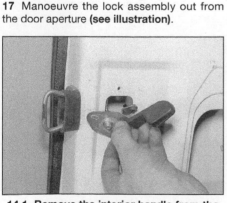

14.1 Remove the interior handle from the rear door

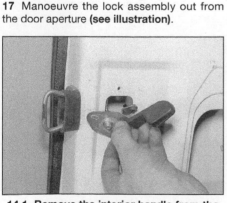

14.3 Undo the bolts from the upper catch

Left rear door interior handle

Removal

1 Undo the two securing bolts and pull the handle slightly away from the door, then detach the upper and lower link rods and withdraw the handle **(see illustration)**.

Refitting

2 Refitting is the reverse of removal.

Left rear door upper catch

Removal

3 Undo the two securing bolts **(see illustration)** and manoeuvre the upper catch off the link rod, then withdraw the catch.

4 To remove the link rod, first unbolt the interior handle and detach it from the link rod. Then detach the link rod from its supporting clip, accessible through a hole about half way between the interior handle and upper catch. Withdraw the link rod through the upper catch aperture, then turn the clip by 90° and withdraw it through the hole **(see illustrations)**.

Refitting

5 Refitting is the reverse of removal.

Left rear door lower catch

Removal

6 Undo the two securing bolts **(see illustration)** and manoeuvre the lower catch off the link rod.

7 To remove the link rod, unbolt the interior handle and detach it from the link rod, then

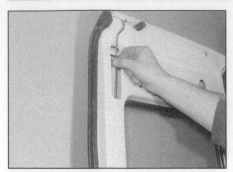

14.4a Withdraw the upper link rod

14.4b Turn the link rod clip by 90° and withdraw it

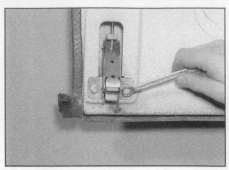

14.6 Undo the bolts from the lower catch

withdraw the link rod through the lower catch aperture.

Refitting

8 Refitting is the reverse of removal.

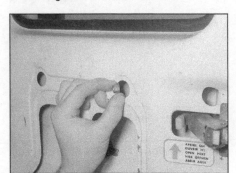

14.9a Undo the two nuts . . .

Right rear door exterior handle

Removal

9 Undo the two nuts with their built-in washers, then withdraw the handle (see illustrations).

Refitting

10 Refitting is the reverse of removal.

Right rear door locking catch and barrel (without central locking)

Note: *The locking catch and barrel on models with central locking have a different design and no specific information is available, except that there is a link rod connecting the locking barrel to the central locking motor.*

Removal

11 Undo the two bolts and remove the

locking catch (see illustration).

12 To remove the lock barrel, pull off the clip and plastic bush, then withdraw the barrel (see illustrations).

Refitting

13 Refitting is the reverse of removal.

Rear door hooks

Removal

14 There are three hooks, one on the left door engaging with the right door lock, and two on the bodywork engaging with the upper and lower left door catches. To remove a hook, undo the single bolt (see illustration).

Refitting

15 Refitting is the reverse of removal.

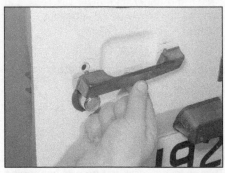

14.9b . . . and withdraw the exterior handle

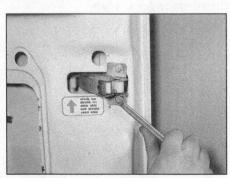

14.11 Undo the two bolts and remove the locking catch

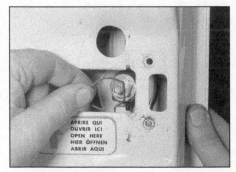

14.12a Remove the clip . . .

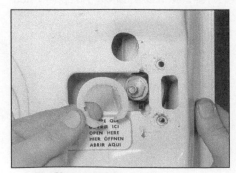

14.12b . . . and the plastic bush . . .

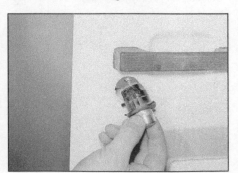

14.12c . . . then withdraw the lock barrel

14.14 Undo the bolt and remove the door hook

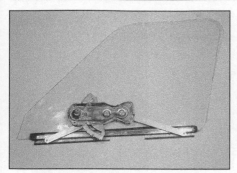

15.1 Window regulator and glass

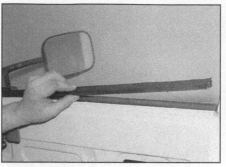

15.3 Remove the inner weatherstrip from below the window

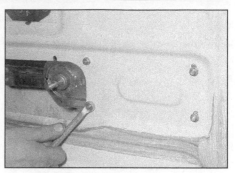

15.4 Undo the four regulator nuts

15 Door window glass and regulator - removal and refitting

Note: *This section covers the front doors only, on models with manual window winders. Models with electric windows have a motor attached to the regulator spindle instead of a door handle, but no specific information is available about removal and refitting. Also there is no information about the sliding side window in the load compartment of the Familiale.*

Removal

1 The window regulator has two arms that slide along a groove in the lower rail of the window glass **(see illustration)**. To remove the regulator and glass, proceed as follows.
2 Remove the window winder handle, the door inner trim panel and the two membranes as described in Section 12.
3 Remove the inner weatherstrip from below the window **(see illustration)**.
4 Undo the four regulator nuts **(see illustration)**, then push the regulator in so that the studs come out of the door.
5 Slide the regulator forward and then back to disengage it from the lower rail on the window glass.
6 Pull the window and lower rail upwards and out of the door **(see illustration)**.
7 Pull the regulator out from the door aperture under the lock mechanism **(see illustration)**.

Refitting

8 Refitting is the reverse of removal, noting the following:
  a) *The two rear bolts on the regulator are in slotted holes for adjustment.*
  b) *The front and rear guide rails are welded into the door and are not adjustable. The small plastic grommet, if it exists at the bottom rear corner of the door, does not lead to an adjustment screw.*

16 Central locking - general

Central locking is available on some later models, so that all the doors can be locked or unlocked using a key in the driver's or passenger's door.

The central locking system consists of a control unit and a number of central locking motors, one for each lock. Each motor is mounted in a door panel and is connected to the locking mechanism by a link rod.

No specific information is available about removing and refitting the central locking system components.

17 Electric windows - general

Electrically operated driver and passenger windows are available on some later models.

Instead of having a window winder, there is a motor inside the door panel, attached to the window regulator spindle. The motors are operated by push-button switches at the left and right side of the heater console.

No specific information is available about removing and refitting the electric window components.

18 Mirrors - removal and refitting

Interior mirror

Removal

1 The interior rear-view mirror is mounted on a base which is bonded to the windscreen using adhesive tape or other special materials. To remove the mirror from the base, slide it upwards **(see illustration)**.

Refitting

2 Refitting the mirror to the base is the reverse of removal. If a new windscreen has been fitted, obtain a new mirror base and bonding kit from your Citroën dealer and follow the instructions supplied. Use masking tape outside the glass to mark the position of the base, 70mm from the top of the windscreen.

Exterior mirror

Removal

3 Undo the three screws and remove the

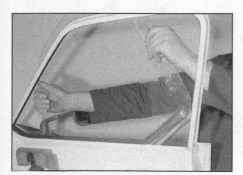

15.6 Remove the window and lower rail from the door

15.7 Withdraw the regulator through the door aperture

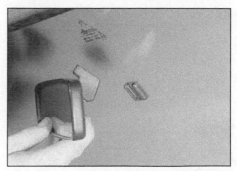

18.1 Slide the mirror upwards and remove it from the base

18.3 Undo the three screws and remove the exterior mirror

20.2 Side trim panel slots and clips

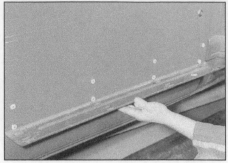

20.3 Remove any broken side trim panel clips

mirror from the door **(see illustration)**. The two upper screws are covered by rubber grommets.

Refitting

4 Refitting is the reverse of removal.

19 Windscreen and fixed windows - general

The windscreen, rear windows and load compartment side windows (if fitted) are bonded in position with special adhesive. Renewal of these windows is a difficult, messy and time-consuming task, which is beyond the scope of the home mechanic. It is difficult, unless one has plenty of practice, to obtain a secure, waterproof fit. Furthermore, the task carries a high risk of breakage. In view of this, you are strongly advised to have this work carried out by one of the many specialist windscreen fitters.

If a new windscreen is fitted, you will also need a new mirror base (see Section 18).

20 Body exterior fittings - removal and refitting

Wheel arch liners

1 The wheel arch liners are secured by a combination of self-tapping screws, and

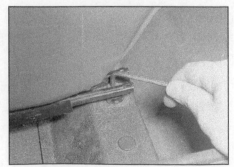

21.1 Undo the seat rail bolts with an Allen key

push-fit clips. Removal is self-evident, and normally the clips can be released by pulling the liner away from the wheel arch.

Side trims

2 The side trims, fitted as standard to the doors and side panels of C15 Champ vans, are secured by plastic clips which fit into slotted holes **(see illustration)**. To remove a door strip, carefully lever it off the upper rear clip which has a vertical slot, then push it sideways to release it from the other clips which have horizontal slots, and withdraw the trim. Carefully remove the load compartment side trim in a similar manner.

3 Insert a small screwdriver into the centre of any broken clips and lift them off, then replace them with new clips **(see illustration)**.

Body trim strips and badges

4 The various body trim strips and badges (except side trims) are held in position with a special adhesive tape. Removal requires the trim/badge to be heated, to soften the adhesive, then cut away from the surface. Due to the high risk of damage to the vehicle paintwork during this operation, it is recommended that this task should be entrusted to a Citroën dealer.

21 Seats - removal and refitting

Note: *This section covers front seats only. No information is available about removal of the rear seat on the Familiale model.*

Removal

1 Slide the seat fully backwards along the rails and undo the two bolts, one at the front of each rail, using an Allen key **(see illustration)**. Recover the special washers which have flat sides so they fit into the rails.

2 Slide the seat fully forwards and undo the two bolts at the back of the rails. Recover the special washers.

3 Remove the seat from the vehicle.

Refitting

4 Refitting is the reverse of removal.

22 Seat belts - removal and refitting

Note: *This section covers only the front seat belts. There are no instructions for removing and refitting the rear seat belts on Familiale models.*

Removal

1 The seat belt has upper and lower anchorage bolts, an inertia loom reel and a hook that fits into the centre stalk **(see illustration)**.

2 The loom reel has a plastic cover in two sections. The lower cover fits around a spacer on the lower anchorage bolt, and the upper cover fits into the top of the lower cover. Undo the screw from the upper cover and remove it by pulling it upwards. Then remove the lower cover by pulling it off the spacer. Slacken the lower anchorage bolt if necessary **(see illustrations)**.

3 Remove the plastic cap from the upper anchorage bolt. Undo the upper and lower anchorage bolts **(see illustrations)**, noting the positions of the washers and spacers, for refitting. Remove the seat belt from the vehicle.

4 If required, undo the bolt from the floor panel and remove the centre stalk **(see illustration)**.

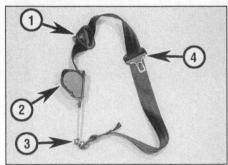

22.1 Seat belt assembly

1 Upper anchorage
2 Inertia loom reel
3 Lower anchorage
4 Hook to centre stalk

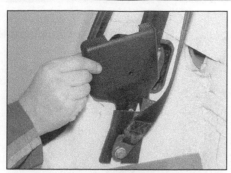

22.2a Remove the upper cover from the inertia loom reel . . .

22.2b . . . then remove the lower cover

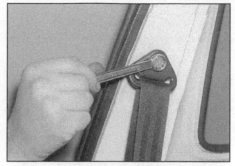

22.3a Undo the upper seat belt anchorage bolt . . .

22.3b . . . and the lower anchorage bolt

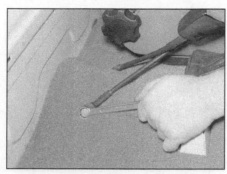

22.4 Undo the bolt and remove the seat belt centre stalk (seat removed)

22.5 The dowel under the inertia loom reel fits into the slotted hole

Refitting

5 Refitting is the reverse of removal, remembering the following points.
a) *Make sure that the belt is untwisted and all the washers and spacers are correctly fitted in their original positions.*
b) *Make sure the dowel under the inertia loom reel fits into the slotted hole in the bodywork (see illustration).*

23 Load compartment bulkhead
- removal and refitting

Removal

1 The bulkhead is fitted behind the front seats on the standard load compartment van.
2 Slacken the three bolts at the bottom of the bulkhead (see illustration).

3 Pull the seat belt to one side and slacken the two bolts at each side of the bulkhead (see illustration).
4 Tilt the bulkhead backwards slightly and lift it out (see illustration)
5 Remove the bulkhead from the vehicle via the rear doors if there are no obstructions such as shelves and other fittings, otherwise remove it via the front doors.

Refitting

6 Refitting is the reverse of removal.

24 Interior trim panels -
general information

Door trim panels

1 Refer to Section 12.

Driver and passenger footwell side panels

2 The speakers are mounted behind the footwell side panels (see Chapter 12).
3 Disconnect the battery negative lead.
4 Remove the appropriate right or left parcel shelf (see Section 26), then undo four screws and remove the panel, taking care not to tension the speaker wiring.
5 Disconnect the wiring from the speaker. If required, undo the four screws and remove the speaker from the panel.

Other interior trim panels

6 The interior trim panels are secured either by screws or by various types of trim fasteners, usually studs or clips. Where press-fit plastic studs are used, it is advisable to use a forked tool to remove them, to avoid damage to the studs and the trim panel (see illustration).

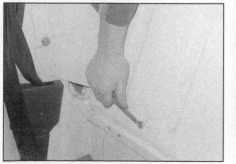

23.2 Slacken the three bolts at the bottom of the bulkhead (seat removed)

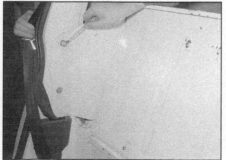

23.3 Slacken the two bolts at each side of the bulkhead

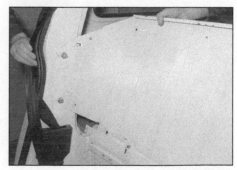

23.4 Tilt the bulkhead backwards slightly and lift it out

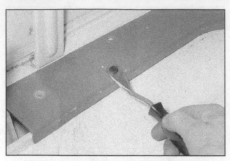

24.6 Using a forked tool to remove a plastic stud on the load compartment wiring harness panel

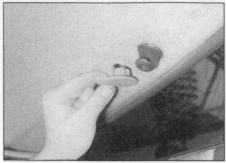

24.11a Remove the aerial grommet . . .

24.11b . . . remove the interior light . . .

24.11c . . . remove the sunvisors . . .

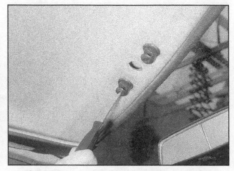

24.11d . . . and the sunvisor central grommets . . .

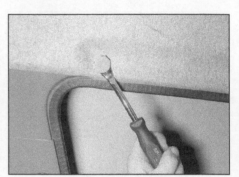

24.11e . . . then undo the plastic clips from around the headlining

7 Check that there are no other panels overlapping the one to be removed. Usually there is a sequence to be followed which will become obvious on close inspection.

8 Remove all the obvious fasteners, such as screws. If a panel will not come free then it is held by hidden clips or fasteners. These are usually situated around the edge of the panel and can be prised up to release them. Note, however, that they can break quite easily so replacements should be available. The best way of releasing such clips, in the absence of the correct type of tool, is to use a large, flat-bladed screwdriver positioned directly beneath the clip.

9 When removing a panel, never use excessive force or the panel may be damaged. Always check that all fasteners have been removed or released before attempting to withdraw a panel.

10 Refitting is the reverse of removal. Secure the fasteners by pressing them firmly into place and ensure that all disturbed components are secured to prevent rattles.

Headlining

11 To remove the headlining, disconnect the battery negative lead and remove the aerial grommet, interior light and sunvisors (including the central grommets), then undo the plastic clips from around the headlining **(see illustrations)**.

Carpets

12 The carpets are secured by screws or various types of clips. Removal and refitting is reasonably straightforward, but can be time-consuming because adjoining trim panels and other components such as the centre console and seats may have to be removed first.

25 Centre console - removal and refitting

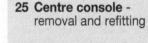

Removal

1 Unscrew the gear lever knob and remove the spacer **(see illustrations)**.

2 Detach the rubber boot from the plastic moulding on the console, releasing it from its clips, then pull the gearstick outer column and boot upwards off the gearstick **(see illustration)**.

3 Undo the four screws on the console **(see illustration)**, then lift the console upwards and off the gearstick, taking care not to tension the radio speaker wires if they are routed through the console, then place the console on the floor.

25.1a Unscrew the gear lever knob . . .

25.1b . . . and remove the spacer

25.2 Pull the gearstick outer column and boot upwards off the gearstick

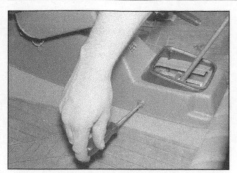

25.3 Undo the four screws on the centre console

26.2 Undo the screws under the left parcel shelf, one at each end

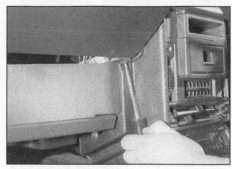

26.3a Undo one screw from the top of the partition panel . . .

4 To remove the console completely, disconnect it from the speaker wires if necessary.

Refitting

5 Refitting is the reverse of removal.

26 Parcel shelf - removal and refitting

Note: *The instructions in this section are for right-hand drive vehicles. The positions are reversed for left-hand drive.*

Left parcel shelf

Removal

1 The left parcel shelf is attached to a vertical partition panel which fits alongside the heater

control panel. To remove the shelf and partition panel as an assembly, proceed as follows.

2 Undo two screws underneath the parcel shelf, one at the left and one at the right **(see illustration)**.

3 Undo one screw at the top of the partition panel. Carefully manoeuvre the assembly off the screw fitting and withdraw it from the vehicle **(see illustrations)**.

4 If required, undo three screws to separate the shelf from the partition panel.

Refitting

5 Refitting is the reverse of removal.

Right parcel shelf

Removal

6 Undo two screws underneath the parcel shelf, one at the left and one at the right **(see illustrations)**.

7 Undo one screw which holds the shelf to the bottom right corner of the steering column lower shroud. Carefully manoeuvre the shelf off the shroud and withdraw it from the vehicle **(see illustrations)**.

Refitting

8 Refitting is the reverse of removal.

27 Steering column lower shroud - removal and refitting

Removal

1 Remove four screws and pull down the lower shroud **(see illustrations)**.

2 For complete removal, pull off the headlamp beam adjuster knob and undo the two screws connecting the adjuster to the

26.3b . . . then remove the shelf and partition panel

26.6a Undo the screws under the right parcel shelf, one at the left . . .

26.6b . . . and one at the right

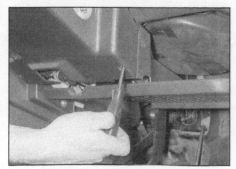

26.7a Undo one screw holding the shelf to the steering column lower shroud . . .

26.7b . . . then remove the shelf

27.1a Remove four screws . . .

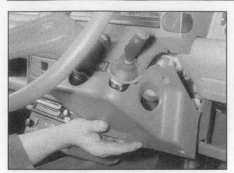

27.1b . . . and pull down the lower shroud

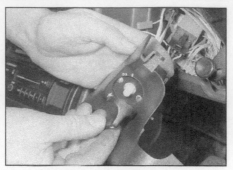

27.2a Pull off the headlamp beam adjuster knob and undo the two screws . . .

27.2b . . . and remove the adjuster from the shroud

shroud **(see illustrations)**. On carburettor petrol models, disconnect the choke cable from the shroud.

Refitting

3 Refitting is a reverse of removal.

28 Facia panel - removal and refitting

Note: *This operation involves disconnecting a large amount of wiring. Make sure all wires and the location of wiring looms are identified for refitting.*

Removal

1 Disconnect the battery negative lead.
2 Remove the steering wheel (see Chapter 10).

3 Remove the steering column lower shroud (see Section 27).
4 Remove the instrument panel (see Chapter 12).
5 Remove any instrument side panels that contains switches, then remove the switches or disconnect the wiring (see Chapter 12).
6 Remove the radio (see Chapter 12) and the heater control panel (see Chapter 3).
7 Remove the ventilation panels from the facia. On each panel, tilt the grille downwards, insert a screwdriver and release the lower clips, then tilt it upwards and release the upper clips **(see illustrations)**. There are four clips on each panel, two at the top and two at the bottom, regardless of whether it is a single or double grille. **Note:** *If, for any reason, the grille becomes detached from the panel, make sure you recover the clip from the edge of the*

grille. It fits on the opposite side to the adjustment knob.
8 Remove the left and right interior lamps from the facia, if fitted **(see illustration)**. Push the lamps out from behind the panel and detach the multi-plugs.
9 Undo four screws from under the facia panel at the following positions, on right-hand drive models **(see illustrations)**.
  a) *One screw at the left of the panel, near the interior lamp;*
  b) *One screw at the left of the heater;*
  c) *One screw at the right of the heater;*
  d) *One screw at the right of the steering column.*
10 Undo the two screws securing the facia to the side panels **(see illustrations)**.
11 Carefully remove the facia and make sure all wiring is detached **(see illustrations)**.

28.7a Undo the lower clip on the ventilation grille panel . . .

28.7b . . . then the upper clip

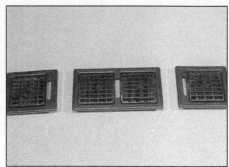

28.7c Ventilation grille panels removed

28.8 Remove the interior lamps from the facia

28.9a Undo one screw at the left of the facia panel . . .

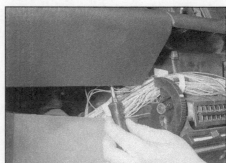

28.9b . . . one screw at the left of the heater . . .

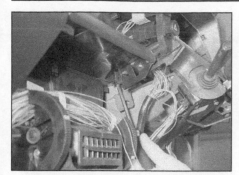

28.9c . . . one screw at the right of the heater . . .

28.9d . . . and one screw at the right of the steering column

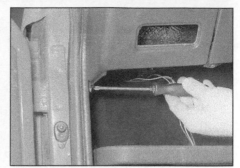

28.10a Undo one screw securing the facia to the left side panel . . .

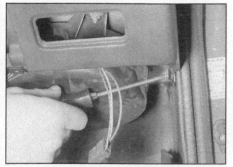

28.10b . . . and one screw from the right side panel

28.11a Remove the facia panel from the vehicle

28.11b Heating and ventilation assembly and ducting with facia removed

Refitting

12 Refitting is the reverse of removal, but note the following points:

a) Ensure that all wiring connections are correctly and securely made. Route the looms as noted during removal.

b) When refitting the steering wheel, make sure it is correctly aligned on the column splines. Use a new nylstop nut and tighten it to the specified torque (see Chapter 10).

c) On completion, test all the switches and controls, including the radio, and make sure that everything works.

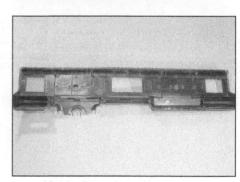

28.11c Rear view of facia panel

Chapter 12
Body electrical systems

Contents

Degrees of difficulty

Easy, suitable for novice with little experience	Fairly easy, suitable for beginner with some experience	Fairly difficult, suitable for competent DIY mechanic 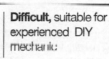	Difficult, suitable for experienced DIY mechanic	Very difficult, suitable for expert DIY or professional

Specifications

System

Type .... 12 volt, negative earth

Fuses

Two fuseboxes on engine compartment scuttle panel with four fuses each, numbered from right to left:

Fuse	Rating (amps)	Circuit(s) protected
1 Green*	10	RH side and tail lamps, switch lighting, RH rear number plate lamp.
2 Red*	16	Direction indicators LH and RH, and warning lamp, windscreen washer, dashboard lighting, lighting for heating controls, blower fan. Warning lamps for brakes, battery charge, fuel gauge and low level, engine oil pressure, preheating, rear screen washer, coolant temperature.
3 Blue*	20	Stop lamps. Engine cooling fans, windscreen wiper, windscreen washer, heated rear screen, horn.
4 Yellow	16	Cigar lighter, interior lamps, radio supply, hazard warning.
5 Green	10	Rear fog lamps and their warning lamp.
6 Red	10	Reversing lights.
7 Blue	30	Central locking, electric front windows (if fitted).
8 Yellow	10	LH side and tail lamps, warning lamp for side and tail lamps. LH rear number plate lamp.

* Some components from fuse 3 depend on a relay that is energised from fuse 2. If components have failed, check both fuses 2 and 3. See Section 3.

One fuse fitted to fly lead close to battery on some models:

Fuse	Rating (amps)	Circuit(s) protected
9	30	Engine cooling fans.

Bulbs

	Wattage
Headlamps (dipped and main beam):	
R2 type bulb	45/40
H4 type bulb	55/60
Side light	4
Direction indicator (front)	21
Direction indicator (rear)	21
Side repeater	4
Tail light	5
Stop light	21
Reversing light	21
Rear fog light	21
Number plate light	5
Interior lights:	
Festoon type	7
Push type	5
Instrument panel and facia illumination (typical)	1.2

Stop-lamp switch

Clearance in front of threaded portion	3.0 mm to 5.0 mm

Torque wrench setting

	Nm	lbf ft
Stop-lamp switch locknut	10	7

---

1 General information and precautions

General Information

The body electrical system consists of all lights, wash/wipe equipment, interior electrical equipment, and associated switches and wiring.

The electrical system is the 12-volt negative earth type. Power to the system is provided by a 12-volt battery which is charged by the alternator. The starting and charging system is described in Chapter 5A. The engine electrical components such as the ignition system on petrol engines and the pre-heating system on diesel engines are described in Chapters 5B and 5C.

This Chapter covers repair and service procedures for the various electrical components not associated with engine or the starting and charging systems.

While some repair procedures are given, the usual course of action is to renew a defective component. The owner whose interest extends beyond mere component renewal should obtain a copy of the *Automotive Electrical and Electronic Systems Manual*, available from the publishers of this Manual.

Precautions

It is necessary to take extra care when working on the electrical system, to avoid damage to semi-conductor devices (diodes and transistors), and to avoid the risk of personal injury. In addition to the precautions given in *"Safety first!"* at the beginning of this manual, observe the following when working on the system:

a) *Always remove rings, watches, etc.*

before working on the electrical system. Even with the battery disconnected, capacitive discharge could occur if a component's live terminal is earthed through a metal object. This could cause a shock or nasty burn.

b) *Do not reverse the battery connections. Components such as the alternator, electronic control units, or any other components having semi-conductor circuitry could be irreparably damaged.*

c) *If the engine is being started using jump leads and a slave battery, connect the batteries positive-to-positive and negative-to-negative (see "Booster battery (jump) starting"). This also applies when connecting a battery charger.*

d) *Never disconnect the battery terminals, the alternator, any electrical wiring or any test instruments when the engine is running.*

e) *Do not allow the engine to turn the alternator when the alternator is not connected.*

f) *Always ensure that the battery negative lead is disconnected when working on the electrical system.*

g) *Before using electric-arc welding equipment on the van, disconnect the battery, alternator and components such as the fuel injection/ignition electronic control unit, to protect them from the risk of damage.*

h) *Some radio/cassette units have a built-in security code, to deter thieves. If the power source to the unit is cut, the anti-theft system will activate. Even if the power source is immediately reconnected, the radio/cassette unit will not function until the correct security code has been entered. Therefore, if you do not know the correct security code for the radio/cassette unit, do not disconnect*

the battery negative terminal, or remove the radio/cassette unit from the vehicle. Refer to "Radio/cassette unit anti-theft system - precaution" in the Reference Section of this manual for further information.

2 Electrical fault-finding - general information

Note: *Refer to the precautions given in "Safety first!" (at the beginning of this manual) and to Section 1 of this Chapter before starting work. The following tests relate to testing of the main electrical circuits, and should not be used to test delicate electronic circuits (such as anti-lock braking systems), particularly where an electronic control module is used.*

A typical electrical circuit consists of an electrical component, any switches, relays, motors, fuses, fusible links or circuit breakers related to that component, and the wiring and connectors that link the component to both the battery and the chassis. To help to pinpoint a problem in an electrical circuit, wiring diagrams are included at the end of this Chapter.

Before attempting to diagnose an electrical fault, first study the appropriate wiring diagram, to obtain a complete understanding of the components included in the particular circuit concerned. The possible sources of a fault can be narrowed down by noting whether other components related to the circuit are operating properly. If several components or circuits fail at one time, the problem is likely to be related to a shared fuse or earth connection.

Electrical problems usually stem from simple causes, such as loose or corroded connections, a faulty earth connection, a blown fuse, a melted fusible link, or a faulty relay (refer to Section 3 for details of testing relays). Visually inspect the condition of all fuses, wires and connections in a problem circuit before testing the components. Use the wiring diagrams to determine which terminal connections will need to be checked, to pinpoint the trouble-spot.

The basic tools required for electrical fault-finding include the following:

a) *A circuit tester or voltmeter (a 12-volt bulb with a set of test leads can also be used for certain tests).*
b) *A self-powered test light (sometimes known as a continuity tester).*
c) *An ohmmeter (to measure resistance).*
d) *A battery.*
e) *A set of test leads.*
f) *A jumper wire, preferably with a circuit breaker or fuse incorporated, which can be used to bypass suspect wires or electrical components.*

Before attempting to locate a problem with test instruments, use the wiring diagram to determine where to make the connections.

To find the source of an intermittent wiring fault (usually due to a poor or dirty connection, or damaged wiring insulation), a "wiggle" test can be performed on the wiring. This involves wiggling the wiring by hand, to see if the fault occurs as the wiring is moved. It should be possible to narrow down the source of the fault to a particular section of wiring. This method of testing can be used in conjunction with any of the tests described in the following sub-sections.

Apart from problems due to poor connections, two basic types of fault can occur in an electrical circuit - open-circuit, or short-circuit.

Open-circuit faults are caused by a break somewhere in the circuit, which prevents current from flowing. An open-circuit fault will prevent a component from working, but will not cause the relevant circuit fuse to blow.

Short-circuit faults are caused by a "short" somewhere in the circuit, which allows the current flowing in the circuit to "escape" along an alternative route, usually to earth. Short-circuit faults are normally caused by a breakdown in wiring insulation, which allows a feed wire to touch either another wire, or an earthed component such as the bodyshell. A short-circuit fault will normally cause the relevant circuit fuse to blow.

Finding an open-circuit

To check for an open-circuit, connect one lead of a circuit tester or voltmeter to either the negative battery terminal or a known good earth.

Connect the other lead to a connector in the circuit being tested, preferably nearest to the battery or fuse.

Switch on the circuit, remembering that some circuits are live only when the ignition switch is moved to a particular position.

If voltage is present (indicated either by the tester bulb lighting or a voltmeter reading, as applicable), this means that the section of the circuit between the relevant connector and the battery is problem-free.

Continue to check the remainder of the circuit in the same fashion.

When a point is reached at which no voltage is present, the problem must lie between that point and the previous test point with voltage. Most problems can be traced to a broken, corroded or loose connection.

Finding a short-circuit

To check for a short-circuit, first disconnect the load(s) from the circuit (loads are the components that draw current from a circuit, such as bulbs, motors, heating elements, etc.).

Remove the relevant fuse from the circuit, and connect a circuit tester or voltmeter to the fuse connections.

Switch on the circuit, remembering that some circuits are live only when the ignition switch is moved to a particular position.

If voltage is present (indicated either by the tester bulb lighting or a voltmeter reading, as applicable), this means that there is a short-circuit.

If no voltage is present, but the fuse still blows with the load(s) connected, this indicates an internal fault in the load(s).

Finding an earth fault

The battery negative terminal is connected to "earth" (the metal of the engine/transmission and the van body), and most systems are wired so that they only receive a positive feed. The current returns through the metal of the van body. This means that the component mounting and the body form part of that circuit. Loose or corroded mountings can therefore cause a range of electrical faults, ranging from total failure of a circuit, to a puzzling partial fault. In particular, lights may shine dimly (especially when another circuit sharing the same earth point is in operation). Motors (e.g. wiper motors or the radiator cooling fan motor) may run slowly, and the operation of one circuit may have an affect on another. Note that on many vehicles, earth straps are used between certain components, such as the engine/transmission and the

body, usually where there is no metal-to-metal contact between components, due to flexible rubber mountings, etc.

To check whether a component is properly earthed, disconnect the battery, and connect one lead of an ohmmeter to a known good earth point. Connect the other lead to the wire or earth connection being tested. The resistance reading should be zero; if not, check the connection as follows.

If an earth connection is thought to be faulty, dismantle the connection, and clean back to bare metal both the bodyshell and the wire terminal or the component earth connection mating surface. Be careful to remove all traces of dirt and corrosion, then use a knife to trim away any paint, so that a clean metal-to-metal joint is made. On reassembly, tighten the joint fasteners securely. If a wire terminal is being refitted, use serrated washers between the terminal and the bodyshell, to ensure a clean and secure connection. When the connection is remade, prevent the onset of corrosion in the future by applying a coat of petroleum jelly or silicone-based grease.

3 Fuses and relays - general

Fuses

1 Fuses are designed to break a circuit when a predetermined current is reached, to protect the components and wiring which could be damaged by excessive current flow. Any excessive current flow will be due to a fault in the circuit, usually a short-circuit (see Section 2).

2 The main fuses are located in two fuse panels in the engine compartment, mounted on the left-hand side of the scuttle panel. The circuits protected by the fuses are given in the Specifications. The fuse positions are colour-coded, according to a coloured dot on the fuseholder (not to be confused with the wire colours which may be different). To remove a fuse, first ensure that the relevant circuit is switched off. Then pull off the cover and remove the relevant fuse from the panel **(see illustrations)**.

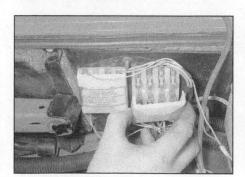

3.2a Pull off the fusebox cover . . .

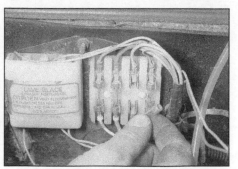

3.2b . . . and remove a fuse (note the coloured dots on the fuseholder)

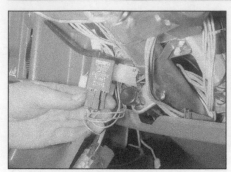

3.7a Flasher unit . . .

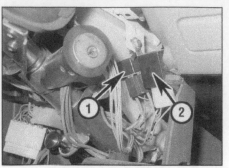

3.7b . . . accessory supply relay (1) and windscreen wiper relay (2)

3.8 Cooling fan relays

3 On some models, a supplementary fuse for the engine cooling fans is fitted to a fly-lead from the battery.

4 Before renewing a blown fuse, trace and rectify the cause, and always use a fuse of the correct rating. Never substitute a fuse of a higher rating, or make temporary repairs using wire or metal foil, as more serious damage or even fire could result.

5 Some components that are powered from fuse 3 are also dependent on an Accessory Supply Relay which is energised from fuse 2 when the ignition is switched on at position "A" or "M". Therefore, if a component from fuse 3 fails to work, it could be either fuse 2 or fuse 3 that has blown.

Relays

6 A relay is an electrically operated switch, used for the following reasons:
  a) A relay can switch a heavy current

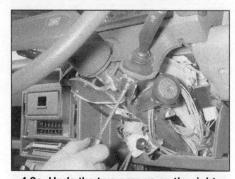

4.3a Undo the two screws on the right-hand stalk switch . . .

remotely from the circuit in which the current is flowing, allowing the use of lighter-gauge wiring and switch contacts.
  b) A relay can receive more than one control input, unlike a mechanical switch.
  c) A relay can have a timer function - for example, the intermittent wiper relay.

7 The following relays are situated above the parcel shelf, next to the steering column **(see illustrations)**. To access them, remove four screws and pull down the steering column lower shroud (see Chapter 11).
  a) Indicator and hazard warning lamp flasher unit.
  b) Accessory supply relay (see paragraph 5).
  c) Front wiper relay.

8 The following relays are mounted at the front left of the engine compartment **(see illustration)** and switch the two engine cooling fans from half power (6V each in series) to full power (12V each in parallel), when the two-stage radiator thermo-switch reaches the temperature for second-stage cooling.
  a) The cooling fan relay.
  b) The cooling fan inverter relay.

9 There may be additional relays for optional accessories, for example rear window wipers.

10 If a circuit or system controlled by a relay develops a fault, and the relay is suspect, operate the system. If the relay is functioning, it should be possible to hear it "click" as it is energised. If this is the case, the fault lies with the components or wiring of the system. If the relay is not being energised, then either the relay is not receiving a main supply or a switching voltage, or the relay itself is faulty.

Testing is by the substitution of a known good unit, but be careful - while some relays are identical in appearance and in operation, others look similar but perform different functions.

11 To remove a relay, first ensure that the relevant circuit is switched off. The relay can then simply be pulled out from the socket, and pushed back into position.

4 Switches - removal and refitting

1 Disconnect the battery negative lead before removing any switch, then reconnect it after refitting the switch.

Steering column stalk switches

Removal

2 Remove four screws and pull down the steering column lower shroud (see Chapter 11).

3 The left and right stalk switches are each secured to a bracket by two screws. Undo the screws and manoeuvre the switch off the bracket, then disconnect the multi-plug **(see illustrations)**.

Refitting

4 Refitting is the reverse of removal.

Instrument side panel switches

Removal

5 Remove the instrument panel as described in Section 11.

4.3b . . . and manoeuvre it out . . .

4.3c . . . then disconnect the multi-plug

4.3d Unscrew and remove the left-hand stalk switch

6 Pull the side panel towards you slightly at the top, then push the lugs upwards to disengage them. Pull the panel a little further towards you and disengage the lower clip, then withdraw the panel **(see illustration)**.

7 Insert a screwdriver behind the switch to release the clips at both sides, then pull out the switch and disconnect the multi-plug **(see illustrations)**.

Refitting

8 Refitting is the reverse of removal.

Heater blower motor switch

Removal

9 Use a small screwdriver to prise off the switch so that it comes off the tabs at the top and bottom **(see illustration)**.

Refitting

10 Refitting is the reverse of removal.

Courtesy lamp switch

Removal

11 The courtesy lamp switches are fitted to the front door pillars so that they turn on the interior lights when the door is opened **(see illustration)**.

12 Open the door, then where necessary prise off the rubber gaiter from the switch.

4.6 Release the clips and withdraw the instrument side panel

13 Remove the securing screw, then withdraw the switch from the door pillar and disconnect the wiring connector.

> **HAYNES HINT** *Tape the wiring to the door pillar, to prevent it falling back into the door pillar. Alternatively, tie a piece of string to the wiring to retrieve it.*

Refitting

14 Refitting is the reverse of removal, ensuring (where applicable) that the rubber gaiter is correctly seated on the switch.

Handbrake warning lamp switch

15 The handbrake warning lamp switch is a push-switch similar to the courtesy lamp switches and is accessed by pulling up the rubber boot from the handbrake.

Stop-lamp switch

Note: *These instructions apply to right-hand drive diesel models with a brake linkage cross-tube. A similar type of switch is mounted above the brake pedal on left-hand drive diesel models. No specific instructions are available about the type of switch fitted to petrol models.*

Removal

16 Remove the left parcel shelf (see Chapter 11).

17 Disconnect the two wiring connectors from the switch **(see illustration)**.

18 Slacken the locknut **(see illustration)**, then unscrew the switch and remove it from the bracket.

Refitting

19 Refitting is the reverse of removal, making sure that the switch is screwed into the correct position and the lamp operates when the brake pedal is depressed. With the brake pedal released, the clearance in front of the threaded portion of the switch should be

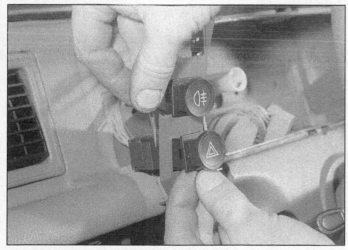

4.7a Insert a screwdriver behind the switch to release the clips at both sides . . .

4.7b . . . then pull out the switch and disconnect the multi-plug

4.9 Prise off the heater blower motor switch so that it comes off the tabs at the top and bottom

4.11 Courtesy lamp switch on door pillar

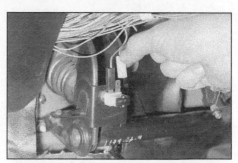

4.17 Undo the connectors from the stop-lamp switch (right-hand drive diesel model with brake linkage cross-tube)

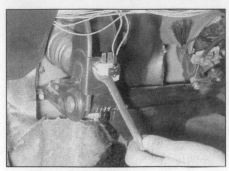

4.18 Slacken the locknut and remove the switch

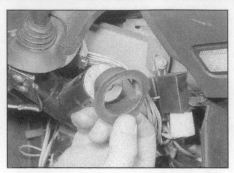

4.21 Remove the rubber cover from the ignition switch

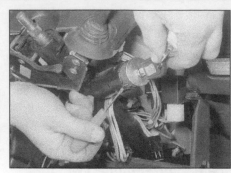

4.22 Release the two lugs and withdraw the switch with the key at the arrow between A and M

according to the Specifications. Tighten the locknut to the specified torque.

Ignition switch/steering column lock

Removal

20 Remove four screws and pull down the steering column lower shroud (see Chapter 11).
21 Remove the rubber cover from the ignition switch **(see illustration)**.
22 Insert the key and turn the switch to the arrow between positions A and M. Use a screwdriver to release the two lugs under the switch, then withdraw the switch **(see illustration)**.

Refitting

23 Refitting is the reverse of removal

---

5 Bulbs (exterior lights) - renewal

General

1 Whenever a bulb is renewed, note the following:

a) *Remember that if the lamp has just been in use, the bulb may be extremely hot.*
b) *Always check the bulb contacts and holder, ensuring that there is a clean metal-to-metal contact between the bulb and its live contacts and earth. Clean off any corrosion or dirt before fitting a new bulb.*
c) *Always ensure that the new bulb is of the correct rating and that the glass envelope is completely clean before fitting. This*

applies particularly to headlamp bulbs. If necessary clean the glass with methylated spirit.

Headlamp (main and dip)

2 Working in the engine compartment, disconnect the wiring connector and remove the rubber cover from behind the headlamp bulb **(see illustrations)**.
3 Release the spring clips and withdraw the bulb from its housing in the reflector **(see illustrations)**.
4 When handling a new bulb, use a tissue or clean cloth to avoid touching the glass with your fingers. Moisture and grease from your skin can cause blackening and rapid failure of this type of bulb.

> **HAYNES HINT** *If the glass of a headlamp bulb is accidentally touched, wipe it clean using methylated spirit.*

5 Refitting is the reverse of removal, making sure that the bulb engages in its locating slots. From time to time, and after every headlamp bulb change, have the beam setting checked by a Citroën dealer.

Front sidelamp

6 Working within the engine compartment, turn the knurled knob and withdraw the bulbholder, then remove the bulb by pushing and turning **(see illustrations)**.
7 Refitting is the reverse of removal.

5.2a Disconnect the headlamp bulb wiring connector . . .

5.2b . . . and remove the rubber cover

5.3a Release the spring clips . . .

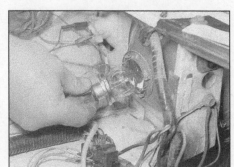

5.3b . . . and withdraw the headlamp bulb

5.6a Withdraw the front sidelamp bulbholder . . .

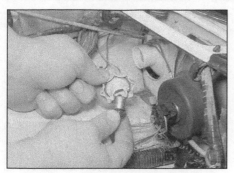

5.6b . . . then remove the bulb

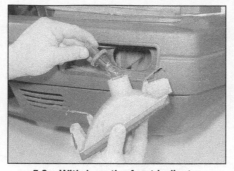

5.9a Withdraw the front indicator bulbholder . . .

5.9b . . . then remove the bulb

Front direction indicators

8 Release the lamp unit from the front bumper, as described in Section 7.
9 Turn the bulbholder and withdraw it, then remove the bulb by pushing and turning **(see illustrations)**.
10 Refitting is the reverse of removal.

Front direction indicator side repeaters

11 Unscrew the lens, then remove the bulb **(see illustrations)**.
12 Refitting is the reverse of removal.

Rear lamps

13 The following bulbs are in the rear lamp unit, from top to bottom.
  a) *Stop and tail lamp (double-filament bulb)*
  b) *Direction indicator*
  c) *Reversing light*
  d) *Fog lamp*

14 Remove the lamp unit as described in Section 7.
15 Push down the central clip and separate the bulbholder panel from the lamp unit, then push and turn the bulbs to remove them **(see illustrations)**.
16 Refitting is the reverse of removal.

Rear number plate lamp

17 Undo the screws and remove the number plate lamp, then remove the bulb.
18 Refitting is the reverse of removal.

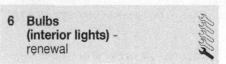

6 Bulbs (interior lights) - renewal

General

1 Refer to Section 5.

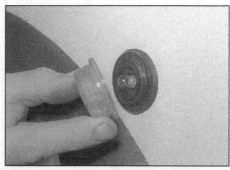

5.11a Unscrew the lens from the indicator side repeater . . .

Overhead lamp

2 Push the lens forward slightly at the back to release the clip, then remove the lens. Remove the festoon-type bulb by pulling it downwards **(see illustrations)**.

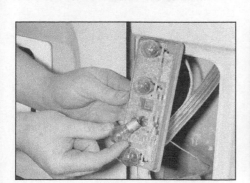

5.11b . . . then remove the bulb

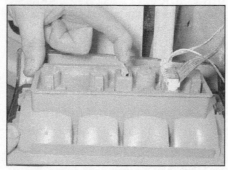

5.15a Push down the central clip on the rear lamp unit . . .

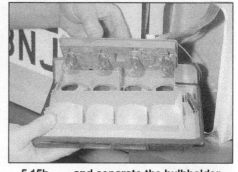

5.15b . . . and separate the bulbholder panel . . .

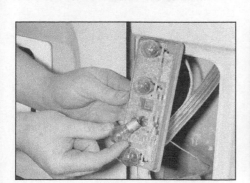

5.15c . . . then remove the bulbs

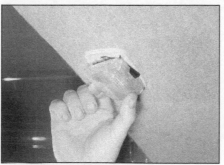

6.2a Release the clip on the overhead lamp and remove the lens . . .

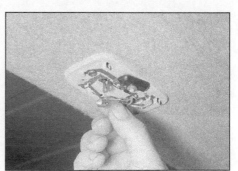

6.2b . . . then remove the bulb

6.4a Release the clip on the load compartment lamp and remove the lens . . .

6.4b . . . then remove the bulb

6.6a Push the facia lamp out from behind the panel . . .

6.6b . . . remove the bulbholder . . .

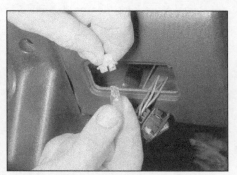

6.6c . . . then remove the bulb from the holder

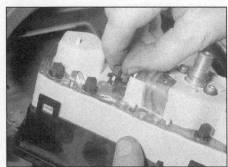

6.9 Remove the bulbholder from the instrument panel

3 Refitting is the reverse of removal. Make sure that the bulb contacts hold the bulb securely, and bend them slightly if necessary.

Load compartment lamp

4 This lamp, fitted to the rear door pillar on some models, is similar to the overhead lamp. Push the lens sideways slightly at the outer edge to release the clip and remove the lens, then remove the bulb **(see illustrations)**.
5 Refitting is the reverse of removal.

Facia panel lamps

6 These lamps are fitted to the left and right facia panels on some models. Push the lamp unit out from behind the panel. Rotate the bulbholder one quarter turn and pull it out, then remove the bulb from the holder **(see illustrations)**.
7 Refitting is the reverse of removal.

Instrument panel illumination and warning lamps

8 Remove the instrument panel as described in Section 11.
9 Twist the relevant bulbholder clockwise and withdraw it from the printed circuit board on the rear of the instrument panel **(see illustration)**. Remove the bulb from the holder, if it has been fitted separately, otherwise the bulb and holder must be renewed as a unit.
10 Refitting is the reverse of removal.

Heater control panel lamps

11 Remove the heater control panel as described in Chapter 3.
12 Remove the bulbs from behind the panel **(see illustration)**.
13 Refitting is the reverse of removal.

7 **Exterior lamp units -** removal and refitting

Headlamp and sidelamp

Removal

1 Remove the stoneguard from the headlamp, if it has been fitted as an accessory **(see illustration)**.
2 Disconnect the headlamp multi-plug and the sidelamp bulbholder.
3 Remove the clip and pull the headlamp off the upper adjuster stud, then pull it off the two lower studs **(see illustrations)**.

Refitting

4 Refitting is the reverse of removal, but

6.12 Bulbs behind heater control panel

7. 1 Remove the stoneguard from the headlamp

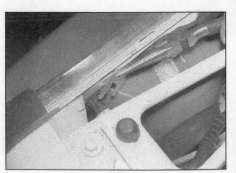

7.3a Remove the clip . . .

7.3b ... and pull the headlamp off the upper adjuster stud ...

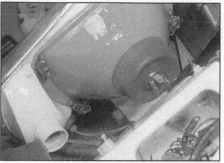

7.3c ... then pull the headlamp off the two lower studs

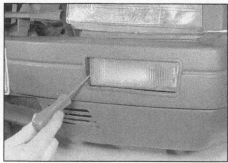

7.5a Release the clip on the front indicator lamp ...

check and if necessary adjust the headlamp beam setting (see Section 8).

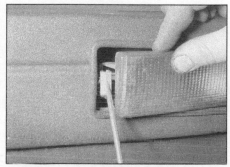

7.5b ... carefully prise out the lamp unit ...

Front direction indicator lamp

Removal

5 Insert a screwdriver and release the clip on the inner edge of the lamp, nearest the number plate, and carefully prise out the lamp unit (see illustrations).
6 Detach the electrical connector.

Refitting

7 Refitting is the reverse of removal. Insert the lamp unit at an angle, with the outer edge first.

Front direction indicator side repeater

Removal

8 Working behind the wheel arch, depress the two tabs and pull out the lamp unit.

Disconnect the two wiring connectors (see illustrations).

Refitting

9 Refitting is the reverse of removal. Make sure the wider tab is at the top.

Rear lamp cluster

Removal

10 Release the clip from the recess inside the van, while supporting the lamp unit from outside, then pull out the lamp unit (see illustrations).
11 Disconnect the two wiring multi-plugs.
12 To access the bulbs, push down the central clip and separate the bulbholder panel from the lamp unit (see Section 5).

7.5c ... and remove it from the bumper

7.8a Depress the tabs and pull out the indicator side repeater lamp ...

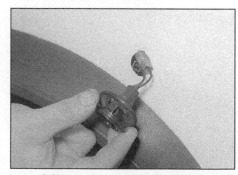

7.8b ... then detach the wiring connectors (lens and bulb removed)

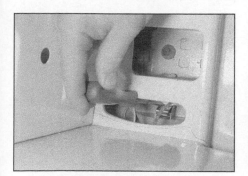

7.10a Release the clip from inside the van ...

7.10b ... while supporting the rear lamp unit from outside ...

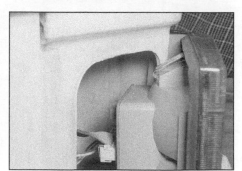

7.10c ... then pull out the lamp unit

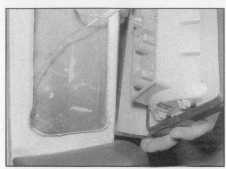

7.13 Make sure the lower clip is correctly engaged when refitting the rear lamp

Refitting

13 Refitting is the reverse of removal. Make sure the lower clip is correctly engaged **(see illustration)**.

Rear number plate lamps

Removal

14 Undo the two screws and remove the rear number plate lamp unit.

Refitting

15 Refitting is the reverse of removal.

8 Headlamp beam alignment - general information

If the headlamps are thought to be out of alignment, then accurate adjustment of their beams is only possible using optical beam

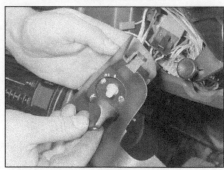

9.5a Pull off the headlamp beam adjuster knob . . .

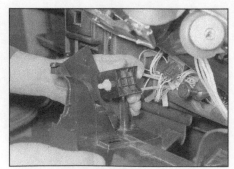

9.5b . . . then undo the two screws and remove the adjuster

setting equipment, and this work should be carried out by a Citroën dealer or suitably-equipped workshop.

For reference, the headlamps can be adjusted from the two adjustment bolts at the rear of each lamp. The lower adjuster, near the grille, controls the horizontal setting. The upper adjuster, near the sidelamp, controls the vertical setting.

Although the headlamps may have been correctly adjusted for normal load, the vehicle may tilt slightly, up or down, depending on the distribution of the load. On some models, the vertical alignment can be adjusted by turning a knob within the vehicle to compensate for load distribution. After loading the vehicle, the knob should be turned so that the centre of the dipped beam hits the road between 30 and 50 metres ahead.

9 Headlamp beam adjuster - removal and refitting

Removal

1 On models fitted with load compensation adjustment, a knob on the steering column lower shroud operates two cables, each connected to a headlamp upper adjustment stud. The cables pass through the bulkhead at the same point as the bonnet release cable.
2 Working within the engine compartment, remove the clips from the upper adjuster studs on both headlamps and pull the headlamps off the studs.
3 Depress the lugs and turn the adjusters by 60° to remove them from the bodywork **(see illustration)**, then unscrew the cables from the adjusters. Detach the cables from any clips in the engine compartment.
4 Working within the passenger compartment, undo the four screws and pull down the steering column lower shroud (see Chapter 11).
5 Pull off the headlamp beam adjuster knob, then undo the two screws and disconnect the adjuster from the shroud **(see illustrations)**.
6 Pull the cables through the bulkhead into the passenger compartment and remove the assembly from the vehicle.

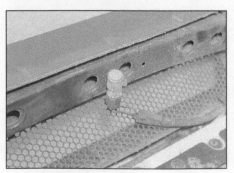

10.2 Headlamp dim-dip resistor

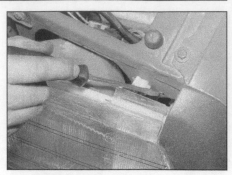

9.3 Depress the lugs and remove headlamp beam adjusters

Refitting

7 Refitting is the reverse of removal. When finished, take the vehicle to a Citroën dealer or suitably-equipped workshop to have the headlamps adjusted for normal load.

10 Dim-dip headlamp system - general information

1 This system, fitted to some models, supplies a low current to the headlamp dipped beam circuit when the sidelights are switched on with the engine running, preventing the vehicle from being driven on the sidelights alone.
2 The current passes through a dim-dip resistor, fitted to a clip behind the front bumper, reducing the dipped beam to one-sixth of its normal value **(see illustration)**.

11 Instrument panel - removal and refitting

Removal

1 Disconnect the battery negative lead.
2 Insert a screwdriver in the top and bottom of the instrument panel and depress the clips **(see illustration)**.
3 Carefully pull the panel away from the facia, then detach the speedometer cable by depressing the plastic lugs on the connector **(see illustration)**.

11.2 Insert a screwdriver and depress the instrument panel clips

4 Disconnect the two multi-plugs and withdraw the panel **(see illustration)**.

5 If required, the bulbs can be removed (see Section 6) and the instrument panel can be dismantled (see Section 12).

Refitting

6 Refitting is the reverse of removal, but ensure that the speedometer cable is not kinked or twisted between the instrument panel and the bulkhead as the panel is refitted.

12 Instrument panel components - removal and refitting

General

1 The instrument panel assembly consists of the instrument cluster housing (from which the instruments can be removed), and a front face with a display panel and clear plastic panel **(see illustration)**.

Removal

2 Remove the instrument panel as described in Section 11.

3 If required, remove the bulbs as described in Section 6.

4 Undo the three clips at the top and bottom of the assembly and separate the instrument cluster from the front face **(see illustrations)**.

5 If required, lever off the clips at the bottom of the face and remove the display panel and the clear plastic panel **(see illustrations)**.

11.3 Carefully pull the panel from the facia and detach the speedometer cable

Fuel gauge

6 Undo three nuts and washers from the

12.1 Instrument panel components

1 Front face
2 Clear plastic panel
3 Display panel
4 Instruments
5 Instrument housing with printed circuit board
6 Screws, nuts and washers

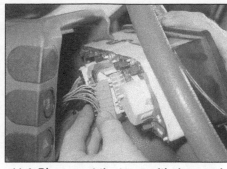
11.4 Disconnect the two multi-plugs and withdraw the panel

printed circuit board and remove the fuel gauge **(see illustrations)**.

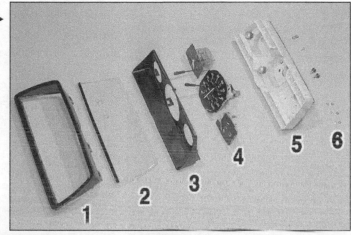

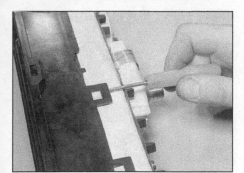
12.4a Undo the three clips at the top and bottom of the assembly . . .

12.4b . . . and separate the instrument cluster from the front face

12.5a Lever off the clips at the bottom of the face . . .

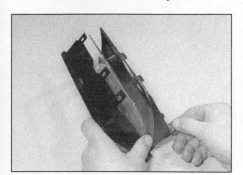
12.5b . . . and remove the display panel and the clear plastic panel

12.6a Undo three nuts and washers from the printed circuit board . . .

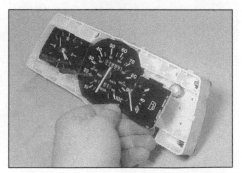
12.6b . . . and remove the fuel gauge

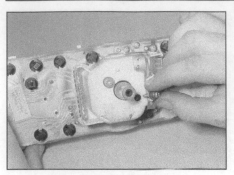

12.7a Undo two screws . . .

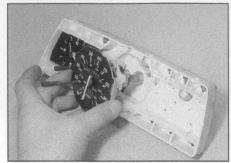

12.7b . . . and remove the speedometer

Speedometer

7 Undo two screws and remove the speedometer **(see illustrations)**.

Clock

8 Undo one screw and two connectors and remove the clock **(see illustrations)**.

Refitting

9 Refitting is the reverse of removal.

13 Cigarette lighter - removal and refitting

Removal

1 Remove the cigarette lighter stub from its metal housing.
2 Insert a screwdriver between the inner metal housing and the outer plastic housing

12.8a Undo one screw and two connectors . . .

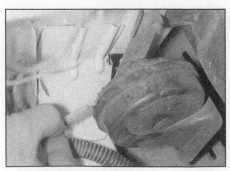

14.4 Disconnect the wiring connector from the horn

(see illustration). Prise out the metal housing slightly, then turn it and pull it out completely.
3 Disconnect the wiring connector.
4 If required, insert a screwdriver and prise out the plastic housing.

Refitting

5 Refitting is the reverse of removal.

14 Horn - removal and refitting

Removal

1 The horn is mounted on a bracket behind the front bumper, to the left of the radiator.
2 Disconnect the battery negative lead.
3 Remove the left headlamp as described in Section 7.

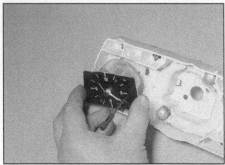

12.8b . . . and remove the clock

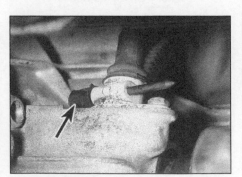

15.3a Pull out the rubber cotter pin (arrowed) . . .

4 Disconnect the wiring connector and undo the nut on the mounting bracket, then remove the horn from the vehicle **(see illustration)**.

Refitting

5 Refitting is the reverse of removal. When finished, take the vehicle to a Citroën dealer or suitably-equipped workshop to have the headlamps aligned.

15 Speedometer cable - removal and refitting

Removal

1 Disconnect the battery negative lead. Remove the instrument panel and disconnect the speedometer cable (see Section 11).
2 The lower end of the cable is connected to the speedometer drive at the rear of the transmission housing, next to the inner end of the right-hand driveshaft. To access it from above, remove the air inlet ducting. Alternatively, to access it from below, jack up the front of the van and support the vehicle securely on axle stands (see *"Jacking and vehicle support"*).
3 Pull out the rubber cotter pin and disconnect the cable from the speedometer drive **(see illustrations)**.
4 The cable passes through the bulkhead via the windscreen wiper motor and linkage compartment. Remove the wheel changing jack and wheelbrace, then undo the four bolts and remove the scuttle panel cover plate (see Section 17).

13.2 Prise out the cigarette lighter inner metal housing

15.3b . . . and disconnect the cable from the speedometer drive in the transmission

16.2a Lift the cover from the wiper arm spindle and undo the nut . . .

16.2b . . . then lift off the wiper arm

17.3 Undo the large plastic sealing nut on the wiper spindle

5 Detach the cable from any clips and withdraw it through the bulkhead. On some models, the cable may be in two parts.

Refitting

6 Refitting is the reverse of removal, but ensure that the cable is not kinked or twisted.

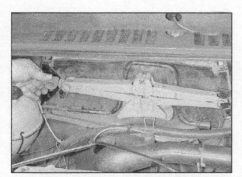

17.4a Open the bonnet and remove the wheel changing jack and wheelbrace . . .

16 Wiper arm - removal and refitting

Windscreen wiper arm

Removal

1 Operate the wiper motor, then switch it off so that the wiper arm returns to the "parked" position.

2 Lift the cover from the wiper arm spindle. Undo the spindle nut and recover the washer, then lift off the wiper arm **(see illustrations)**. If necessary, lever the arm off the spindle using a large, flat-bladed screwdriver.

> **HAYNES HiNT**
> *Stick a piece of tape on the windscreen, along the edge of the wiper blade, to use as an alignment aid on refitting.*

Refitting

3 Refitting is the reverse of removal. Ensure that the wiper arm and spindle splines are clean and dry, and align the blade with the tape on the windscreen.

Rear window wiper arm

4 Rear window wipers are fitted as an option on some models. No specific instructions are available for removal and refitting, but the design of the wiper arm is similar to the windscreen wiper.

17 Wiper motor and linkage - removal and refitting

Windscreen wiper motor and linkage

Removal

1 Disconnect the battery negative lead.

2 Remove the windscreen wiper arm (see Section 16).

3 Undo the large plastic sealing nut on the spindle **(see illustration)**.

4 Open the bonnet and remove the wheel changing jack and wheelbrace, then undo the four bolts and remove the scuttle panel cover plate **(see illustrations)**.

5 Disconnect the multi-plug and withdraw the cable through the scuttle panel. Undo the cable clip if necessary **(see illustrations)**.

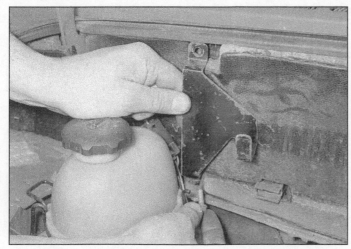

17.4b . . . undo four bolts from the wheel jack mounting brackets . . .

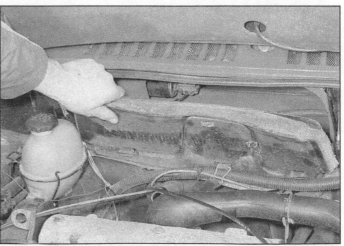

17.4c . . . and remove the scuttle panel cover plate

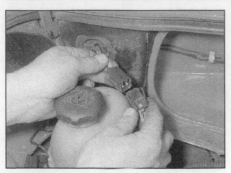

17.5a Disconnect the multi-plug and withdraw the cable through the scuttle panel

17.5b Undo the cable clip

6 Undo one vertical bolt securing the linkage to the upper bodywork and two horizontal bolts securing it to the bulkhead, then remove the wiper motor and linkage from the vehicle **(see illustrations)**.

7 To separate the motor from the linkage, undo the motor spindle nut and the three flange bolts **(see illustration)**.

Refitting

8 Refitting is the reverse of removal.

Rear window wiper motor

9 Rear window wipers are fitted as an option on some models. The motors are fitted inside a panel on each of the rear doors. No specific instructions are available for removal and refitting.

---

18 Windscreen washer system components - general, removal and refitting

General

1 The windscreen washer system consists of a fluid reservoir, pump and washer nozzle. The reservoir is mounted at the left of the engine compartment rear scuttle panel. There are a number of different designs with the pump in a variety of locations as follows:

a) *Mounted on a bracket on the scuttle panel and connected to the reservoir by a hose (see illustration).*
b) *Mounted inside the reservoir.*
c) *Mounted externally on the reservoir and secured by a sealing grommet.*

Washer fluid reservoir

Removal

2 On models with the pump mounted on a bracket on the scuttle panel, disconnect the hose connecting the pump to the reservoir.

3 On models with the pump fitted inside the reservoir, turn the cap and withdraw the pump.

4 On models with the pump mounted externally on the reservoir, disconnect the battery negative lead, then disconnect the electrical connectors from the pump. Disconnect the fluid hose and be prepared for spillage.

5 Undo the wire clip and remove the reservoir from the scuttle panel.

6 On models with the pump mounted externally on the reservoir, remove the pump if necessary by carefully prising it off from its sealing grommet.

Refitting

7 Refitting is the reverse of removal. If an externally mounted pump has been removed from the reservoir, check the condition of the sealing grommet and renew it if necessary.

Washer pump

Removal

Note: *On models with the pump mounted externally on the fluid reservoir, it is recomm-*

17.6a Undo one vertical bolt . . .

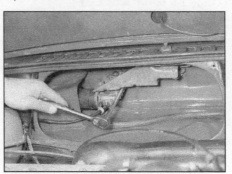

17.6b . . . and two horizontal bolts . . .

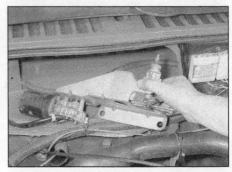

17.6c . . . then remove the wiper motor and linkage

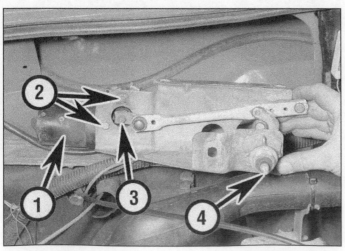

17.7 Wiper motor and linkage assembly

1 *Motor*
2 *Flange bolts*
3 *Motor spindle nut*
4 *Wiper arm spindle*

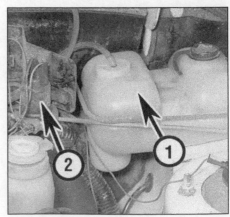

18.1 Windscreen washer reservoir (1) and pump (2)

20.5 Insert the release tools and withdraw the radio

20.6a Disconnect the aerial . . .

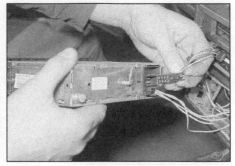

20.6b . . . and the wiring connectors from behind the radio

ended that the reservoir is at least half empty before renewal, to minimise fluid loss.

8 Disconnect the battery negative lead, then disconnect the electrical connectors from the pump.
9 Disconnect the fluid hose(s) from the pump.
10 On models with a pump fitted to a bracket on the scuttle panel, undo the nuts and bolts and remove the pump from the bracket.
11 On models with a pump fitted inside the reservoir, turn the cap and remove the pump.
12 On models with a pump fitted externally on the reservoir, remove the pump by carefully prising it off from its sealing grommet.

Refitting

13 Refitting is the reverse of removal. In the case of an external pump mounted on the reservoir, check the condition of the sealing grommet and renew it if necessary.

Washer nozzle

Removal

14 Carefully prise the washer nozzle from the bonnet, taking care not to damage the paintwork, and disconnect the fluid hose.

Refitting

15 Refitting is the reverse of removal. Adjust the aim of the washer jet, if necessary, by using a pin in the nozzle.

19 Rear window/headlamp washer system components - general

Rear window washers and headlamp washers are fitted as optional extras to some models.
Rear window washers have a separate reservoir and pump mounted on the load compartment side panel.
Control of the headlamp washers is coupled to the windscreen washer, although there may be a separate reservoir and pump.
No specific information is available about removal and refitting, although in principle it is the same as removing and refitting the windscreen washer system.

20 Radio/cassette player - general, removal and refitting

Note: *Refer to " Radio/cassette unit anti-theft system" at the end of this manual, before proceeding.*

General

1 A radio, or radio/cassette player is fitted as standard on most models. All models are fitted with standard "pre-radio equipment", regardless of whether or not a radio was fitted during production. This equipment consists of a roof-mounted aerial and coaxial aerial cable, normal interference suppression and radio connection wiring harnesses including speaker wiring.
2 On models without a radio fitted, the wiring is accessed by removing the storage pocket at the bottom of the heater control panel. On early models, the wiring consists of a number of separate leads, and on later models there are two multi-plugs. All models have an aerial cable. The wiring configuration varies according to model, and details are given in your Owner's Handbook. If in doubt about the type of radio that can be fitted, consult your Citroën dealer.
3 The speakers are mounted in the footwell side panels on the driver and passenger side.

Removal

4 Disconnect the battery negative lead.
5 The radio is secured by clips which can be released using special tools consisting of two pieces of wire, each bent into a "U" shape. These tools may have been provided with the vehicle when new, otherwise they can be obtained from your Citroën dealer or any audio accessory outlet. Insert the tools into the holes and withdraw the radio **(see illustration)**.
6 Disconnect the aerial and the wiring connectors from behind the radio and withdraw the unit from the vehicle **(see illustrations)**.

Refitting

7 Refitting is the reverse of removal. Check that the wiring is positioned correctly behind the radio, to allow space for the unit to be pushed back into position, and make sure that the radio is fully engaged with its retaining clips. Enter the access code if the radio is coded for anti-theft protection.

21 Speakers - removal and refitting

Removal

1 The speakers are mounted behind the footwell side trim panels on the driver and passenger side.
2 Disconnect the battery negative lead.
3 Remove the appropriate right or left parcel shelf (see Chapter 11).
4 Undo four screws and remove the panel, taking care not to tension the speaker wiring. Alternatively, just undo three screws and swivel the panel on the upper screw so that the speaker becomes accessible.
5 Disconnect the wiring from speaker.
6 Undo four screws and remove the speaker from the panel **(see illustration)**.

Refitting

7 Refitting is the reverse of removal.

21.6 Remove the speaker from behind the footwell side trim panel

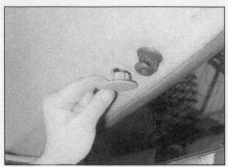

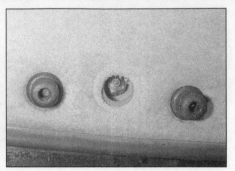

22.2a Remove the plastic cap from the headlining . . .

22.2b . . . then undo the nut from under the aerial mounting

22 Radio aerial - removal and refitting

Removal

1 The aerial is fitted to the roof and can be unscrewed from its mounting. It should always be removed before entering a carwash.

2 To remove the aerial mounting, prise off the plastic cap from the headlining, then undo the nut using a socket **(see illustrations)**. Detach the co-axial cable from the aerial spindle and withdraw the mounting.

3 To remove the co-axial cable, remove the radio (see Section 20) and tie a piece of string to the end of the cable. The string must be at least as long as the cable, so that it occupies the body channels when the cable is withdrawn. The cable passes through the body pillar on either the left or right side. Remove the parcel shelf or side trim panel as appropriate, to gain access to the cable. Remove the sunvisor and pull down the headlining to expose the cable where it enters the body pillar. Carefully withdraw the cable from the top, leaving the string in place for refitting, then detach the cable from the string.

Refitting

4 Refitting is the reverse of removal. If the co-axial cable has been removed, tie the string to the end of the cable and use it to feed the cable through the body channels.

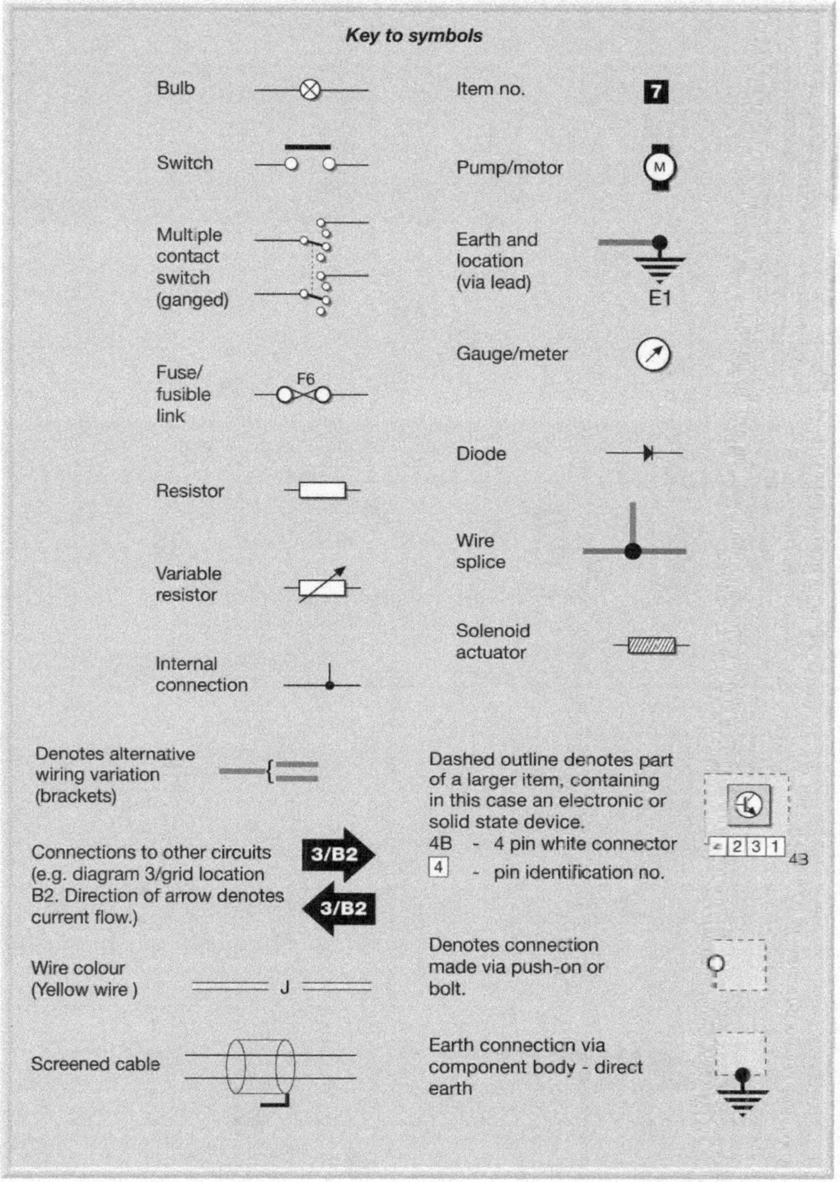
Key to symbols
Bulb
Switch
Multiple contact switch (ganged)
Fuse/ fusible link
F6
Resistor
Variable resistor
Internal connection
Item no.
7
Pump/motor
M
Earth and location (via lead)
E1
Gauge/meter
Diode
Wire splice
Solenoid actuator
Denotes alternative wiring variation (brackets)
Connections to other circuits (e.g. diagram 3/grid location B2. Direction of arrow denotes current flow.)
3/B2
3/B2
Wire colour (Yellow wire)
J
Screened cable
Dashed outline denotes part of a larger item, containing in this case an electronic or solid state device.
4B - 4 pin white connector
4 - pin identification no.
2 3 1
43
Denotes connection made via push-on or bolt.
Earth connection via component body - direct earth

Standard terminal identification (typical)

15	Ignition switch 'ignition' position
30	Battery +ve
31	Earth
50	Ignition switch 'start' position
85	Relay winding input
86	Relay winding earth
87	Relay output
87a	Relay output

Typical fuse box (bulkhead)

Fuse	*Rating*	*Circuit protected*
F1	10A	RH side and tail lights, switch illumination and sidelight warning light
F2	16A	LH and RH direction indicator, accessory supply relay, instrument illumination, heater lights, heater blower, warning lights and gauges
F3	20A	Stop lights, engine cooling fan, wash/wipe and horn
F4	16A	Cigar lighter, interior lighting, radio supply, clock and hazard warning
F5	10A	Rear foglight
F6	10A	Reversing light
F7	-	-
F8	10A	LH side and tail lights, and number plate light

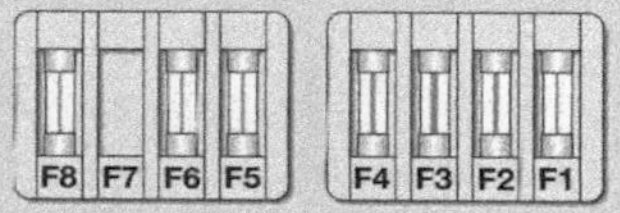
F8 F7 F6 F5
F4 F3 F2 F1

Additional fuse (near battery)

Fuse	*Rating*	*Circuit protected*
F9	30A	Engine cooling fan

Earth locations

E1	RH front engine bay, below headlight
E2	LH front engine bay, below headlight
E3	Behind dashboard on bulkhead
E4	By RH tail light, on load area floor
E5	By LH tail light, on load area floor

H31100

Diagram 1 : Information for wiring diagrams - typical

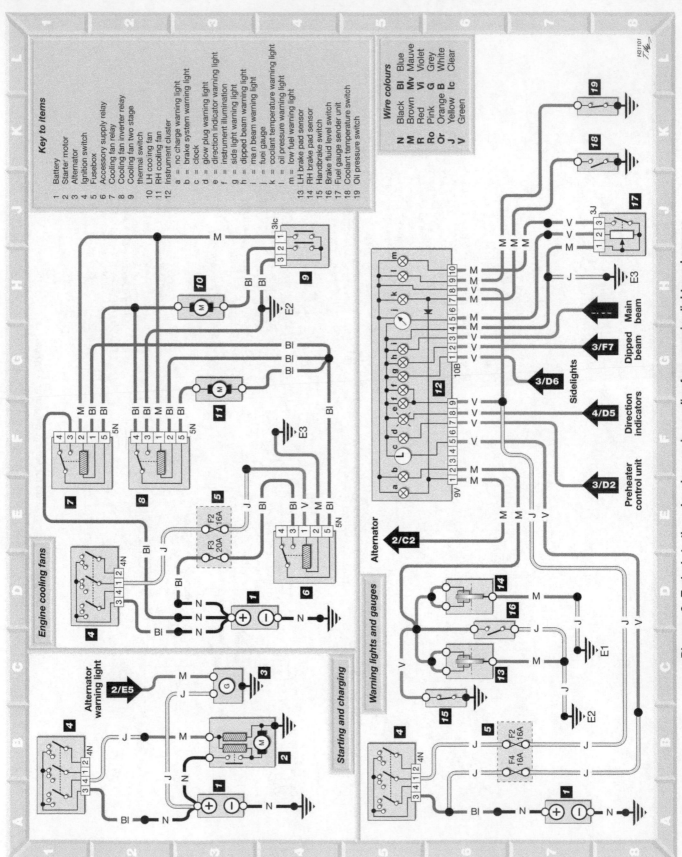

Key to items

1 Battery
2 Starter motor
3 Alternator
4 Ignition switch
5 Fusebox
6 Accessory supply relay
7 Cooling fan relay
8 Cooling fan inverter relay
9 Cooling fan two stage thermal switch
10 LH cooling fan
11 RH cooling fan
12 Instrument cluster
  a = nc charge warning light
  b = brake system warning light
  c = clock
  d = glow plug warning light
  e = direction indicator warning light
  f = instrument illumination
  g = side light warning light
  h = dipped beam warning light
  i = main beam warning light
  j = fuel gauge
  k = coolant temperature warning light
  l = oil pressure warning light
  m = low fuel warning light
13 LH brake pad sensor
14 RH brake pad sensor
15 Handbrake switch
16 Brake fluid level switch
17 Fuel gauge sender unit
18 Coolant temperature switch
19 Oil pressure switch

Wire colours

N	Black	Bl	Blue
R	Brown	Mv	Mauve
Ro	Red	Vi	Violet
Or	Pink	G	Grey
J	Orange	B	White
	Yellow	Ic	Clear
V	Green		

Engine cooling fans

Alternator warning light

Starting and charging

Warning lights and gauges

Alternator

Main beam

Dipped beam 3/F7

Sidelights 3/D6

Direction indicators 4/D5

Preheater control unit 3/D2

2/C2

2/E5

Diagram 2 : Typical starting, charging, engine cooling fans, warning lights and gauges

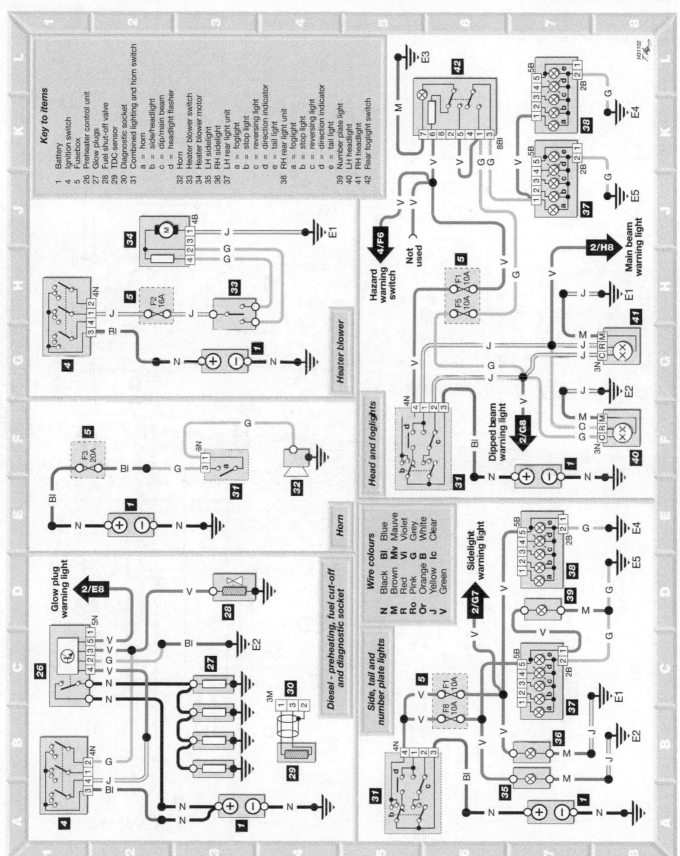

Key to items

1 Battery
4 Ignition switch
5 Fusebox
26 Preheater control unit
27 Glow plugs
28 Fuel shut-off valve
29 TDC sensor
30 Diagnostic socket
31 Combined lighting and horn switch
   a = horn
   b = side/headlight
   c = dip/main beam
   d = headlight flasher
32 Horn
33 Heater blower switch
34 Heater blower motor
35 LH sidelight
36 RH sidelight
37 LH rear light unit
   a = foglight
   b = stop light
   c = reversing light
   d = direction indicator
   e = tail light
38 RH rear light unit
   a = foglight
   b = stop light
   c = reversing light
   d = direction indicator
   e = tail light
39 Number plate light
40 LH headlight
41 RH headlight
42 Rear foglight switch

Wire colours

N	Black	Bl	Blue
M	Brown	Mv	Mauve
R	Red	Vi	Violet
Ro	Pink	G	Grey
Or	Orange	B	White
J	Yellow	Ic	Clear
V	Green		

Heater blower

Horn

Diesel - preheating, fuel cut-off and diagnostic socket

Head and foglights

Side, tail and number plate lights

Glow plug warning light

Hazard warning switch

Not used

Main beam warning light

Dipped beam warning light

Sidelight warning light

Diagram 3 : Typical Diesel engine electrics, horn, heater blower and exterior lighting

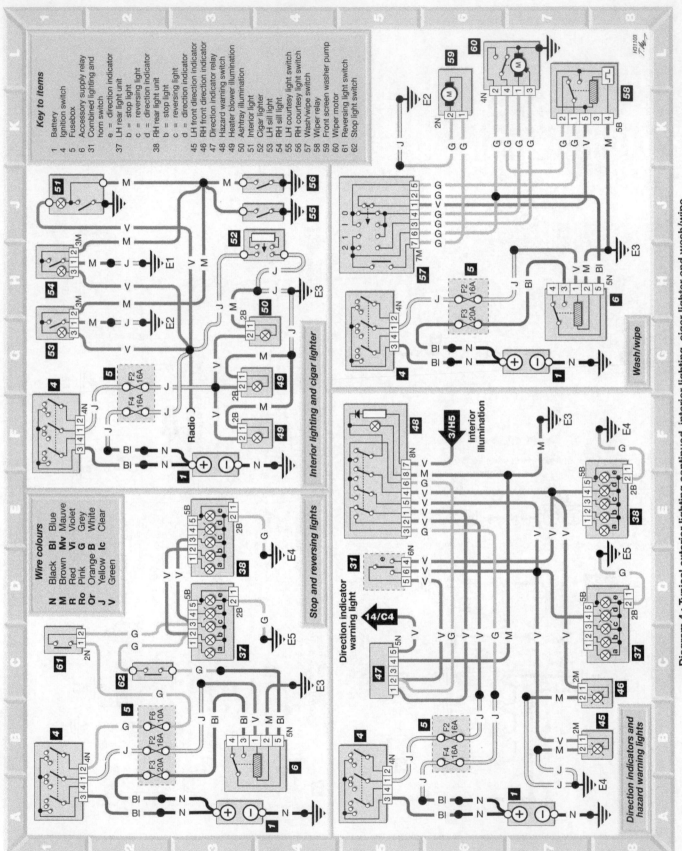

Key to items

1 Battery
4 Ignition switch
5 Fusebox
6 Accessory supply relay
31 Combined lighting and horn switch
37 LH rear light unit
   e = direction indicator
   b = stop light
   c = reversing light
   d = direction indicator
38 RH rear light unit
   b = stop light
   c = reversing light
   d = direction indicator
45 LH front direction indicator
46 RH front direction indicator
47 Direction indicator relay
48 Hazard warning switch
49 Heater blower illumination
50 Ashtray illumination
51 Interior light
52 Cigar lighter
53 LH sill light
54 RH sill light
55 LH courtesy light switch
56 RH courtesy light switch
57 Wash/wipe switch
58 Wiper relay
59 Front screen washer pump
60 Wiper motor
61 Reversing light switch
62 Stop light switch

Wire colours

N	Black	**Bl**	Blue
M	Brown	**Mv**	Mauve
R	Red	**Vi**	Violet
Ro	Pink	**G**	Grey
Or	Orange	**B**	White
J	Yellow	**Ic**	Clear
V	Green		

Interior lighting and cigar lighter

Stop and reversing lights

Wash/wipe

Interior illumination

Direction indicator warning light

Direction indicators and hazard warning lights

Diagram 4 : Typical exterior lighting continued, interior lighting, cigar lighter and wash/wipe

H31103

Dimensions and weights

Note: *All figures are approximate, and may vary according to model. Refer to manufacturer's data for exact figures.*

Dimensions

Overall length	3995 mm
Overall width:	
Including mirrors	1890 mm
Excluding mirrors	1636 mm
Overall height (unladen)	1805 mm
Wheelbase	2740 mm
Track width:	
Front:	
Early petrol models	1302 mm
Diesel models and later petrol models	1360 mm
Rear	1340 mm

Weights

	Petrol models, 1124 cc*	Diesel models
475 kg payload:		
Unladen weight	860 kg	Not applicable
Maximum gross vehicle weight	1335 kg	Not applicable
Maximum weight on front axle	700 kg	Not applicable
Maximum weight on rear axle	710 kg	Not applicable
Maximum gross train weight	1870 kg	Not applicable
Maximum towing weight (unbraked trailer)	430 kg	Not applicable
600 kg payload:		
Unladen weight	865 kg	945 kg
Maximum gross vehicle weight	1465 kg	1545 kg
Maximum weight on front axle	700 kg	740 kg
Maximum weight on rear axle	830 kg	830 kg
Maximum gross train weight	1990 kg	2165 kg
Maximum towing weight (unbraked trailer)	430 kg	470 kg
765 kg payload:		
Unladen weight	875 kg	945 kg
Maximum gross vehicle weight	1640 kg	1710 kg
Maximum weight on front axle	720 kg	765 kg
Maximum weight on rear axle	945 kg	990 kg
Maximum gross train weight	2165 kg	2330 kg
Maximum towing weight (unbraked trailer)	435 kg	470 kg

* No figures available for 954 cc petrol models. Refer to manufacturer's data.

All models

Maximum towing weight (braked trailer)	800 kg**
Maximum trailer nose weight	45 kg
Maximum roof rack load	80 kg

** Without exceeding maximum gross train weight.

Length (distance)

Inches (in)	x 25.4	= Millimetres (mm)	x 0.0394	= Inches (in)	
Feet (ft)	x 0.305	= Metres (m)	x 3.281	= Feet (ft)	
Miles	x 1.609	= Kilometres (km)	x 0.621	= Miles	

Volume (capacity)

Cubic inches (cu in; in³)	x 16.387	= Cubic centimetres (cc; cm³)	x 0.061	= Cubic inches (cu in; in³)	
Imperial pints (Imp pt)	x 0.568	= Litres (l)	x 1.76	= Imperial pints (Imp pt)	
Imperial quarts (Imp qt)	x 1.137	= Litres (l)	x 0.88	= Imperial quarts (Imp qt)	
Imperial quarts (Imp qt)	x 1.201	= US quarts (US qt)	x 0.833	= Imperial quarts (Imp qt)	
US quarts (US qt)	x 0.946	= Litres (l)	x 1.057	= US quarts (US qt)	
Imperial gallons (Imp gal)	x 4.546	= Litres (l)	x 0.22	= Imperial gallons (Imp gal)	
Imperial gallons (Imp gal)	x 1.201	= US gallons (US gal)	x 0.833	= Imperial gallons (Imp gal)	
US gallons (US gal)	x 3.785	= Litres (l)	x 0.264	= US gallons (US gal)	

Mass (weight)

Ounces (oz)	x 28.35	= Grams (g)	x 0.035	= Ounces (oz)	
Pounds (lb)	x 0.454	= Kilograms (kg)	x 2.205	= Pounds (lb)	

Force

Ounces-force (ozf; oz)	x 0.278	= Newtons (N)	x 3.6	= Ounces-force (ozf; oz)	
Pounds-force (lbf; lb)	x 4.448	= Newtons (N)	x 0.225	= Pounds-force (lbf; lb)	
Newtons (N)	x 0.1	= Kilograms-force (kgf; kg)	x 9.81	= Newtons (N)	

Pressure

Pounds-force per square inch (psi; lbf/in²; lb/in²)	x 0.070	= Kilograms-force per square centimetre (kgf/cm²; kg/cm²)	x 14.223	= Pounds-force per square inch (psi; lbf/in²; lb/in²)	
Pounds-force per square inch (psi; lbf/in²; lb/in²)	x 0.068	= Atmospheres (atm)	x 14.696	= Pounds-force per square inch (psi; lbf/in²; lb/in²)	
Pounds-force per square inch (psi; lbf/in²; lb/in²)	x 0.069	= Bars	x 14.5	= Pounds-force per square inch (psi; lbf/in²; lb/in²)	
Pounds-force per square inch (psi; lbf/in²; lb/in²)	x 6.895	= Kilopascals (kPa)	x 0.145	= Pounds-force per square inch (psi; lbf/in²; lb/in²)	
Kilopascals (kPa)	x 0.01	= Kilograms-force per square centimetre (kgf/cm²; kg/cm²)	x 98.1	= Kilopascals (kPa)	
Millibar (mbar)	x 100	= Pascals (Pa)	x 0.01	= Millibar (mbar)	
Millibar (mbar)	x 0.0145	= Pounds-force per square inch (psi; lbf/in²; lb/in²)	x 68.947	= Millibar (mbar)	
Millibar (mbar)	x 0.75	= Millimetres of mercury (mmHg)	x 1.333	= Millibar (mbar)	
Millibar (mbar)	x 0.401	= Inches of water (inH₂O)	x 2.491	= Millibar (mbar)	
Millimetres of mercury (mmHg)	x 0.535	= Inches of water (inH₂O)	x 1.868	= Millimetres of mercury (mmHg)	
Inches of water (inH₂O)	x 0.036	= Pounds-force per square inch (psi; lbf/in²; lb/in²)	x 27.68	= Inches of water (inH₂O)	

Torque (moment of force)

Pounds-force inches (lbf in; lb in)	x 1.152	= Kilograms-force centimetre (kgf cm; kg cm)	x 0.868	= Pounds-force inches (lbf in; lb in)	
Pounds-force inches (lbf in; lb in)	x 0.113	= Newton metres (Nm)	x 8.85	= Pounds-force inches (lbf in; lb in)	
Pounds-force inches (lbf in; lb in)	x 0.083	= Pounds-force feet (lbf ft; lb ft)	x 12	= Pounds-force inches (lbf in; lb in)	
Pounds-force feet (lbf ft; lb ft)	x 0.138	= Kilograms-force metres (kgf m; kg m)	x 7.233	= Pounds-force feet (lbf ft; lb ft)	
Pounds-force feet (lbf ft; lb ft)	x 1.356	= Newton metres (Nm)	x 0.738	= Pounds-force feet (lbf ft; lb ft)	
Newton metres (Nm)	x 0.102	= Kilograms-force metres (kgf m; kg m)	x 9.804	= Newton metres (Nm)	

Power

Horsepower (hp)	x 745.7	= Watts (W)	x 0.0013	= Horsepower (hp)	

Velocity (speed)

Miles per hour (miles/hr; mph)	x 1.609	= Kilometres per hour (km/hr; kph)	x 0.621	= Miles per hour (miles/hr; mph)	

Fuel consumption*

Miles per gallon, Imperial (mpg)	x 0.354	= Kilometres per litre (km/l)	x 2.825	= Miles per gallon, Imperial (mpg)	
Miles per gallon, US (mpg)	x 0.425	= Kilometres per litre (km/l)	x 2.352	= Miles per gallon, US (mpg)	

Temperature

Degrees Fahrenheit = (°C x 1.8) + 32

Degrees Celsius (Degrees Centigrade; °C) = (°F - 32) x 0.56

It is common practice to convert from miles per gallon (mpg) to litres/100 kilometres (l/100km), where mpg x l/100 km = 282

Spare parts are available from many sources, including maker's appointed garages, accessory shops, and motor factors. To be sure of obtaining the correct parts, it will sometimes be necessary to quote the vehicle identification number. If possible, it can also be useful to take the old parts along for positive identification. Items such as starter motors and alternators may be available under a service exchange scheme - any parts returned should always be clean.

Our advice regarding spare part sources is as follows.

Officially-appointed garages

This is the best source of parts which are peculiar to your vehicle, and which are not otherwise generally available (e.g. badges, interior trim, certain body panels, etc). It is also the only place at which you should buy parts if the vehicle is still under warranty.

Accessory shops

These are very good places to buy materials and components needed for the maintenance of your vehicle (oil, air and fuel filters, spark plugs, light bulbs, drivebelts, oils and greases, brake pads, touch-up paint, etc). Components of this nature sold by a reputable shop are of the same standard as those used by the vehicle manufacturer.

Besides components, these shops also sell tools and general accessories, usually have convenient opening hours, charge lower prices, and can often be found not far from home. Some accessory shops have parts counters where the components needed for almost any repair job can be purchased or ordered.

Motor factors

Good factors will stock all the more important components which wear out comparatively quickly, and can sometimes supply individual components needed for the overhaul of a larger assembly (e.g. brake seals and hydraulic parts, bearing shells, pistons, valves, alternator brushes). They may also handle work such as cylinder block reboring, crankshaft regrinding and balancing, etc.

Tyre and exhaust specialists

These outlets may be independent, or members of a local or national chain. They frequently offer competitive prices when compared with a main dealer or local garage, but it will pay to obtain several quotes before making a decision. When researching prices, also ask what "extras" may be added - for instance, fitting a new valve and balancing the wheel are both commonly charged on top of the price of a new tyre.

Other sources

Beware of parts or materials obtained from market stalls, car boot sales or similar outlets. Such items are not invariably sub-standard, but there is little chance of compensation if they do prove unsatisfactory. In the case of safety-critical components such as brake pads, there is the risk not only of financial loss but also of an accident causing injury or death.

Second-hand components or assemblies obtained from a car breaker can be a good buy in some circumstances, but this sort of purchase is best made by the experienced DIY mechanic.

Vehicle identification

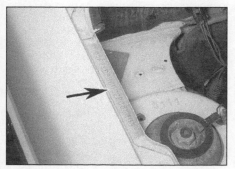

VIN plate (1) and replacement parts number (2)

Modifications are a continuing and unpublicised process in vehicle manufacture, quite apart from major model changes. Spare parts manuals and lists are compiled upon a numerical basis, the individual vehicle identification numbers being essential to correct identification of the component concerned.

When ordering spare parts, always give as much information as possible. Quote the vehicle model, year of manufacture, VIN number, engine number and replacement parts number as appropriate.

The *Vehicle Identification Number (VIN)* plate is located in the engine compartment, behind the right-hand suspension strut. The plate carries the VIN and vehicle weight information. The VIN number is also stamped in the gutter, where the bonnet fits over the right-hand wing (see illustrations).

The *replacement parts number* is associated with the production date of the vehicle and is painted on the right-hand suspension strut turret (see illustration, previous paragraph).

The *engine number* is stamped on a plate on the front of the cylinder block. On petrol engines, the plate is at the left-hand end of the front face, and the first three characters give the engine code, for example 'H1A' (see illustrations).

Other *identification numbers* or codes are stamped on major items such as the gearbox, etc.

VIN number (arrowed) in gutter on right-hand wing

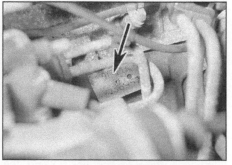

Engine number on the plate (arrowed) on the front of the cylinder block - petrol engine . . .

. . . and diesel engine

Whenever servicing, repair or overhaul work is carried out on the car or its components, observe the following procedures and instructions. This will assist in carrying out the operation efficiently and to a professional standard of workmanship.

Joint mating faces and gaskets

When separating components at their mating faces, never insert screwdrivers or similar implements into the joint between the faces in order to prise them apart. This can cause severe damage which results in oil leaks, coolant leaks, etc upon reassembly. Separation is usually achieved by tapping along the joint with a soft-faced hammer in order to break the seal. However, note that this method may not be suitable where dowels are used for component location.

Where a gasket is used between the mating faces of two components, a new one must be fitted on reassembly; fit it dry unless otherwise stated in the repair procedure. Make sure that the mating faces are clean and dry, with all traces of old gasket removed. When cleaning a joint face, use a tool which is unlikely to score or damage the face, and remove any burrs or nicks with an oilstone or fine file.

Make sure that tapped holes are cleaned with a pipe cleaner, and keep them free of jointing compound, if this is being used, unless specifically instructed otherwise.

Ensure that all orifices, channels or pipes are clear, and blow through them, preferably using compressed air.

Oil seals

Oil seals can be removed by levering them out with a wide flat-bladed screwdriver or similar implement. Alternatively, a number of self-tapping screws may be screwed into the seal, and these used as a purchase for pliers or some similar device in order to pull the seal free.

Whenever an oil seal is removed from its working location, either individually or as part of an assembly, it should be renewed.

The very fine sealing lip of the seal is easily damaged, and will not seal if the surface it contacts is not completely clean and free from scratches, nicks or grooves. If the original sealing surface of the component cannot be restored, and the manufacturer has not made provision for slight relocation of the seal relative to the sealing surface, the component should be renewed.

Protect the lips of the seal from any surface which may damage them in the course of fitting. Use tape or a conical sleeve where possible. Lubricate the seal lips with oil before fitting and, on dual-lipped seals, fill the space between the lips with grease.

Unless otherwise stated, oil seals must be fitted with their sealing lips toward the lubricant to be sealed.

Use a tubular drift or block of wood of the appropriate size to install the seal and, if the seal housing is shouldered, drive the seal down to the shoulder. If the seal housing is unshouldered, the seal should be fitted with its face flush with the housing top face (unless otherwise instructed).

Screw threads and fastenings

Seized nuts, bolts and screws are quite a common occurrence where corrosion has set in, and the use of penetrating oil or releasing fluid will often overcome this problem if the offending item is soaked for a while before attempting to release it. The use of an impact driver may also provide a means of releasing such stubborn fastening devices, when used in conjunction with the appropriate screwdriver bit or socket. If none of these methods works, it may be necessary to resort to the careful application of heat, or the use of a hacksaw or nut splitter device.

Studs are usually removed by locking two nuts together on the threaded part, and then using a spanner on the lower nut to unscrew the stud. Studs or bolts which have broken off below the surface of the component in which they are mounted can sometimes be removed using a stud extractor. Always ensure that a blind tapped hole is completely free from oil, grease, water or other fluid before installing the bolt or stud. Failure to do this could cause the housing to crack due to the hydraulic action of the bolt or stud as it is screwed in.

When tightening a castellated nut to accept a split pin, tighten the nut to the specified torque, where applicable, and then tighten further to the next split pin hole. Never slacken the nut to align the split pin hole, unless stated in the repair procedure.

When checking or retightening a nut or bolt to a specified torque setting, slacken the nut or bolt by a quarter of a turn, and then retighten to the specified setting. However, this should not be attempted where angular tightening has been used.

For some screw fastenings, notably cylinder head bolts or nuts, torque wrench settings are no longer specified for the latter stages of tightening, "angle-tightening" being called up instead. Typically, a fairly low torque wrench setting will be applied to the bolts/nuts in the correct sequence, followed by one or more stages of tightening through specified angles.

Locknuts, locktabs and washers

Any fastening which will rotate against a component or housing during tightening should always have a washer between it and the relevant component or housing.

Spring or split washers should always be renewed when they are used to lock a critical component such as a big-end bearing retaining bolt or nut. Locktabs which are folded over to retain a nut or bolt should always be renewed.

Self-locking nuts can be re-used in non-critical areas, providing resistance can be felt when the locking portion passes over the bolt or stud thread. However, it should be noted that self-locking stiffnuts tend to lose their effectiveness after long periods of use, and should then be renewed as a matter of course.

Split pins must always be replaced with new ones of the correct size for the hole.

When thread-locking compound is found on the threads of a fastener which is to be re-used, it should be cleaned off with a wire brush and solvent, and fresh compound applied on reassembly.

Special tools

Some repair procedures in this manual entail the use of special tools such as a press, two or three-legged pullers, spring compressors, etc. Wherever possible, suitable readily-available alternatives to the manufacturer's special tools are described, and are shown in use. In some instances, where no alternative is possible, it has been necessary to resort to the use of a manufacturer's tool, and this has been done for reasons of safety as well as the efficient completion of the repair operation. Unless you are highly-skilled and have a thorough understanding of the procedures described, never attempt to bypass the use of any special tool when the procedure described specifies its use. Not only is there a very great risk of personal injury, but expensive damage could be caused to the components involved.

Environmental considerations

When disposing of used engine oil, brake fluid, antifreeze, etc, give due consideration to any detrimental environmental effects. Do not, for instance, pour any of the above liquids down drains into the general sewage system, or onto the ground to soak away. Many local council refuse tips provide a facility for waste oil disposal, as do some garages. If none of these facilities are available, consult your local Environmental Health Department, or the National Rivers Authority, for further advice.

With the universal tightening-up of legislation regarding the emission of environmentally-harmful substances from motor vehicles, most vehicles have tamperproof devices fitted to the main adjustment points of the fuel system. These devices are primarily designed to prevent unqualified persons from adjusting the fuel/air mixture, with the chance of a consequent increase in toxic emissions. If such devices are found during servicing or overhaul, they should, wherever possible, be renewed or refitted in accordance with the manufacturer's requirements or current legislation.

OIL CARE

FOLLOW THE CODE

OIL BANK LINE
0800 66 33 66

Note: It is antisocial and illegal to dump oil down the drain. To find the location of your local oil recycling bank, call this number free.

The jack supplied with the vehicle tool kit should only be used for changing the roadwheels - see *"Wheel changing"* at the front of this manual. When carrying out any other kind of work, raise the vehicle using a hydraulic (or "trolley") jack, and always supplement the jack with axle stands positioned under the vehicle jacking points. Use a shaped block of wood on the jack head or axle stand if necessary, to prevent denting or scratching the underside of the sill. The jacking points are between two horizontal flanges in the lower sill - there are two jacking points on each side of the vehicle, one behind the front wheel, and one in front of the rear wheel **(see illustration)**.

To raise the front of the vehicle, locate a beam transversely under the front subframe, near the inner ends of the track control arms. Otherwise, raise one side at a time using axle stands positioned behind each front wheel. **Do not** jack the vehicle under the sump, or any of the steering or suspension components.

To raise the rear of the vehicle, position the jack head centrally under the rear crossmember, making sure that it is clear of the brake pressure limiter and any hydraulic hoses. Otherwise, raise one side at a time using axle stands positioned in front of each rear wheel. **Do not** attempt to raise the vehicle with the jack positioned underneath the spare wheel, as the vehicle floor will almost certainly be damaged.

Never work under, around, or near a raised vehicle, unless it is adequately supported in at least two places.

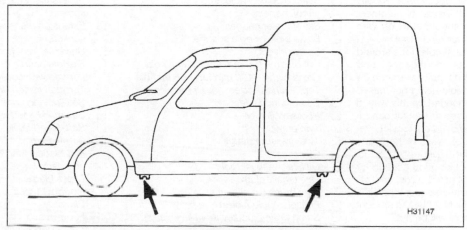

Jacking points (arrowed)

Radio/cassette unit anti-theft system - precaution

The radio/cassette unit fitted by Citroën may be equipped with a built-in security code, to deter thieves. If the power source to the unit is cut, the anti-theft system will activate. Even if the power source is immediately reconnected, the unit will not function until the correct security code has been entered. Therefore, if you do not know the correct code, DO NOT disconnect the battery negative lead, or remove the radio/cassette unit from the vehicle. The exact procedure for reprogramming a unit which has been disconnected from its power supply varies from model to model. Consult the radio booklet which should have been supplied with the vehicle for specific details. On production of proof of ownership, a Citroën dealer (or possibly a vehicle audio entertainment specialist) should be able to enter the code.

Introduction

A selection of good tools is a fundamental requirement for anyone contemplating the maintenance and repair of a motor vehicle. For the owner who does not possess any, their purchase will prove a considerable expense, offsetting some of the savings made by doing-it-yourself. However, provided that the tools purchased meet the relevant national safety standards and are of good quality, they will last for many years and prove an extremely worthwhile investment.

To help the average owner to decide which tools are needed to carry out the various tasks detailed in this manual, we have compiled three lists of tools under the following headings: *Maintenance and minor repair, Repair and overhaul*, and *Special*. Newcomers to practical mechanics should start off with the *Maintenance and minor repair* tool kit, and confine themselves to the simpler jobs around the vehicle. Then, as confidence and experience grow, more difficult tasks can be undertaken, with extra tools being purchased as, and when, they are needed. In this way, a *Maintenance and minor repair* tool kit can be built up into a *Repair and overhaul* tool kit over a considerable period of time, without any major cash outlays. The experienced do-it-yourselfer will have a tool kit good enough for most repair and overhaul procedures, and will add tools from the *Special* category when it is felt that the expense is justified by the amount of use to which these tools will be put.

Maintenance and minor repair tool kit

The tools given in this list should be considered as a minimum requirement if routine maintenance, servicing and minor repair operations are to be undertaken. We recommend the purchase of combination spanners (ring one end, open-ended the other); although more expensive than open-ended ones, they do give the advantages of both types of spanner.

☐ *Combination spanners:*
 Metric - 8 to 19 mm inclusive
☐ *Adjustable spanner - 35 mm jaw (approx.)*
☐ *Spark plug spanner (with rubber insert) - petrol models*
☐ *Spark plug gap adjustment tool - petrol models*
☐ *Set of feeler gauges*
☐ *Brake bleed nipple spanner*
☐ *Screwdrivers:*
 Flat blade - 100 mm long x 6 mm dia
 Cross blade - 100 mm long x 6 mm dia
 Torx - various sizes (not all vehicles)
☐ *Combination pliers*
☐ *Hacksaw (junior)*
☐ *Tyre pump*
☐ *Tyre pressure gauge*
☐ *Oil can*
☐ *Oil filter removal tool*
☐ *Fine emery cloth*
☐ *Wire brush (small)*
☐ *Funnel (medium size)*
☐ *Sump drain plug key (not all vehicles)*

Repair and overhaul tool kit

These tools are virtually essential for anyone undertaking any major repairs to a motor vehicle, and are additional to those given in the *Maintenance and minor repair* list. Included in this list is a comprehensive set of sockets. Although these are expensive, they will be found invaluable as they are so versatile - particularly if various drives are included in the set. We recommend the half-inch square-drive type, as this can be used with most proprietary torque wrenches.

The tools in this list will sometimes need to be supplemented by tools from the *Special* list:

☐ *Sockets (or box spanners) to cover range in previous list (including Torx sockets)*
☐ *Reversible ratchet drive (for use with sockets)*
☐ *Extension piece, 250 mm (for use with sockets)*
☐ *Universal joint (for use with sockets)*
☐ *Flexible handle or sliding T "breaker bar" (for use with sockets)*
☐ *Torque wrench (for use with sockets)*
☐ *Self-locking grips*
☐ *Ball pein hammer*
☐ *Soft-faced mallet (plastic or rubber)*
☐ *Screwdrivers:*
 Flat blade - long & sturdy, short (chubby), and narrow (electrician's) types
 Cross blade – long & sturdy, and short (chubby) types
☐ *Pliers:*
 Long-nosed
 Side cutters (electrician's)
 Circlip (internal and external)
☐ *Cold chisel - 25 mm*
☐ *Scriber*
☐ *Scraper*
☐ *Centre-punch*
☐ *Pin punch*
☐ *Hacksaw*
☐ *Brake hose clamp*
☐ *Brake/clutch bleeding kit*
☐ *Selection of twist drills*
☐ *Steel rule/straight-edge*
☐ *Allen keys (inc. splined/Torx type)*
☐ *Selection of files*
☐ *Wire brush*
☐ *Axle stands*
☐ *Jack (strong trolley or hydraulic type)*
☐ *Light with extension lead*
☐ *Universal electrical multi-meter*

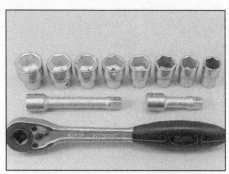

Sockets and reversible ratchet drive

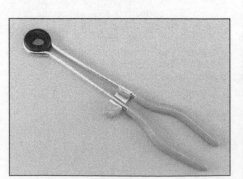

Brake bleeding kit

Torx key, socket and bit

Hose clamp

Angular-tightening gauge

Special tools

The tools in this list are those which are not used regularly, are expensive to buy, or which need to be used in accordance with their manufacturers' instructions. Unless relatively difficult mechanical jobs are undertaken frequently, it will not be economic to buy many of these tools. Where this is the case, you could consider clubbing together with friends (or joining a motorists' club) to make a joint purchase, or borrowing the tools against a deposit from a local garage or tool hire specialist. It is worth noting that many of the larger DIY superstores now carry a large range of special tools for hire at modest rates.

The following list contains only those tools and instruments freely available to the public, and not those special tools produced by the vehicle manufacturer specifically for its dealer network. You will find occasional references to these manufacturers' special tools in the text of this manual. Generally, an alternative method of doing the job without the vehicle manufacturers' special tool is given. However, sometimes there is no alternative to using them. Where this is the case and the relevant tool cannot be bought or borrowed, you will have to entrust the work to a dealer.

- ☐ Angular-tightening gauge
- ☐ Valve spring compressor
- ☐ Valve grinding tool
- ☐ Piston ring compressor
- ☐ Piston ring removal/installation tool
- ☐ Cylinder bore hone
- ☐ Balljoint separator
- ☐ Coil spring compressors (where applicable)
- ☐ Two/three-legged hub and bearing puller
- ☐ Impact screwdriver
- ☐ Micrometer and/or vernier calipers
- ☐ Dial gauge
- ☐ Stroboscopic timing light
- ☐ Dwell angle meter/tachometer
- ☐ Fault code reader
- ☐ Cylinder compression gauge
- ☐ Hand-operated vacuum pump and gauge
- ☐ Clutch plate alignment set
- ☐ Brake shoe steady spring cup removal tool
- ☐ Bush and bearing removal/installation set
- ☐ Stud extractors
- ☐ Tap and die set
- ☐ Lifting tackle
- ☐ Trolley jack

Buying tools

Reputable motor accessory shops and superstores often offer excellent quality tools at discount prices, so it pays to shop around.

Remember, you don't have to buy the most expensive items on the shelf, but it is always advisable to steer clear of the very cheap tools. Beware of 'bargains' offered on market stalls or at car boot sales. There are plenty of good tools around at reasonable prices, but always aim to purchase items which meet the relevant national safety standards. If in doubt, ask the proprietor or manager of the shop for advice before making a purchase.

Care and maintenance of tools

Having purchased a reasonable tool kit, it is necessary to keep the tools in a clean and serviceable condition. After use, always wipe off any dirt, grease and metal particles using a clean, dry cloth, before putting the tools away. Never leave them lying around after they have been used. A simple tool rack on the garage or workshop wall for items such as screwdrivers and pliers is a good idea. Store all normal spanners and sockets in a metal box. Any measuring instruments, gauges, meters, etc, must be carefully stored where they cannot be damaged or become rusty.

Take a little care when tools are used. Hammer heads inevitably become marked, and screwdrivers lose the keen edge on their blades from time to time. A little timely attention with emery cloth or a file will soon restore items like this to a good finish.

Working facilities

Not to be forgotten when discussing tools is the workshop itself. If anything more than routine maintenance is to be carried out, a suitable working area becomes essential.

It is appreciated that many an owner-mechanic is forced by circumstances to remove an engine or similar item without the benefit of a garage or workshop. Having done this, any repairs should always be done under the cover of a roof.

Wherever possible, any dismantling should be done on a clean, flat workbench or table at a suitable working height.

Any workbench needs a vice; one with a jaw opening of 100 mm is suitable for most jobs. As mentioned previously, some clean dry storage space is also required for tools, as well as for any lubricants, cleaning fluids, touch-up paints etc, which become necessary.

Another item which may be required, and which has a much more general usage, is an electric drill with a chuck capacity of at least 8 mm. This, together with a good range of twist drills, is virtually essential for fitting accessories.

Last, but not least, always keep a supply of old newspapers and clean, lint-free rags available, and try to keep any working area as clean as possible.

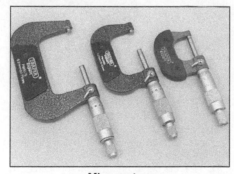

Micrometers

Dial test indicator ("dial gauge")

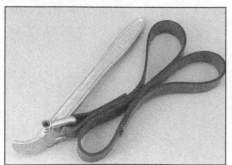

Strap wrench

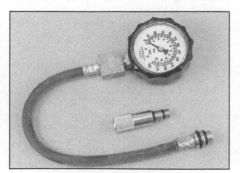

Compression tester

Fault code reader

This is a guide to getting your vehicle through the MOT test. Obviously it will not be possible to examine the vehicle to the same standard as the professional MOT tester. However, working through the following checks will enable you to identify any problem areas before submitting the vehicle for the test.

Where a testable component is in borderline condition, the tester has discretion in deciding whether to pass or fail it. The basis of such discretion is whether the tester would be happy for a close relative or friend to use the vehicle with the component in that condition. If the vehicle presented is clean and evidently well cared for, the tester may be more inclined to pass a borderline component than if the vehicle is scruffy and apparently neglected.

It has only been possible to summarise the test requirements here, based on the regulations in force at the time of printing. Test standards are becoming increasingly stringent, although there are some exemptions for older vehicles. For full details obtain a copy of the Haynes publication Pass the MOT! (available from stockists of Haynes manuals).

An assistant will be needed to help carry out some of these checks.

The checks have been sub-divided into four categories, as follows:

1 Checks carried out **FROM THE DRIVER'S SEAT**

2 Checks carried out **WITH THE VEHICLE ON THE GROUND**

3 Checks carried out **WITH THE VEHICLE RAISED AND THE WHEELS FREE TO TURN**

4 Checks carried out on **YOUR VEHICLE'S EXHAUST EMISSION SYSTEM**

---

1 Checks carried out **FROM THE DRIVER'S SEAT**

Handbrake

☐ Test the operation of the handbrake. Excessive travel (too many clicks) indicates incorrect brake or cable adjustment.
☐ Check that the handbrake cannot be released by tapping the lever sideways. Check the security of the lever mountings.

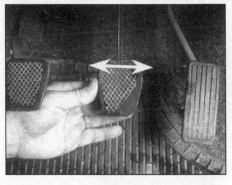

Footbrake

☐ Depress the brake pedal and check that it does not creep down to the floor, indicating a master cylinder fault. Release the pedal, wait a few seconds, then depress it again. If the pedal travels nearly to the floor before firm resistance is felt, brake adjustment or repair is necessary. If the pedal feels spongy, there is air in the hydraulic system which must be removed by bleeding.

☐ Check that the brake pedal is secure and in good condition. Check also for signs of fluid leaks on the pedal, floor or carpets, which would indicate failed seals in the brake master cylinder.
☐ Check the servo unit (when applicable) by operating the brake pedal several times, then keeping the pedal depressed and starting the engine. As the engine starts, the pedal will move down slightly. If not, the vacuum hose or the servo itself may be faulty.

Steering wheel and column

☐ Examine the steering wheel for fractures or looseness of the hub, spokes or rim.
☐ Move the steering wheel from side to side and then up and down. Check that the steering wheel is not loose on the column, indicating wear or a loose retaining nut. Continue moving the steering wheel as before, but also turn it slightly from left to right.
☐ Check that the steering wheel is not loose on the column, and that there is no abnormal

movement of the steering wheel, indicating wear in the column support bearings or couplings.

Windscreen and mirrors

☐ The windscreen must be free of cracks or other significant damage within the driver's field of view. (Small stone chips are acceptable.) Rear view mirrors must be secure, intact, and capable of being adjusted.

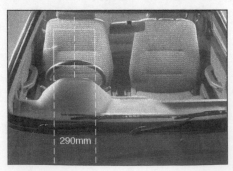

290mm

Seat belts and seats

Note: *The following checks are applicable to all seat belts, front and rear.*

☐ Examine the webbing of all the belts (including rear belts if fitted) for cuts, serious fraying or deterioration. Fasten and unfasten each belt to check the buckles. If applicable, check the retracting mechanism. Check the security of all seat belt mountings accessible from inside the vehicle.

☐ The front seats themselves must be securely attached and the backrests must lock in the upright position.

Doors

☐ Both front doors must be able to be opened and closed from outside and inside, and must latch securely when closed.

2 Checks carried out WITH THE VEHICLE ON THE GROUND

Vehicle identification

☐ Number plates must be in good condition, secure and legible, with letters and numbers correctly spaced – spacing at (A) should be twice that at (B).

☐ The VIN plate and/or homologation plate must be legible.

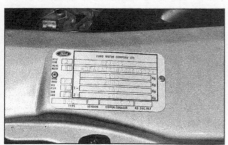

Electrical equipment

☐ Switch on the ignition and check the operation of the horn.

☐ Check the windscreen washers and wipers, examining the wiper blades; renew damaged or perished blades. Also check the operation of the stop-lights.

☐ Check the operation of the sidelights and number plate lights. The lenses and reflectors must be secure, clean and undamaged.

☐ Check the operation and alignment of the headlights. The headlight reflectors must not be tarnished and the lenses must be undamaged.

☐ Switch on the ignition and check the operation of the direction indicators (including the instrument panel tell-tale) and the hazard warning lights. Operation of the sidelights and stop-lights must not affect the indicators - if it does, the cause is usually a bad earth at the rear light cluster.

☐ Check the operation of the rear foglight(s), including the warning light on the instrument panel or in the switch.

Footbrake

☐ Examine the master cylinder, brake pipes and servo unit for leaks, loose mountings, corrosion or other damage.

☐ The fluid reservoir must be secure and the fluid level must be between the upper (**A**) and lower (**B**) markings.

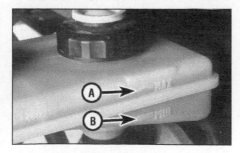

☐ Inspect both front brake flexible hoses for cracks or deterioration of the rubber. Turn the steering from lock to lock, and ensure that the hoses do not contact the wheel, tyre, or any part of the steering or suspension mechanism. With the brake pedal firmly depressed, check the hoses for bulges or leaks under pressure.

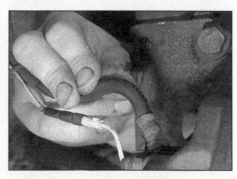

Steering and suspension

☐ Have your assistant turn the steering wheel from side to side slightly, up to the point where the steering gear just begins to transmit this movement to the roadwheels. Check for excessive free play between the steering wheel and the steering gear, indicating wear or insecurity of the steering column joints, the column-to-steering gear coupling, or the steering gear itself.

☐ Have your assistant turn the steering wheel more vigorously in each direction, so that the roadwheels just begin to turn. As this is done, examine all the steering joints, linkages, fittings and attachments. Renew any component that shows signs of wear or damage. On vehicles with power steering, check the security and condition of the steering pump, drivebelt and hoses.

☐ Check that the vehicle is standing level, and at approximately the correct ride height.

Shock absorbers

☐ Depress each corner of the vehicle in turn, then release it. The vehicle should rise and then settle in its normal position. If the vehicle continues to rise and fall, the shock absorber is defective. A shock absorber which has seized will also cause the vehicle to fail.

Exhaust system

☐ Start the engine. With your assistant holding a rag over the tailpipe, check the entire system for leaks. Repair or renew leaking sections.

3 Checks carried out
WITH THE VEHICLE RAISED AND THE WHEELS FREE TO TURN

Jack up the front and rear of the vehicle, and securely support it on axle stands. Position the stands clear of the suspension assemblies. Ensure that the wheels are clear of the ground and that the steering can be turned from lock to lock.

Steering mechanism

☐ Have your assistant turn the steering from lock to lock. Check that the steering turns smoothly, and that no part of the steering mechanism, including a wheel or tyre, fouls any brake hose or pipe or any part of the body structure.
☐ Examine the steering rack rubber gaiters for damage or insecurity of the retaining clips. If power steering is fitted, check for signs of damage or leakage of the fluid hoses, pipes or connections. Also check for excessive stiffness or binding of the steering, a missing split pin or locking device, or severe corrosion of the body structure within 30 cm of any steering component attachment point.

Front and rear suspension and wheel bearings

☐ Starting at the front right-hand side, grasp the roadwheel at the 3 o'clock and 9 o'clock positions and shake it vigorously. Check for free play or insecurity at the wheel bearings, suspension balljoints, or suspension mountings, pivots and attachments.
☐ Now grasp the wheel at the 12 o'clock and 6 o'clock positions and repeat the previous inspection. Spin the wheel, and check for roughness or tightness of the front wheel bearing.

☐ If excess free play is suspected at a component pivot point, this can be confirmed by using a large screwdriver or similar tool and levering between the mounting and the component attachment. This will confirm whether the wear is in the pivot bush, its retaining bolt, or in the mounting itself (the bolt holes can often become elongated).

☐ Carry out all the above checks at the other front wheel, and then at both rear wheels.

Springs and shock absorbers

☐ Examine the suspension struts (when applicable) for serious fluid leakage, corrosion, or damage to the casing. Also check the security of the mounting points.
☐ If coil springs are fitted, check that the spring ends locate in their seats, and that the spring is not corroded, cracked or broken.
☐ If leaf springs are fitted, check that all leaves are intact, that the axle is securely attached to each spring, and that there is no deterioration of the spring eye mountings, bushes, and shackles.

☐ The same general checks apply to vehicles fitted with other suspension types, such as torsion bars, hydraulic displacer units, etc. Ensure that all mountings and attachments are secure, that there are no signs of excessive wear, corrosion or damage, and (on hydraulic types) that there are no fluid leaks or damaged pipes.
☐ Inspect the shock absorbers for signs of serious fluid leakage. Check for wear of the mounting bushes or attachments, or damage to the body of the unit.

Driveshafts (fwd vehicles only)

☐ Rotate each front wheel in turn and inspect the constant velocity joint gaiters for splits or damage. Also check that each driveshaft is straight and undamaged.

Braking system

☐ If possible without dismantling, check brake pad wear and disc condition. Ensure that the friction lining material has not worn excessively, (A) and that the discs are not fractured, pitted, scored or badly worn (B).

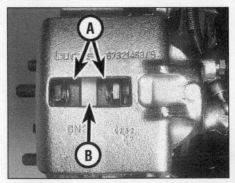

☐ Examine all the rigid brake pipes underneath the vehicle, and the flexible hose(s) at the rear. Look for corrosion, chafing or insecurity of the pipes, and for signs of bulging under pressure, chafing, splits or deterioration of the flexible hoses.
☐ Look for signs of fluid leaks at the brake calipers or on the brake backplates. Repair or renew leaking components.
☐ Slowly spin each wheel, while your assistant depresses and releases the footbrake. Ensure that each brake is operating and does not bind when the pedal is released.

□ Examine the handbrake mechanism, checking for frayed or broken cables, excessive corrosion, or wear or insecurity of the linkage. Check that the mechanism works on each relevant wheel, and releases fully, without binding.

□ It is not possible to test brake efficiency without special equipment, but a road test can be carried out later to check that the vehicle pulls up in a straight line.

Fuel and exhaust systems

□ Inspect the fuel tank (including the filler cap), fuel pipes, hoses and unions. All components must be secure and free from leaks.

□ Examine the exhaust system over its entire length, checking for any damaged, broken or missing mountings, security of the retaining clamps and rust or corrosion.

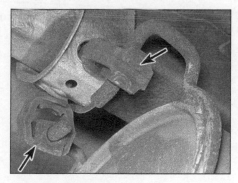

Wheels and tyres

□ Examine the sidewalls and tread area of each tyre in turn. Check for cuts, tears, lumps, bulges, separation of the tread, and exposure of the ply or cord due to wear or damage. Check that the tyre bead is correctly seated on the wheel rim, that the valve is sound and

properly seated, and that the wheel is not distorted or damaged.

□ Check that the tyres are of the correct size for the vehicle, that they are of the same size and type on each axle, and that the pressures are correct.

□ Check the tyre tread depth. The legal minimum at the time of writing is 1.6 mm over at least three-quarters of the tread width. Abnormal tread wear may indicate incorrect front wheel alignment.

Body corrosion

□ Check the condition of the entire vehicle structure for signs of corrosion in load-bearing areas. (These include chassis box sections, side sills, cross-members, pillars, and all suspension, steering, braking system and seat belt mountings and anchorages.) Any corrosion which has seriously reduced the thickness of a load-bearing area is likely to cause the vehicle to fail. In this case professional repairs are likely to be needed.

□ Damage or corrosion which causes sharp or otherwise dangerous edges to be exposed will also cause the vehicle to fail.

4 Checks carried out on YOUR VEHICLE'S EXHAUST EMISSION SYSTEM

Petrol models

□ Have the engine at normal operating temperature, and make sure that it is in good tune (ignition system in good order, air filter element clean, etc).

□ Before any measurements are carried out, raise the engine speed to around 2500 rpm, and hold it at this speed for 20 seconds.

Allow the engine speed to return to idle, and watch for smoke emissions from the exhaust tailpipe. If the idle speed is obviously much too high, or if dense blue or clearly-visible black smoke comes from the tailpipe for more than 5 seconds, the vehicle will fail. As a rule of thumb, blue smoke signifies oil being burnt (engine wear) while black smoke signifies unburnt fuel (dirty air cleaner element, or other carburettor or fuel system fault).

□ An exhaust gas analyser capable of measuring carbon monoxide (CO) and hydrocarbons (HC) is now needed. If such an instrument cannot be hired or borrowed, a local garage may agree to perform the check for a small fee.

CO emissions (mixture)

□ At the time of writing, the maximum CO level at idle is 3.5% for vehicles first used after August 1986 and 4.5% for older vehicles. From January 1996 a much tighter limit (around 0.5%) applies to catalyst-equipped vehicles first used from August 1992. If the CO level cannot be reduced far enough to pass the test (and the fuel and ignition systems are otherwise in good condition) then the carburettor is badly worn, or there is some problem in the fuel injection system or catalytic converter (as applicable).

HC emissions

□ With the CO emissions within limits, HC emissions must be no more than 1200 ppm (parts per million). If the vehicle fails this test at idle, it can be re-tested at around 2000 rpm; if the HC level is then 1200 ppm or less, this counts as a pass.

□ Excessive HC emissions can be caused by oil being burnt, but they are more likely to be due to unburnt fuel.

Diesel models

□ The only emission test applicable to Diesel engines is the measuring of exhaust smoke density. The test involves accelerating the engine several times to its maximum unloaded speed.

Note: *It is of the utmost importance that the engine timing belt is in good condition before the test is carried out.*

□ Excessive smoke can be caused by a dirty air cleaner element. Otherwise, professional advice may be needed to find the cause.

Engine

- ☐ Engine fails to rotate when attempting to start
- ☐ Engine rotates, but will not start
- ☐ Engine difficult to start when cold
- ☐ Engine difficult to start when hot
- ☐ Starter motor noisy or excessively-rough in engagement
- ☐ Engine starts, but stops immediately
- ☐ Engine idles erratically
- ☐ Engine misfires at idle speed
- ☐ Engine misfires throughout the driving speed range
- ☐ Engine hesitates on acceleration
- ☐ Engine stalls
- ☐ Engine lacks power
- ☐ Engine backfires
- ☐ Oil pressure warning light illuminated with engine running
- ☐ Oil consumption excessive
- ☐ Engine runs-on after switching off
- ☐ Engine noises

Cooling system

- ☐ Overheating
- ☐ Overcooling
- ☐ External coolant leakage
- ☐ Internal coolant leakage
- ☐ Corrosion

Fuel and exhaust systems

- ☐ Excessive fuel consumption
- ☐ Fuel leakage and/or fuel odour
- ☐ Excessive noise or fumes from exhaust system

Clutch

- ☐ Pedal travels to floor - no pressure or very little resistance
- ☐ Clutch fails to disengage (unable to select gears)
- ☐ Clutch slips (engine speed increases, with no increase in vehicle speed)
- ☐ Judder as clutch is engaged
- ☐ Noise when depressing or releasing clutch pedal

Manual transmission

- ☐ Noisy in neutral with engine running
- ☐ Noisy in one particular gear
- ☐ Difficulty engaging gears
- ☐ Jumps out of gear
- ☐ Vibration
- ☐ Lubricant leaks

Driveshafts

- ☐ Clicking or knocking noise on turns (at slow speed on full-lock)
- ☐ Vibration when accelerating or decelerating

Braking system

- ☐ Poor brake performance
- ☐ Vehicle pulls to one side under braking
- ☐ Noise (grinding or high-pitched squeal) when brakes applied
- ☐ Excessive brake pedal travel
- ☐ Brake pedal feels spongy when depressed
- ☐ Excessive brake pedal effort required to stop vehicle
- ☐ Judder felt through brake pedal or steering wheel when braking
- ☐ Brakes binding
- ☐ Rear wheels locking under normal braking

Suspension and steering

- ☐ Vehicle pulls to one side
- ☐ Wheel wobble and vibration
- ☐ Excessive pitching and/or rolling round corners, or during braking
- ☐ Wandering or general instability
- ☐ Excessively stiff steering
- ☐ Excessive play in steering
- ☐ Lack of power assistance (where applicable)
- ☐ Tyre wear excessive

Electrical system

- ☐ Battery will not hold a charge for more than a few days
- ☐ Ignition/no-charge warning light stays on with engine running
- ☐ Ignition/no-charge warning light fails to come on
- ☐ Lights inoperative
- ☐ Instrument readings inaccurate or erratic
- ☐ Horn inoperative, or unsatisfactory in operation
- ☐ Windscreen/rear window wipers inoperative, or unsatisfactory in operation
- ☐ Windscreen/rear window washers inoperative, or unsatisfactory in operation
- ☐ Electric windows inoperative, or unsatisfactory in operation
- ☐ Central locking system inoperative, or unsatisfactory in operation

Introduction

The vehicle owner who does his or her own maintenance according to the recommended service schedules should not have to use this section of the manual very often. Modern component reliability is such that, provided those items subject to wear or deterioration are inspected or renewed at the specified intervals, sudden failure is comparatively rare. Faults do not usually just happen as a result of sudden failure, but develop over a period of time. Major mechanical failures in particular are usually preceded by characteristic symptoms over hundreds or even thousands of miles. Those components which do occasionally fail without warning are often small and easily carried in the vehicle.

With any fault-finding, the first step is to decide where to begin investigations.

Sometimes this is obvious, but on other occasions, a little detective work will be necessary. The owner who makes half a dozen haphazard adjustments or replacements may be successful in curing a fault (or its symptoms), but will be none the wiser if the fault recurs, and ultimately may have spent more time and money than was necessary. A calm and logical approach will be found to be more satisfactory in the long run. Always take into account any warning signs or abnormalities that may have been noticed in the period preceding the fault - power loss, high or low gauge readings, unusual smells, etc - and remember that failure of components such as fuses or spark plugs may only be pointers to some underlying fault.

The pages which follow provide an easy-reference guide to the more common problems which may occur during the operation of the vehicle. These problems and their possible causes are grouped under headings denoting various components or systems, such as Engine, Cooling system, etc. The Chapter and/or Section which deals with the problem is also shown in brackets. Whatever the fault, certain basic principles apply. These are as follows:

Verify the fault. This is simply a matter of being sure that you know what the symptoms are before starting work. This is particularly important if you are investigating a fault for someone else, who may not have described it very accurately.

Don't overlook the obvious. For example, if

the vehicle won't start, is there petrol in the tank? (Don't take anyone else's word on this particular point, and don't trust the fuel gauge either!). If an electrical fault is indicated, look for loose or broken wires before digging out the test gear. Establish what work (if any) has recently been carried out - any job which has not been done properly will sometimes lead to puzzling side-effects.

Cure the disease, not the symptom.

Substituting a flat battery with a fully-charged one will get you off the hard shoulder, but if the underlying cause is not attended to, the new battery will go the same way. Similarly, changing oil-fouled spark plugs (petrol models) for a new set will get you moving again, but remember that the reason for the fouling (if it wasn't simply an incorrect grade of plug) will have to be established and corrected.

Don't take anything for granted. Particularly, don't forget that a "new" component may itself be defective (especially if it's been rattling around in the boot for months), and don't leave components out of a fault diagnosis sequence just because they are new or recently-fitted. When you do finally diagnose a difficult fault, you'll probably realise that all the evidence was there from the start.

1 Engine

Engine fails to rotate when attempting to start

- [ ] Battery terminal connections loose or corroded (*"Weekly checks"*).
- [ ] Battery discharged or faulty (Chapter 5A).
- [ ] Broken, loose or disconnected wiring in the starting circuit (Chapter 5A).
- [ ] Defective starter solenoid or switch (Chapter 5A).
- [ ] Defective starter motor (Chapter 5A).
- [ ] Starter pinion or flywheel ring gear teeth loose or broken (Chapters 2A, 2B and 5A).
- [ ] Engine earth strap broken or disconnected (Chapter 5A).

Engine rotates, but will not start

- [ ] Fuel tank empty.
- [ ] Battery discharged (engine rotates slowly) (Chapter 5A).
- [ ] Battery terminal connections loose or corroded (*"Weekly checks"*).
- [ ] Ignition components damp or damaged - petrol models (Chapters 1A and 5B).
- [ ] Broken, loose or disconnected wiring in the ignition circuit - petrol models (Chapters 1A and 5B).
- [ ] Worn, faulty or incorrectly-gapped spark plugs - petrol models (Chapter 1A).
- [ ] Preheating system faulty - diesel models (Chapter 5C).
- [ ] Choke mechanism incorrectly adjusted, worn or sticking - carburettor petrol models (Chapter 4A).
- [ ] Fuel injection system fault - fuel-injected petrol models (Chapter 4B).
- [ ] Stop solenoid faulty - diesel models (Chapter 4C).
- [ ] Air in fuel system - diesel models (Chapter 4C).
- [ ] Major mechanical failure (e.g. camshaft drive) (Chapter 2A, 2B or 2C).
- [ ] Inertia switch in operation - reset (*"Roadside repairs"*).

Engine difficult to start when cold

- [ ] Battery discharged (Chapter 5A).
- [ ] Battery terminal connections loose or corroded (*"Weekly checks"*).
- [ ] Worn, faulty or incorrectly-gapped spark plugs - petrol models (Chapter 1A).
- [ ] Preheating system faulty - diesel models (Chapter 5C).
- [ ] Choke mechanism incorrectly adjusted, worn or sticking - carburettor petrol models (Chapter 4A).
- [ ] Injection pump timing incorrect - diesel models (Chapter 4C)
- [ ] Fuel injection system fault - fuel-injected petrol models (Chapter 4B).
- [ ] Other ignition system fault - petrol models (Chapters 1A and 5B).
- [ ] Fast idle valve incorrectly adjusted - diesel models (Chapter 4C).
- [ ] Low cylinder compressions (Chapter 2A, 2B or 2C).
- [ ] Engine management system in "limp-home" mode (engine check warning light illuminated) - fuel-injected petrol models (Chapter 4B).

Engine difficult to start when hot

- [ ] Air filter element dirty or clogged (Chapter 1A or 1B).
- [ ] Choke mechanism incorrectly adjusted, worn or sticking - carburettor petrol models (Chapter 4A).
- [ ] Fuel injection system fault - fuel-injected petrol models (Chapter 4B).
- [ ] Low cylinder compressions (Chapter 2A or 2B).

Starter motor noisy or excessively-rough in engagement

- [ ] Starter pinion or flywheel ring gear teeth loose or broken (Chapters 2A, 2B and 5A).
- [ ] Starter motor mounting bolts loose or missing (Chapter 5A).
- [ ] Starter motor internal components worn or damaged (Chapter 5A).

Engine starts, but stops immediately

- [ ] Loose or faulty electrical connections in the ignition circuit - petrol models (Chapters 1A and 5B).
- [ ] Vacuum leak at the carburettor/throttle body or inlet manifold - petrol models (Chapter 4A or 4B).
- [ ] Blocked carburettor jet(s) or internal passages - carburettor petrol models (Chapter 4A).
- [ ] Blocked injector/fuel injection system fault - fuel-injected petrol models (Chapter 4B).

Engine idles erratically

- [ ] Air filter element clogged (Chapter 1A or 1B).
- [ ] Vacuum leak at the carburettor/throttle body, inlet manifold or associated hoses - petrol models (Chapter 4A or 4B).
- [ ] Worn, faulty or incorrectly-gapped spark plugs - petrol models (Chapter 1A).
- [ ] Uneven or low cylinder compressions (Chapter 2A or 2B).
- [ ] Camshaft lobes worn (Chapter 2A or 2B).
- [ ] Timing belt incorrectly tensioned (Chapter 2A or 2B).
- [ ] Blocked carburettor jet(s) or internal passages - carburettor petrol models (Chapter 4A).
- [ ] Blocked injector/fuel injection system fault - fuel-injected petrol models (Chapter 4B).
- [ ] Engine management system in "limp-home" mode (engine check warning light illuminated) - fuel-injected petrol models (Chapter 4B).
- [ ] Faulty injector(s) - diesel models (Chapter 4C).

Engine (continued)

Engine misfires at idle speed

☐ Worn, faulty or incorrectly-gapped spark plugs - petrol models (Chapter 1A).
☐ Faulty spark plug HT leads - petrol models (Chapter 1A).
☐ Vacuum leak at the carburettor/throttle body, inlet manifold or associated hoses - petrol models (Chapter 4A or 4B).
☐ Blocked carburettor jet(s) or internal passages - carburettor petrol models (Chapter 4A).
☐ Blocked injector/fuel injection system fault - fuel-injected petrol models (Chapter 4B).
☐ Faulty injector(s) - diesel models (Chapter 4C).
☐ Uneven or low cylinder compressions (Chapter 2A or 2B).
☐ Disconnected, leaking, or perished crankcase ventilation hoses (Chapter 4D).

Engine misfires throughout the driving speed range

☐ Fuel filter choked (Chapter 1A or 1B).
☐ Fuel pump faulty, or delivery pressure low - petrol models (Chapter 4A or 4B).
☐ Fuel tank vent blocked, or fuel pipes restricted (Chapter 4A, 4B or 4C).
☐ Vacuum leak at the carburettor/throttle body, inlet manifold or associated hoses - petrol models (Chapter 4A or 4B).
☐ Worn, faulty or incorrectly-gapped spark plugs - petrol models (Chapter 1A).
☐ Faulty spark plug HT leads - petrol models (Chapter 1A).
☐ Faulty injector(s) - diesel models (Chapter 4C).
☐ Faulty ignition coil - petrol models (Chapter 5B).
☐ Uneven or low cylinder compressions (Chapter 2A or 2B).
☐ Blocked carburettor jet(s) or internal passages - carburettor petrol models (Chapter 4A).
☐ Blocked injector/fuel injection system fault - fuel-injected petrol models (Chapter 4B).

Engine hesitates on acceleration

☐ Worn, faulty or incorrectly-gapped spark plugs - petrol models (Chapter 1A).
☐ Vacuum leak at the carburettor/throttle body, inlet manifold or associated hoses - petrol models (Chapter 4A or 4B).
☐ Blocked carburettor jet(s) or internal passages - carburettor petrol models (Chapter 4A).
☐ Blocked injector/fuel injection system fault - fuel-injected petrol models (Chapter 4B).
☐ Faulty injector(s) - diesel models (Chapter 4C).

Engine stalls

☐ Vacuum leak at the carburettor/throttle body, inlet manifold or associated hoses - petrol models (Chapter 4A or 4B).
☐ Fuel filter choked (Chapter 1A or 1B).
☐ Fuel pump faulty, or delivery pressure low - petrol models (Chapter 4A or 4B).
☐ Fuel tank vent blocked, or fuel pipes restricted (Chapter 4A, 4B or 4C).
☐ Blocked carburettor jet(s) or internal passages - carburettor petrol models (Chapter 4A).
☐ Blocked injector/fuel injection system fault - fuel-injected petrol models (Chapter 4B).
☐ Faulty injector(s) - diesel models (Chapter 4C).
☐ Perished diaphragm in hand-priming pump - diesel models (Chapter 4C)
☐ Engine management system in "limp-home" mode (engine check warning light illuminated) - fuel-injected petrol models (Chapter 4B).

Engine lacks power

☐ Timing belt incorrectly fitted or tensioned (Chapter 2A or 2B).
☐ Fuel filter choked (Chapter 1A or 1B).
☐ Fuel pump faulty, or delivery pressure low - petrol models (Chapter 4A or 4B).
☐ Uneven or low cylinder compressions (Chapter 2A or 2B).
☐ Worn, faulty or incorrectly-gapped spark plugs - petrol models (Chapter 1A).
☐ Vacuum leak at the carburettor/throttle body, inlet manifold or associated hoses - petrol models (Chapter 4A or 4B).
☐ Blocked carburettor jet(s) or internal passages - carburettor petrol models (Chapter 4A).
☐ Blocked injector/fuel injection system fault - fuel-injected petrol models (Chapter 4B).
☐ Faulty injector(s) - diesel models (Chapter 4C).
☐ Injection pump timing incorrect - diesel models (Chapter 4C).
☐ Brakes binding (Chapters 1 and 9).
☐ Clutch slipping (Chapter 6).

Engine backfires

☐ Timing belt incorrectly fitted or tensioned - petrol models (Chapter 2A).
☐ Vacuum leak at the carburettor/throttle body, inlet manifold or associated hoses - petrol models (Chapter 4A or 4B).
☐ Blocked carburettor jet(s) or internal passages - carburettor petrol models (Chapter 4A).
☐ Blocked injector/fuel injection system fault - fuel-injected petrol models (Chapter 4B).

Oil pressure warning light illuminated with engine running

☐ Low oil level, or incorrect oil grade ("Weekly checks").
☐ Faulty oil pressure warning light switch (Chapter 5A).
☐ Worn engine bearings and/or oil pump (Chapter 2).
☐ High engine operating temperature (Chapter 3).
☐ Oil pick-up strainer clogged (Chapter 2A or 2B).

Oil consumption excessive

☐ External leakage (standing or running) ("Roadside repairs" and relevant Part of Chapter 2).
☐ New or overhauled engine not yet run-in (Chapter 2C).
☐ Engine oil incorrect grade or poor quality ("Lubricants and fluids").
☐ Oil level too high ("Weekly checks").
☐ Defective engine breather system (relevant Part of Chapter 2).
☐ Oil leakage from ancillary component (diesel engine vacuum pump etc.) (Chapters 2 or 9).
☐ Oil leaking into coolant (Chapters 2C and 3).
☐ Oil leaking into injection pump (Chapters 2C and 4C).
☐ Leaking cylinder head gasket (relevant Part of Chapter 2).
☐ Air cleaner dirty (Chapter 1A or 1B).
☐ Cylinder bores glazed (Chapter 2C).
☐ Piston rings broken or worn (Chapter 2C).
☐ Pistons and/or bores worn (Chapter 2C).
☐ Valve stems or guides worn (Chapter 2C).
☐ Valve stem oil seals worn (Chapter 2C).

Engine runs-on after switching off

☐ Excessive carbon build-up in engine - petrol models (Chapter 2C).
☐ High engine operating temperature - carburettor petrol models (Chapter 3).
☐ Faulty stop solenoid - diesel models (Chapter 4C).

Engine (continued)

Engine noises

Pre-ignition (pinking) or knocking during acceleration or under load

- ☐ Ignition timing incorrect/ignition system fault - petrol models (Chapters 1A and 5B).
- ☐ Incorrect grade of spark plug - petrol models (Chapter 1A).
- ☐ Incorrect grade of fuel - petrol models (Chapter 4A or 4B).
- ☐ Vacuum leak at the carburettor/throttle body, inlet manifold or associated hoses - petrol models (Chapter 4A or 4B).
- ☐ Excessive carbon build-up in engine - petrol models (Chapter 2C).
- ☐ Blocked carburettor jet(s) or internal passages - carburettor petrol models (Chapter 4A).
- ☐ Blocked injector/fuel injection system fault - fuel-injected petrol models (Chapter 4B).

Whistling or wheezing noises

- ☐ Leaking inlet manifold or carburettor/throttle body gasket - petrol models (Chapter 4A or 4B).

- ☐ Leaking exhaust manifold gasket or pipe-to-manifold joint (relevant Part of Chapter 4).
- ☐ Leaking vacuum hose (relevant Part of Chapter 4 or Chapter 9).
- ☐ Blowing cylinder head gasket (Chapter 2A or 2B).

Tapping or rattling noises

- ☐ Worn valve gear or camshaft (Chapter 2A or 2B).
- ☐ Ancillary component fault (coolant pump, alternator, etc) (Chapters 3, 5A, etc).

Knocking or thumping noises

- ☐ Worn big-end bearings (regular heavy knocking, perhaps less under load) (Chapter 2C).
- ☐ Worn main bearings (rumbling and knocking, perhaps worsening under load) (Chapter 2C).
- ☐ Piston slap (most noticeable when cold) - engine worn (Chapter 2C).
- ☐ Ancillary component fault (coolant pump, alternator, etc) (Chapters 3, 5A, etc).

2 Cooling system

Overheating

- ☐ Insufficient coolant in system ("Weekly checks").
- ☐ Thermostat faulty (Chapter 3).
- ☐ Radiator core blocked, or grille restricted (Chapter 3).
- ☐ Electric cooling fan or thermoswitch faulty (Chapter 3).
- ☐ Pressure cap faulty (Chapter 3).
- ☐ Ignition timing incorrect/ignition system fault - petrol models (Chapters 1A and 5B).
- ☐ Inaccurate temperature gauge sender unit (Chapter 3).
- ☐ Airlock in cooling system (Chapter 1A or 1B).

Overcooling

- ☐ Thermostat faulty (Chapter 3).
- ☐ Inaccurate temperature gauge sender unit (Chapter 3).

External coolant leakage

- ☐ Deteriorated or damaged hoses or hose clips (Chapter 1A or 1B).
- ☐ Radiator core or heater matrix leaking (Chapter 3).
- ☐ Pressure cap faulty (Chapter 3).
- ☐ Water pump seal leaking (Chapter 3).
- ☐ Boiling due to overheating (Chapter 3).
- ☐ Core plug leaking (Chapter 2C).

Internal coolant leakage

- ☐ Leaking cylinder head gasket (Chapter 2A or 2B).
- ☐ Cracked cylinder head or cylinder bore (relevant Part of Chapter 2).

Corrosion

- ☐ Infrequent draining and flushing (Chapter 1A or 1B).
- ☐ Incorrect coolant mixture or inappropriate coolant type ("Weekly checks").

3 Fuel and exhaust systems

Excessive fuel consumption

- ☐ Air filter element dirty or clogged (Chapter 1A or 1B).
- ☐ Choke mechanism incorrectly adjusted, worn or sticking - carburettor petrol models (Chapter 4A).
- ☐ Fuel injection system fault - fuel injected petrol models (Chapter 4B).
- ☐ Faulty injector(s) - diesel models (Chapter 4C).
- ☐ Ignition timing incorrect/ignition system fault - petrol models (Chapters 1A and 5B).
- ☐ Tyres under-inflated ("Weekly checks").

Fuel leakage and/or fuel odour

- ☐ Damaged or corroded fuel tank, pipes or connections (relevant Part of Chapter 4).
- ☐ Carburettor float chamber flooding (float height incorrect) - carburettor petrol models (Chapter 4A).

Excessive noise or fumes from exhaust system

- ☐ Leaking exhaust system or manifold joints (relevant Part of Chapters 1 and 4).
- ☐ Leaking, corroded or damaged silencers or pipe (relevant Part of Chapters 1 and 4).
- ☐ Broken mountings causing body or suspension contact (Chapter 1A or 1B).

4 Clutch

Pedal travels to floor - no pressure or very little resistance

- [ ] Broken clutch cable (Chapter 6).
- [ ] Incorrect clutch cable adjustment (Chapter 6).
- [ ] Broken clutch release bearing or fork (Chapter 6).
- [ ] Broken diaphragm spring in clutch pressure plate (Chapter 6).

Clutch fails to disengage (unable to select gears)

- [ ] Incorrect clutch cable adjustment (Chapter 6).
- [ ] Clutch disc sticking on gearbox input shaft splines (Chapter 6).
- [ ] Clutch disc sticking to flywheel or pressure plate (Chapter 6).
- [ ] Faulty pressure plate assembly (Chapter 6).
- [ ] Clutch release mechanism worn or incorrectly assembled (Chapter 6).

Clutch slips (engine speed increases, with no increase in vehicle speed)

- [ ] Incorrect clutch cable adjustment (Chapter 6).
- [ ] Clutch disc linings excessively worn (Chapter 6).
- [ ] Clutch disc linings contaminated with oil or grease (Chapter 6).
- [ ] Faulty pressure plate or weak diaphragm spring (Chapter 6).

Judder as clutch is engaged

- [ ] Clutch disc linings contaminated with oil or grease (Chapter 6).
- [ ] Clutch disc linings excessively worn (Chapter 6).
- [ ] Clutch cable sticking or frayed (Chapter 6).
- [ ] Faulty or distorted pressure plate or diaphragm spring (Chapter 6).
- [ ] Worn or loose engine or gearbox mountings (Chapter 2A or 2B).
- [ ] Clutch disc hub or gearbox input shaft splines worn (Chapter 6).

Noise when depressing or releasing clutch pedal

- [ ] Worn clutch release bearing (Chapter 6).
- [ ] Worn or dry clutch pedal bushes (Chapter 6).
- [ ] Faulty pressure plate assembly (Chapter 6).
- [ ] Pressure plate diaphragm spring broken (Chapter 6).
- [ ] Broken clutch disc cushioning springs (Chapter 6).

5 Manual transmission

Noisy in neutral with engine running

- [ ] Input shaft bearings worn (noise apparent with clutch pedal released, but not when depressed) (Chapter 7).*
- [ ] Clutch release bearing worn (noise apparent with clutch pedal depressed, possibly less when released) (Chapter 6).

Noisy in one particular gear

- [ ] Worn, damaged or chipped gear teeth (Chapter 7).*

Difficulty engaging gears

- [ ] Clutch fault (Chapter 6).
- [ ] Worn or damaged gear linkage (Chapter 7).
- [ ] Incorrectly-adjusted gear linkage (Chapter 7).
- [ ] Worn synchroniser units (Chapter 7).*

Jumps out of gear

- [ ] Worn or damaged gear linkage (Chapter 7).
- [ ] Incorrectly-adjusted gear linkage (Chapter 7).
- [ ] Worn synchroniser units (Chapter 7).*
- [ ] Worn selector forks (Chapter 7).*

Vibration

- [ ] Lack of oil (Chapter 1A or 1B).
- [ ] Worn bearings (Chapter 7).*

Lubricant leaks

- [ ] Leaking differential output oil seal (Chapter 7).
- [ ] Leaking housing joint (Chapter 7).*
- [ ] Leaking input shaft oil seal (Chapter 7).*

Although the corrective action necessary to remedy the symptoms described is beyond the scope of the home mechanic, the above information should be helpful in isolating the cause of the condition, so that the owner can communicate clearly with a professional mechanic.

6 Driveshafts

Clicking or knocking noise on turns (at slow speed on full-lock)

- [ ] Lack of constant velocity joint lubricant, possibly due to damaged gaiter (Chapter 8).
- [ ] Worn outer constant velocity joint (Chapter 8).

Vibration when accelerating or decelerating

- [ ] Worn inner constant velocity joint (Chapter 8).
- [ ] Bent or distorted driveshaft (Chapter 8).

7 Braking system

Note: *Before assuming that a brake problem exists, make sure that the tyres are in good condition and correctly inflated, that the front wheel alignment is correct, and that the vehicle is not loaded with weight in an unequal manner.*

Poor brake performance

- ☐ Air in hydraulic system - bleed the brakes (Chapter 9).
- ☐ Brake fluid contaminated - change fluid (Chapters 1 and 9).
- ☐ Seized or partially-seized brake caliper piston(s) (Chapter 9).
- ☐ Brake pads incorrectly fitted (Chapters 1 and 9).
- ☐ Incorrect grade of brake pads fitted (Chapters 1 and 9).
- ☐ Brake pads "glazed" - replace (Chapter 9).
- ☐ Automatic adjustment mechanism on rear brake shoes seized (Chapter 9).
- ☐ Brake pads or linings contaminated with oil or brake fluid (Chapters 1 and 9).

Vehicle pulls to one side under braking

- ☐ Worn, defective, damaged or contaminated brake pads on one side (Chapters 1 and 9).
- ☐ Seized or partially-seized front brake caliper or rear wheel cylinder piston (Chapters 1 and 9).
- ☐ A mixture of brake pad/shoe lining materials fitted between sides (Chapters 1 and 9).
- ☐ Front brake caliper or rear brake backplate mounting bolts loose (Chapter 9).
- ☐ Worn or damaged steering or suspension components (Chapters 1 and 10).

Noise (grinding or high-pitched squeal) when brakes applied

- ☐ Brake pad friction lining material worn down to metal backing (Chapters 1 and 9).
- ☐ Excessive corrosion of brake disc or drum (may be apparent after the vehicle has been standing for some time) (Chapters 1 and 9).
- ☐ Foreign object (stone chipping, etc) trapped between brake disc and shield (Chapters 1 and 9).

Excessive brake pedal travel

- ☐ Inoperative rear brake self-adjust mechanism (Chapter 9).
- ☐ Faulty master cylinder (Chapter 9).
- ☐ Faulty vacuum servo unit (Chapter 9).
- ☐ Faulty brake vacuum pump - diesel models (Chapter 9).
- ☐ Disconnected, damaged or insecure brake servo vacuum hose (Chapters 1 and 9).
- ☐ Air in hydraulic system - bleed the brakes (Chapter 9).
- ☐ Brake fluid contaminated - change fluid (Chapters 1 and 9).
- ☐ Brake fluid leak - check especially fluid unions and bleed screws (Chapters 1 and 9).

Brake pedal feels spongy when depressed

- ☐ Air in hydraulic system (Chapters 1 and 9).
- ☐ Deteriorated flexible rubber brake hoses (Chapters 1 and 9).
- ☐ Master cylinder mounting nuts loose (Chapter 9).
- ☐ Faulty master cylinder (Chapter 9).

Excessive brake pedal effort required to stop vehicle

- ☐ Faulty vacuum servo unit (Chapter 9).
- ☐ Faulty brake vacuum pump - diesel models (Chapter 9).
- ☐ Disconnected, damaged or insecure brake servo vacuum hose (Chapters 1 and 9).
- ☐ Primary or secondary hydraulic circuit failure (Chapter 9).
- ☐ Seized brake caliper or wheel cylinder piston(s) (Chapter 9).
- ☐ Brake pads or brake shoes incorrectly fitted (Chapters 1 and 9).
- ☐ Incorrect grade of brake pads or brake shoes fitted (Chapters 1 and 9).
- ☐ Brake pads or brake shoe linings contaminated (Chapters 1 and 9).
- ☐ Brake pedal-to-servo linkage binding - right-hand-drive models (Chapter 9).

Judder felt through brake pedal or steering wheel when braking

- ☐ Excessive run-out or distortion of front discs or rear drums (Chapters 1 and 9).
- ☐ Brake pad or brake shoe linings worn (Chapters 1 and 9).
- ☐ Front brake caliper or rear brake backplate mounting bolts loose (Chapter 9).
- ☐ Wear in suspension or steering components or mountings (Chapters 1 and 10).

Brakes binding

- ☐ Seized brake caliper or wheel cylinder piston (Chapter 9).
- ☐ Incorrectly-adjusted handbrake mechanism (Chapter 1).
- ☐ Faulty master cylinder (Chapter 9).

Rear wheels locking under normal braking

- ☐ Rear brake shoe linings contaminated (Chapters 1 and 9).
- ☐ Faulty brake pressure limiter (Chapter 9).

8 Suspension and steering

Note: *Before diagnosing suspension or steering faults, be sure that the trouble is not due to incorrect tyre pressures, mixtures of tyre types, or binding brakes.*

Vehicle pulls to one side

☐ Defective tyre (*"Weekly checks"*).
☐ Excessive wear in suspension or steering components (Chapters 1 and 10).
☐ Incorrect front wheel alignment (Chapter 10).
☐ Accident damage to steering or suspension components (Chapter 1A or 1B).

Wheel wobble and vibration

☐ Front roadwheels out of balance (vibration felt mainly through the steering wheel) (*"Weekly checks"*).
☐ Rear roadwheels out of balance (vibration felt throughout the vehicle) (*"Weekly checks"*).
☐ Roadwheels damaged or distorted (*"Weekly checks"*).
☐ Faulty or damaged tyre (*"Weekly checks"*).
☐ Worn steering or suspension joints, bushes or components (Chapters 1 and 10).
☐ Wheel bolts loose.

Excessive pitching and/or rolling round corners, or during braking

☐ Defective shock absorbers (Chapters 1 and 10).
☐ Broken or weak coil spring and/or suspension component (Chapters 1 and 10).
☐ Worn or damaged anti-roll bar or mountings (Chapter 10).

Wandering or general instability

☐ Incorrect front wheel alignment (Chapter 10).
☐ Worn steering or suspension joints, bushes or components (Chapters 1 and 10).
☐ Roadwheels out of balance (*"Weekly checks"*).
☐ Faulty or damaged tyre (*"Weekly checks"*).
☐ Wheel bolts loose.
☐ Defective shock absorbers (Chapters 1 and 10).

Excessively stiff steering

☐ Lack of steering gear lubricant (Chapter 10).
☐ Seized track-rod end balljoint or suspension balljoint (Chapters 1 and 10).
☐ Incorrect front wheel alignment (Chapter 10).
☐ Steering rack or column bent or damaged (Chapter 10).

Excessive play in steering

☐ Worn steering column intermediate shaft universal joint (Chapter 10).
☐ Worn steering track-rod end balljoints (Chapters 1 and 10).
☐ Worn rack-and-pinion steering gear (Chapter 10).
☐ Worn steering or suspension joints, bushes or components (Chapters 1 and 10).

Lack of power assistance (where applicable)

☐ Broken or incorrectly-adjusted power steering pump drivebelt (Chapter 1B).
☐ Incorrect power steering fluid level (*"Weekly checks"*).
☐ Air in power steering system - bleed (Chapter 10).
☐ Power steering fluid leak (Chapter 10).
☐ Faulty power steering pump (Chapter 10).
☐ Faulty rack-and-pinion steering gear (Chapter 10).

Tyre wear excessive

Tyres worn on inside or outside edges

☐ Tyres under-inflated (wear on both edges) (*"Weekly checks"*).
☐ Incorrect camber or castor angles (wear on one edge only) (Chapter 10).
☐ Worn steering or suspension joints, bushes or components (Chapters 1 and 10).
☐ Excessively-hard cornering.
☐ Accident damage.

Tyre treads exhibit feathered edges

☐ Incorrect toe setting (Chapter 10).

Tyres worn in centre of tread

☐ Tyres over-inflated (*"Weekly checks"*).

Tyres worn on inside and outside edges

☐ Tyres under-inflated (*"Weekly checks"*).

Tyres worn unevenly

☐ Tyres/wheels out of balance (*"Weekly checks"*).
☐ Bent or damaged wheel rim, or badly fitted tyre (*"Weekly checks"*).
☐ Worn shock absorbers (Chapters 1 and 10).
☐ Faulty tyre (*"Weekly checks"*).

9 Electrical system

Note: *For problems associated with the starting system, refer to the faults listed under "Engine" earlier in this Section.*

Battery will not hold a charge for more than a few days

☐ Battery defective internally (Chapter 5A).
☐ Battery electrolyte level low - where applicable (*"Weekly checks"*).
☐ Battery terminal connections loose or corroded (*"Weekly checks"*).
☐ Auxiliary drivebelt worn or incorrectly adjusted (Chapter 1A or 1B).
☐ Alternator not charging at correct output (Chapter 5A).
☐ Alternator or voltage regulator faulty (Chapter 5A).
☐ Short-circuit causing continual battery drain (Chapters 5A and 12).

Ignition/no-charge warning light stays on with engine running

☐ Auxiliary drivebelt broken, worn, or incorrectly adjusted (Chapter 1A or 1B).
☐ Alternator brushes worn, sticking, or dirty (Chapter 5A).
☐ Alternator brush springs weak or broken (Chapter 5A).
☐ Internal fault in alternator or voltage regulator (Chapter 5A).
☐ Broken, disconnected, or loose wiring in charging circuit (Chapter 5A).

Ignition/no-charge warning light fails to come on

☐ Warning light bulb blown (Chapter 12).
☐ Broken, disconnected, or loose wiring in warning light circuit (Chapter 12).
☐ Alternator faulty (Chapter 5A).

Electrical system (continued)

Lights inoperative

- [ ] Bulb blown (Chapter 12).
- [ ] Corrosion of bulb or bulbholder contacts (Chapter 12).
- [ ] Blown fuse (Chapter 12).
- [ ] Faulty relay (Chapter 12).
- [ ] Broken, loose, or disconnected wiring (Chapter 12).
- [ ] Faulty switch (Chapter 12).

Instrument readings inaccurate or erratic

Fuel or temperature gauges give no reading

- [ ] Faulty gauge sender unit (Chapters 3 and 4A, 4B or 4C).
- [ ] Wiring open-circuit (Chapter 12).
- [ ] Faulty gauge (Chapter 12).

Fuel or temperature gauges give continuous maximum reading

- [ ] Faulty gauge sender unit (Chapters 3 and 4A, 4B or 4C).
- [ ] Wiring short-circuit (Chapter 12).
- [ ] Faulty gauge (Chapter 12).

Horn inoperative, or unsatisfactory in operation

Horn operates all the time

- [ ] Horn push either earthed or stuck down (Chapter 12).
- [ ] Horn wire to horn push earthed (Chapter 12).

Horn fails to operate

- [ ] Blown fuse (Chapter 12).
- [ ] Wiring connections loose, broken or disconnected (Chapter 12).
- [ ] Faulty horn (Chapter 12).

Horn emits intermittent or unsatisfactory sound

- [ ] Wiring connections loose (Chapter 12).
- [ ] Horn mountings loose (Chapter 12).
- [ ] Faulty horn (Chapter 12).

Windscreen/rear window wipers inoperative, or unsatisfactory in operation

Wipers fail to operate, or operate very slowly

- [ ] Wiper blades stuck to screen, or linkage seized or binding ("Weekly checks" and Chapter 12).
- [ ] Blown fuse (Chapter 12).
- [ ] Wiring connections loose, broken or disconnected (Chapter 12).
- [ ] Faulty relay (Chapter 12).
- [ ] Faulty wiper motor (Chapter 12).

Wiper blades sweep over too large or too small an area of the glass

- [ ] Wiper arms incorrectly positioned on spindles (Chapter 12).
- [ ] Excessive wear of wiper linkage (Chapter 12).
- [ ] Wiper motor or linkage mountings loose or insecure (Chapter 12).

Wiper blades fail to clean the glass effectively

- [ ] Wiper blade rubbers worn or perished ("Weekly checks").
- [ ] Wiper arm tension springs broken, or arm pivots seized (Chapter 12).
- [ ] Insufficient windscreen washer additive to adequately remove road film ("Weekly checks").

Windscreen/rear window washers inoperative, or unsatisfactory in operation

One or more washer jets inoperative

- [ ] Blocked washer jet.
- [ ] Disconnected, kinked or restricted fluid hose (Chapter 12).
- [ ] Insufficient fluid in washer reservoir ("Weekly checks").

Washer pump fails to operate

- [ ] Broken or disconnected wiring or connections (Chapter 12).
- [ ] Blown fuse (Chapter 12).
- [ ] Faulty washer switch (Chapter 12).
- [ ] Faulty washer pump (Chapter 12).

Electric windows inoperative, or unsatisfactory in operation

Window glass will only move in one direction

- [ ] Faulty switch (Chapter 12).

Window glass slow to move

- [ ] Regulator seized or damaged, or in need of lubrication (Chapter 11).
- [ ] Door internal components or trim fouling regulator (Chapter 11).
- [ ] Faulty motor (Chapter 11).

Window glass fails to move

- [ ] Blown fuse (Chapter 12).
- [ ] Faulty switch or relay (Chapter 12).
- [ ] Broken or disconnected wiring or connections (Chapter 12).
- [ ] Faulty motor (Chapter 11).

Central locking system inoperative, or unsatisfactory in operation

Complete system failure

- [ ] Blown fuse (Chapter 12).
- [ ] Faulty relay (Chapter 12).
- [ ] Broken or disconnected wiring or connections (Chapter 12).
- [ ] Faulty control unit (Chapter 11).

Latch locks but will not unlock, or unlocks but will not lock

- [ ] Broken or disconnected latch operating rods or levers (Chapter 11).
- [ ] Faulty relay (Chapter 12).
- [ ] Faulty control unit (Chapter 11).

One solenoid/motor fails to operate

- [ ] Broken or disconnected wiring or connections (Chapter 12).
- [ ] Faulty solenoid/motor (Chapter 11).
- [ ] Broken, binding or disconnected latch operating rods or levers (Chapter 11).
- [ ] Fault in door latch (Chapter 11).

A

ABS (Anti-lock brake system) A system, usually electronically controlled, that senses incipient wheel lockup during braking and relieves hydraulic pressure at wheels that are about to skid.

Air bag An inflatable bag hidden in the steering wheel (driver's side) or the dash or glovebox (passenger side). In a head-on collision, the bags inflate, preventing the driver and front passenger from being thrown forward into the steering wheel or windscreen.

Air cleaner A metal or plastic housing, containing a filter element, which removes dust and dirt from the air being drawn into the engine.

Air filter element The actual filter in an air cleaner system, usually manufactured from pleated paper and requiring renewal at regular intervals.

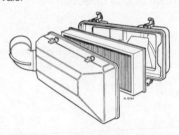

Air filter

Allen key A hexagonal wrench which fits into a recessed hexagonal hole.

Alligator clip A long-nosed spring-loaded metal clip with meshing teeth. Used to make temporary electrical connections.

Alternator A component in the electrical system which converts mechanical energy from a drivebelt into electrical energy to charge the battery and to operate the starting system, ignition system and electrical accessories.

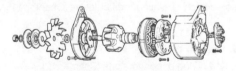

Alternator (exploded view)

Ampere (amp) A unit of measurement for the flow of electric current. One amp is the amount of current produced by one volt acting through a resistance of one ohm.

Anaerobic sealer A substance used to prevent bolts and screws from loosening. Anaerobic means that it does not require oxygen for activation. The Loctite brand is widely used.

Antifreeze A substance (usually ethylene glycol) mixed with water, and added to a vehicle's cooling system, to prevent freezing of the coolant in winter. Antifreeze also contains chemicals to inhibit corrosion and the formation of rust and other deposits that would tend to clog the radiator and coolant passages and reduce cooling efficiency.

Anti-seize compound A coating that reduces the risk of seizing on fasteners that are subjected to high temperatures, such as exhaust manifold bolts and nuts.

Anti-seize compound

Asbestos A natural fibrous mineral with great heat resistance, commonly used in the composition of brake friction materials. Asbestos is a health hazard and the dust created by brake systems should never be inhaled or ingested.

Axle A shaft on which a wheel revolves, or which revolves with a wheel. Also, a solid beam that connects the two wheels at one end of the vehicle. An axle which also transmits power to the wheels is known as a live axle.

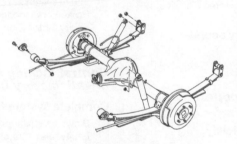

Axle assembly

Axleshaft A single rotating shaft, on either side of the differential, which delivers power from the final drive assembly to the drive wheels. Also called a driveshaft or a halfshaft.

B

Ball bearing An anti-friction bearing consisting of a hardened inner and outer race with hardened steel balls between two races.

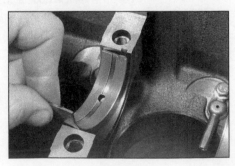

Bearing

Bearing The curved surface on a shaft or in a bore, or the part assembled into either, that permits relative motion between them with minimum wear and friction.

Big-end bearing The bearing in the end of the connecting rod that's attached to the crankshaft.

Bleed nipple A valve on a brake wheel cylinder, caliper or other hydraulic component that is opened to purge the hydraulic system of air. Also called a bleed screw.

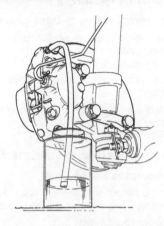

Brake bleeding

Brake bleeding Procedure for removing air from lines of a hydraulic brake system.

Brake disc The component of a disc brake that rotates with the wheels.

Brake drum The component of a drum brake that rotates with the wheels.

Brake linings The friction material which contacts the brake disc or drum to retard the vehicle's speed. The linings are bonded or riveted to the brake pads or shoes.

Brake pads The replaceable friction pads that pinch the brake disc when the brakes are applied. Brake pads consist of a friction material bonded or riveted to a rigid backing plate.

Brake shoe The crescent-shaped carrier to which the brake linings are mounted and which forces the lining against the rotating drum during braking.

Braking systems For more information on braking systems, consult the *Haynes Automotive Brake Manual*.

Breaker bar A long socket wrench handle providing greater leverage.

Bulkhead The insulated partition between the engine and the passenger compartment.

C

Caliper The non-rotating part of a disc-brake assembly that straddles the disc and carries the brake pads. The caliper also contains the hydraulic components that cause the pads to pinch the disc when the brakes are applied. A caliper is also a measuring tool that can be set to measure inside or outside dimensions of an object.

Camshaft A rotating shaft on which a series of cam lobes operate the valve mechanisms. The camshaft may be driven by gears, by sprockets and chain or by sprockets and a belt.

Canister A container in an evaporative emission control system; contains activated charcoal granules to trap vapours from the fuel system.

Canister

Carburettor A device which mixes fuel with air in the proper proportions to provide a desired power output from a spark ignition internal combustion engine.

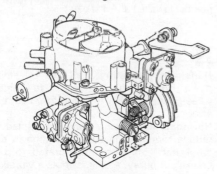

Carburettor

Castellated Resembling the parapets along the top of a castle wall. For example, a castellated balljoint stud nut.

Castellated nut

Castor In wheel alignment, the backward or forward tilt of the steering axis. Castor is positive when the steering axis is inclined rearward at the top.

Catalytic converter A silencer-like device in the exhaust system which converts certain pollutants in the exhaust gases into less harmful substances.

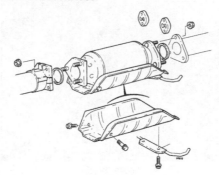

Catalytic converter

Circlip A ring-shaped clip used to prevent endwise movement of cylindrical parts and shafts. An internal circlip is installed in a groove in a housing; an external circlip fits into a groove on the outside of a cylindrical piece such as a shaft.

Clearance The amount of space between two parts. For example, between a piston and a cylinder, between a bearing and a journal, etc.

Coil spring A spiral of elastic steel found in various sizes throughout a vehicle, for example as a springing medium in the suspension and in the valve train.

Compression Reduction in volume, and increase in pressure and temperature, of a gas, caused by squeezing it into a smaller space.

Compression ratio The relationship between cylinder volume when the piston is at top dead centre and cylinder volume when the piston is at bottom dead centre.

Constant velocity (CV) joint A type of universal joint that cancels out vibrations caused by driving power being transmitted through an angle.

Core plug A disc or cup-shaped metal device inserted in a hole in a casting through which core was removed when the casting was formed. Also known as a freeze plug or expansion plug.

Crankcase The lower part of the engine block in which the crankshaft rotates.

Crankshaft The main rotating member, or shaft, running the length of the crankcase, with offset "throws" to which the connecting rods are attached.

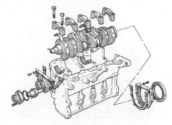

Crankshaft assembly

Crocodile clip See Alligator clip

D

Diagnostic code Code numbers obtained by accessing the diagnostic mode of an engine management computer. This code can be used to determine the area in the system where a malfunction may be located.

Disc brake A brake design incorporating a rotating disc onto which brake pads are squeezed. The resulting friction converts the energy of a moving vehicle into heat.

Double-overhead cam (DOHC) An engine that uses two overhead camshafts, usually one for the intake valves and one for the exhaust valves.

Drivebelt(s) The belt(s) used to drive accessories such as the alternator, water pump, power steering pump, air conditioning compressor, etc. off the crankshaft pulley.

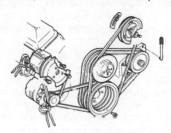

Accessory drivebelts

Driveshaft Any shaft used to transmit motion. Commonly used when referring to the axleshafts on a front wheel drive vehicle.

Driveshaft

Drum brake A type of brake using a drum-shaped metal cylinder attached to the inner surface of the wheel. When the brake pedal is pressed, curved brake shoes with friction linings press against the inside of the drum to slow or stop the vehicle.

Drum brake assembly

E

EGR valve A valve used to introduce exhaust gases into the intake air stream.

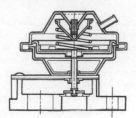

EGR valve

Electronic control unit (ECU) A computer which controls (for instance) ignition and fuel injection systems, or an anti-lock braking system. For more information refer to the *Haynes Automotive Electrical and Electronic Systems Manual.*

Electronic Fuel Injection (EFI) A computer controlled fuel system that distributes fuel through an injector located in each intake port of the engine.

Emergency brake A braking system, independent of the main hydraulic system, that can be used to slow or stop the vehicle if the primary brakes fail, or to hold the vehicle stationary even though the brake pedal isn't depressed. It usually consists of a hand lever that actuates either front or rear brakes mechanically through a series of cables and linkages. Also known as a handbrake or parking brake.

Endfloat The amount of lengthwise movement between two parts. As applied to a crankshaft, the distance that the crankshaft can move forward and back in the cylinder block.

Engine management system (EMS) A computer controlled system which manages the fuel injection and the ignition systems in an integrated fashion.

Exhaust manifold A part with several passages through which exhaust gases leave the engine combustion chambers and enter the exhaust pipe.

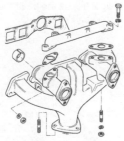

Exhaust manifold

F

Fan clutch A viscous (fluid) drive coupling device which permits variable engine fan speeds in relation to engine speeds.

Feeler blade A thin strip or blade of hardened steel, ground to an exact thickness, used to check or measure clearances between parts.

Feeler blade

Firing order The order in which the engine cylinders fire, or deliver their power strokes, beginning with the number one cylinder.

Flywheel A heavy spinning wheel in which energy is absorbed and stored by means of momentum. On cars, the flywheel is attached to the crankshaft to smooth out firing impulses.

Free play The amount of travel before any action takes place. The "looseness" in a linkage, or an assembly of parts, between the initial application of force and actual movement. For example, the distance the brake pedal moves before the pistons in the master cylinder are actuated.

Fuse An electrical device which protects a circuit against accidental overload. The typical fuse contains a soft piece of metal which is calibrated to melt at a predetermined current flow (expressed as amps) and break the circuit.

Fusible link A circuit protection device consisting of a conductor surrounded by heat-resistant insulation. The conductor is smaller than the wire it protects, so it acts as the weakest link in the circuit. Unlike a blown fuse, a failed fusible link must frequently be cut from the wire for replacement.

G

Gap The distance the spark must travel in jumping from the centre electrode to the side

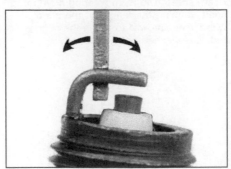

Adjusting spark plug gap

electrode in a spark plug. Also refers to the spacing between the points in a contact breaker assembly in a conventional points-type ignition, or to the distance between the reluctor or rotor and the pickup coil in an electronic ignition.

Gasket Any thin, soft material - usually cork, cardboard, asbestos or soft metal - installed between two metal surfaces to ensure a good seal. For instance, the cylinder head gasket seals the joint between the block and the cylinder head.

Gasket

Gauge An instrument panel display used to monitor engine conditions. A gauge with a movable pointer on a dial or a fixed scale is an analogue gauge. A gauge with a numerical readout is called a digital gauge.

H

Halfshaft A rotating shaft that transmits power from the final drive unit to a drive wheel, usually when referring to a live rear axle.

Harmonic balancer A device designed to reduce torsion or twisting vibration in the crankshaft. May be incorporated in the crankshaft pulley. Also known as a vibration damper.

Hone An abrasive tool for correcting small irregularities or differences in diameter in an engine cylinder, brake cylinder, etc.

Hydraulic tappet A tappet that utilises hydraulic pressure from the engine's lubrication system to maintain zero clearance (constant contact with both camshaft and valve stem). Automatically adjusts to variation in valve stem length. Hydraulic tappets also reduce valve noise.

I

Ignition timing The moment at which the spark plug fires, usually expressed in the number of crankshaft degrees before the piston reaches the top of its stroke.

Inlet manifold A tube or housing with passages through which flows the air-fuel mixture (carburettor vehicles and vehicles with throttle body injection) or air only (port fuel-injected vehicles) to the port openings in the cylinder head.

J

Jump start Starting the engine of a vehicle with a discharged or weak battery by attaching jump leads from the weak battery to a charged or helper battery.

L

Load Sensing Proportioning Valve (LSPV) A brake hydraulic system control valve that works like a proportioning valve, but also takes into consideration the amount of weight carried by the rear axle.

Locknut A nut used to lock an adjustment nut, or other threaded component, in place. For example, a locknut is employed to keep the adjusting nut on the rocker arm in position.

Lockwasher A form of washer designed to prevent an attaching nut from working loose.

M

MacPherson strut A type of front suspension system devised by Earle MacPherson at Ford of England. In its original form, a simple lateral link with the anti-roll bar creates the lower control arm. A long strut - an integral coil spring and shock absorber - is mounted between the body and the steering knuckle. Many modern so-called MacPherson strut systems use a conventional lower A-arm and don't rely on the anti-roll bar for location.

Multimeter An electrical test instrument with the capability to measure voltage, current and resistance.

N

NOx Oxides of Nitrogen. A common toxic pollutant emitted by petrol and diesel engines at higher temperatures.

O

Ohm The unit of electrical resistance. One volt applied to a resistance of one ohm will produce a current of one amp.

Ohmmeter An instrument for measuring electrical resistance.

O-ring A type of sealing ring made of a special rubber-like material; in use, the O-ring is compressed into a groove to provide the sealing action.

O-ring

Overhead cam (ohc) engine An engine with the camshaft(s) located on top of the cylinder head(s).

Overhead valve (ohv) engine An engine with the valves located in the cylinder head, but with the camshaft located in the engine block.

Oxygen sensor A device installed in the engine exhaust manifold, which senses the oxygen content in the exhaust and converts this information into an electric current. Also called a Lambda sensor.

P

Phillips screw A type of screw head having a cross instead of a slot for a corresponding type of screwdriver.

Plastigage A thin strip of plastic thread, available in different sizes, used for measuring clearances. For example, a strip of Plastigage is laid across a bearing journal. The parts are assembled and dismantled; the width of the crushed strip indicates the clearance between journal and bearing.

Plastigage

Propeller shaft The long hollow tube with universal joints at both ends that carries power from the transmission to the differential on front-engined rear wheel drive vehicles.

Proportioning valve A hydraulic control valve which limits the amount of pressure to the rear brakes during panic stops to prevent wheel lock-up.

R

Rack-and-pinion steering A steering system with a pinion gear on the end of the steering shaft that mates with a rack (think of a geared wheel opened up and laid flat). When the steering wheel is turned, the pinion turns, moving the rack to the left or right. This movement is transmitted through the track rods to the steering arms at the wheels.

Radiator A liquid-to-air heat transfer device designed to reduce the temperature of the coolant in an internal combustion engine cooling system.

Refrigerant Any substance used as a heat transfer agent in an air-conditioning system. R-12 has been the principle refrigerant for many years; recently, however, manufacturers have begun using R-134a, a non-CFC substance that is considered less harmful to the ozone in the upper atmosphere.

Rocker arm A lever arm that rocks on a shaft or pivots on a stud. In an overhead valve engine, the rocker arm converts the upward movement of the pushrod into a downward movement to open a valve.

Rotor In a distributor, the rotating device inside the cap that connects the centre electrode and the outer terminals as it turns, distributing the high voltage from the coil secondary winding to the proper spark plug. Also, that part of an alternator which rotates inside the stator. Also, the rotating assembly of a turbocharger, including the compressor wheel, shaft and turbine wheel.

Runout The amount of wobble (in-and-out movement) of a gear or wheel as it's rotated. The amount a shaft rotates "out-of-true." The out-of-round condition of a rotating part.

S

Sealant A liquid or paste used to prevent leakage at a joint. Sometimes used in conjunction with a gasket.

Sealed beam lamp An older headlight design which integrates the reflector, lens and filaments into a hermetically-sealed one-piece unit. When a filament burns out or the lens cracks, the entire unit is simply replaced.

Serpentine drivebelt A single, long, wide accessory drivebelt that's used on some newer vehicles to drive all the accessories, instead of a series of smaller, shorter belts. Serpentine drivebelts are usually tensioned by an automatic tensioner.

Serpentine drivebelt

Shim Thin spacer, commonly used to adjust the clearance or relative positions between two parts. For example, shims inserted into or under bucket tappets control valve clearances. Clearance is adjusted by changing the thickness of the shim.

Slide hammer A special puller that screws into or hooks onto a component such as a shaft or bearing; a heavy sliding handle on the shaft bottoms against the end of the shaft to knock the component free.

Sprocket A tooth or projection on the periphery of a wheel, shaped to engage with a chain or drivebelt. Commonly used to refer to the sprocket wheel itself.

Glossary of technical terms

Starter inhibitor switch On vehicles with an automatic transmission, a switch that prevents starting if the vehicle is not in Neutral or Park.

Strut See MacPherson strut.

T

Tappet A cylindrical component which transmits motion from the cam to the valve stem, either directly or via a pushrod and rocker arm. Also called a cam follower.

Thermostat A heat-controlled valve that regulates the flow of coolant between the cylinder block and the radiator, so maintaining optimum engine operating temperature. A thermostat is also used in some air cleaners in which the temperature is regulated.

Thrust bearing The bearing in the clutch assembly that is moved in to the release levers by clutch pedal action to disengage the clutch. Also referred to as a release bearing.

Timing belt A toothed belt which drives the camshaft. Serious engine damage may result if it breaks in service.

Timing chain A chain which drives the camshaft.

Toe-in The amount the front wheels are closer together at the front than at the rear. On rear wheel drive vehicles, a slight amount of toe-in is usually specified to keep the front wheels running parallel on the road by offsetting other forces that tend to spread the wheels apart.

Toe-out The amount the front wheels are closer together at the rear than at the front. On front wheel drive vehicles, a slight amount of toe-out is usually specified.

Tools For full information on choosing and using tools, refer to the *Haynes Automotive Tools Manual.*

Tracer A stripe of a second colour applied to a wire insulator to distinguish that wire from another one with the same colour insulator.

Tune-up A process of accurate and careful adjustments and parts replacement to obtain the best possible engine performance.

Turbocharger A centrifugal device, driven by exhaust gases, that pressurises the intake air. Normally used to increase the power output from a given engine displacement, but can also be used primarily to reduce exhaust emissions (as on VW's "Umwelt" Diesel engine).

U

Universal joint or U-joint A double-pivoted connection for transmitting power from a driving to a driven shaft through an angle. A U-joint consists of two Y-shaped yokes and a cross-shaped member called the spider.

V

Valve A device through which the flow of liquid, gas, vacuum, or loose material in bulk may be started, stopped, or regulated by a movable part that opens, shuts, or partially obstructs one or more ports or passageways. A valve is also the movable part of such a device.

Valve clearance The clearance between the valve tip (the end of the valve stem) and the rocker arm or tappet. The valve clearance is measured when the valve is closed.

Vernier caliper A precision measuring instrument that measures inside and outside dimensions. Not quite as accurate as a micrometer, but more convenient.

Viscosity The thickness of a liquid or its resistance to flow.

Volt A unit for expressing electrical "pressure" in a circuit. One volt that will produce a current of one ampere through a resistance of one ohm.

W

Welding Various processes used to join metal items by heating the areas to be joined to a molten state and fusing them together. For more information refer to the *Haynes Automotive Welding Manual.*

Wiring diagram A drawing portraying the components and wires in a vehicle's electrical system, using standardised symbols. For more information refer to the *Haynes Automotive Electrical and Electronic Systems Manual.*

Note: *References throughout this index are in the form - "Chapter number" • "Page number"*

Preserving Our Motoring Heritage

< The Model J Duesenberg Derham Tourster. Only eight of these magnificent cars were ever built – this is the only example to be found outside the United States of America

Almost every car you've ever loved, loathed or desired is gathered under one roof at the Haynes Motor Museum. Over 300 immaculately presented cars and motorbikes represent every aspect of our motoring heritage, from elegant reminders of bygone days, such as the superb Model J Duesenberg to curiosities like the bug-eyed BMW Isetta. There are also many old friends and flames. Perhaps you remember the 1959 Ford Popular that you did your courting in? The magnificent 'Red Collection' is a spectacle of classic sports cars including AC, Alfa Romeo, Austin Healey, Ferrari, Lamborghini, Maserati, MG, Riley, Porsche and Triumph.

A Perfect Day Out

Each and every vehicle at the Haynes Motor Museum has played its part in the history and culture of Motoring. Today, they make a wonderful spectacle and a great day out for all the family. Bring the kids, bring Mum and Dad, but above all bring your camera to capture those golden memories for ever. You will also find an impressive array of motoring memorabilia, a comfortable 70 seat video cinema and one of the most extensive transport book shops in Britain. The Pit Stop Cafe serves everything from a cup of tea to wholesome, home-made meals or, if you prefer, you can enjoy the large picnic area nestled in the beautiful rural surroundings of Somerset.

> John Haynes O.B.E., Founder and Chairman of the museum at the wheel of a Haynes Light 12.

< Graham Hill's Lola Cosworth Formula 1 car next to a 1934 Riley Sports.

The Museum is situated on the A359 Yeovil to Frome road at Sparkford, just off the A303 in Somerset. It is about 40 miles south of Bristol, and 25 minutes drive from the M5 intersection at Taunton.
Open 9.30am - 5.30pm (10.00am - 4.00pm Winter) 7 days a week, *except Christmas Day, Boxing Day and New Years Day*
Special rates available for schools, coach parties and outings Charitable Trust No. 292048